Research in Science Education in Europe

Research in Science Education in Europe:
Current Issues and Themes

Edited by

Geoff Welford
Jonathan Osborne
Phil Scott

 The Falmer Press

(A member of the Taylor & Francis Group)
London • Washington, D.C.

UK   The Falmer Press, 1 Gunpowder Square, London, EC4A 3DE
USA   The Falmer Press, Taylor & Francis Inc., 1900 Frost Road, Suite 101,
Bristol, PA 19007

First published in 1996

**A catalogue record for this book is available from the British
Library**

**Library of Congress Cataloging-in-Publication Data are available on
request**

ISBN 0 7507 0547 7 paper

Jacket design by Caroline Archer

Typeset in 10/12pt Times by
Graphicraft Typesetters Ltd., Hong Kong.

*Printed in Great Britain by Biddles Ltd., Guildford and King's Lynn on
paper which has a specified pH value on final paper manufacture of not
less than 7.5 and is therefore 'acid free'.*

Contents

Editorial   1

Part I Teaching and Learning

Introduction   5
Phil Scott

1 Developing New Teaching Sequences in Science: The Example
   of 'Gases and Their Properties'   7
   Björn Andersson and Frank Bach

2 Exploring the Use of Analogy in the Teaching of Heat,
   Temperature and Thermal Equilibrium   22
   Michael Arnold and Robin Millar

3 Developing Scientific Concepts in the Primary Classroom:
   Teaching about Electric Circuits   36
   Hilary Asoko

4 Teaching Electricity by Help of a Water Analogy
   (How to Cope with the Need for Conceptual Change)   50
   Hannelore Schwedes and Wolff-Gerhard Dudeck

5 Students' Computer-based Problem-solving in Electricity:
   Strategies and Collaborative Talk   64
   René Amigues

6 Group Concept Mapping, Language and Children's
   Learning in Primary Science   74
   Steve Sizmur

7 Using a Picture Language to Teach about Processes
   of Change   85
   Richard Boohan

8 Construction of Prototypical Situations in Teaching the
   Concept of Energy   100
   Andrée Tiberghien

9 Decision-making on Science-related Social Issues: The Case
   of Garbage in Physical Science — A Problem-posing Approach   115
   Koos Kortland

Contents

10 Adolescent Decision-making, by Individuals and Groups,
about Science-related Societal Issues 126
Mary Ratcliffe

Part II Developing and Understanding Models in Science Education

Introduction 141
John Gilbert

11 The Scientific Model as a Form of Speech 143
Clive Sutton

12 Modelling the Growth of Scientific Knowledge 153
Richard A. Duschl and Sibel Erduran

13 Mental Modelling 166
Reinders Duit and Shawn Glynn

14 Texts and Contexts: Framing Modelling in the Primary
Science Classroom 177
Carolyn Boulter and John Gilbert

Part III Approaches to Science Instruction

Introduction 189
Geoff Welford

15 Knowledge and Action: Students' Understanding of the
Procedures of Scientific Enquiry 191
Robin Millar and Fred Lubben

16 What Changes in Conceptual Change?: From Ideas to Theories 200
Montserrat Benlloch and Juan Ignacio Pozo

17 Development of Pupils' Ideas of the Particulate Nature of Matter:
Long-term Research Project 212
Michael Lichtfeldt

18 Biotechnology and Genetic Engineering: Student Knowledge
and Attitudes: Implications for Teaching Controversial Issues
and the Public Understanding of Science 229
Roger Lock

19 Progression in Pupils' Understanding of Combustion 243
Rod Watson and Justin Dillon

20 Students' Conceptions of Quantum Physics 254
Azam Mashhadi

Part IV   Students' and Teachers' Perceptions of the Nature of Science

Introduction   267
Geoff Welford

21   Students' Understanding of the Nature of Science   269
John Leach

22   Probing Teachers' Views of the Nature of Science: How Should
We Do It and Where Should We Be Looking?   283
Mick Nott and Jerry Wellington

Part V   Social Interactions in Science Classrooms

Introduction   295
Geoff Welford

23   Gender Effects in Science Classrooms   297
Elizabeth Whitelegg

24   Geology: A Science, a Teacher, or a Course?: How Students
Construct the Image of Geological Disciplines and That of
Their Teachers   312
Alfredo Bezzi

25   Social Interactions and Personal Meaning Making in Secondary
Science Classrooms   325
Phil Scott

26   Primary Teachers' Cosmologies: The Case of the 'Universe'   337
Vassiliki Spiliotopoulou and George Ioannidis

27   The Perceived Value of Classroom Research as an Element of an
Initial Teacher Education Course for Science Teachers   351
Bob Campbell and Judith Ramsden

Part VI   Language and Imagery

Introduction   359
Geoff Welford

28   A Comparison of Some Linguistic Variables in
Fifteen Science Texts   361
Gottfried Merzyn

29   Imagery of Information: Prerequisites and Consequences   370
Elke Sumfleth, Hans-Dieter Körner and Andrea Gnoyke

Contents

Part VII Science Education Research in Europe

Introduction   381
Geoff Welford

30 Science Education Research in Europe and Scientific Culture:
   What Can Be Done?   383
   Joan Solomon

31 First Experiences with European PhD Summer Schools on
   Research in Science Education   394
   Piet Lijnse

32 Science Education Research in Europe: Some Reflections for the
   Future Association   399
   Svein Sjøberg

List of Contributors   405

Index   409

Editorial

This volume offers a selection of papers presented at the first European Conference on Research in Science Education held at the University of Leeds April 7–11, 1995. The conference was organized jointly by the University of Leeds and King's College, University of London and sponsored by the British Council, Leeds City Council, the Nuffield Foundation, Taylor and Francis (Publishers), the Wellcome Centre for Medical Science and the Association for Science Education. Over five busy days more than 160 participants were treated to 126 presentations through papers, poster sessions and symposia. This conference provided the opportunity for the first meeting of the embryonic European Science Education Research Association. In two specially convened sessions the Constitution was developed, amended and approved and the first officers of the Association, including the President, Professor Dimitris Psillos, University of Thessaloniki, and secretary, Dr Philip Adey, King's College London, were elected. In addition to the working programme of the conference, three receptions were provided by Taylor and Francis, the British Council and Leeds City Council and other social occasions provided a valuable opportunity to meet colleagues and discuss current issues in science education.

The idea for such a conference started life in Utrecht when a small group of science educators from around Europe took their 'Wouldn't it be nice if . . .' musings that one stage further. They initiated the process of turning into reality their dreams of establishing a forum for the exchange of outcomes, strategies and ideas for science education research in Europe. An International Advisory Group, comprising Dr Sylvia Caravita, Professor John Gilbert, Professor Wynne Harlen, Professor Piet Lijnse, Professor Arthur Lucas, Dr Svein Sjøberg and Dr Andrée Tiberghien, was formed, soon to be followed by a Local Planning Committee — consisting of colleagues from King's College and Leeds University and led by Professor Rosalind Driver.

Fortunately, early in the planning process, Falmer Press (Taylor and Francis) agreed to publish the proceedings. Whilst assuring a professional standard of publication and the opportunity to reach a very wide audience, a limit of 150,000 words inevitably curtailed the number of papers that could be published. Consequently, editing this collection of papers from among the eighty-nine received has been a challenging experience. The pleasure of receiving so many papers for consideration was masked by the realization that only about thirty papers, less than a quarter of all those presented at the Conference, could be published.

All of the papers published in this selection have been refereed. The referees were drawn from the international advisory group and colleagues on the planning committee. One of the immediate problems we were aware of was that many papers were limited and constrained by the requirement to be written in English. To compensate, referees were asked, first and foremost, to pay particular attention to the quality of the research and the ideas expressed in the paper when evaluating its merits. As editors, we have accepted the responsibility for attempting to enhance, where necessary, the quality of the English and the clarity of expression.

Our principal concern has been to select a range of papers of the highest quality. Inevitably this has been balanced with a need to select a range of papers that are representative of the issues and themes of the conference and to achieve a degree of geographical representation. In passing it is interesting to note that, of the final collection of papers, nearly half have been drawn from those presented as posters.

The papers in this volume have been arranged in a set of major themes which were an emergent property of the conference. These are 'Teaching and Learning', 'Developing and Understanding Models in Science Education', 'Approaches to Science Instruction', 'Students' and Teachers' Perceptions of the Nature of Science', 'Social Interactions in Science Classrooms', 'Language and Imagery' and 'Science Education Research in Europe'. As with all categorizations, there is room for argument as to the particular placing of one paper or another, but these areas have provided us with a way of highlighting the strengths of science education research currently obtaining in Europe.

Presenters at symposia were urged to submit one joint paper, but tight submission deadlines precluded, in some cases, collaborative redrafting between colleagues in different countries and individual papers were offered. The application by the referees and editors of the criterion of quality has meant therefore that, in some cases, several papers from one symposia have been included and in other cases none. This has provided a source of imbalance in the selection.

In general, however, we feel that the selection of papers is broadly representative of the conference with two exceptions. Firstly, a large number of papers were presented in the area of children's understanding of scientific concepts, yet only a few were rated highly by the referees. Perhaps the message writ large is that the pursuit of more of the same is now unacceptable as it has little to add to the large volume of research which is well documented in several books (Driver, Guesne, E. and Tiberghien, 1985; Driver, Squires, Rushworth and Wood-Robinson, 1994; Osborne and Freyberg, 1985) and for which there is a very extensive bibliography (Pfundt and Duit, 1994). Hence researchers seeking to conduct further work in this area must offer imaginative, creative and novel approaches that will generate fresh insight into our understanding of the growth of scientific knowledge in the child. Secondly, papers on assessment in science, policy determination and implementation, and on the procedural aspects of science were poorly represented, both at the Conference and consequently in this collation of papers.

Our major lament is that there is an unfortunate preponderance of contributors from the UK. In part, this is a natural reflection of the geographical location of the conference, the large number of people attending from the UK and the high proportion of them who submitted their papers for this publication. One would hope that the next conference to be held in Italy will draw a more representative cross-section of participants from Europe. Other possible explanations may lie in the fact that in some countries, research in science education itself is a relatively new field, whereas in the UK and some other countries, it has a more established history. In addition, being a new organization and the first conference, lines of communication about the conference were often reliant on informal and personal links and many individuals heard about the conference too late to submit abstracts or even attend as the number of available places was constrained by limited accommodation and the conference was well over-subscribed. Hopefully, the new association will remedy the problems in the future and seek to encourage the establishment and recognition of the work in the field encompassed within science education.

More importantly, we hope that these proceedings, the Association and this conference will help to foster broader understandings of each others' work throughout Europe. Running through many of the presentations and much of the discussion was a clear difference in perspectives. As Sjøberg, for example, shows in his chapter in this volume, even the title science education is problematic in terms of how it is interpreted by researchers from different countries. However, such richness and diversity must be valued for it is important that no one vocabulary becomes discursively hegemonic. We all have much to learn from our different cultures, traditions and approaches to science education. We hope that readers will find the papers in this volume a useful contribution to our knowledge and understanding of science education research.

References

DRIVER, R., GUESNE, E. and TIBERGHIEN, A.E. (Ed) (1985) *Children's Ideas in Science*, Milton Keynes, Open University Press.

DRIVER, R., SQUIRES, A., RUSHWORTH, P. and WOOD-ROBINSON, V. (1994) *Making Sense of Secondary Science*, London, Routledge.

OSBORNE, R. and FREYBERG, P. (Ed) (1985) *Learning in Science*, London, Heinemann.

PFUNDT, H. and DUIT, R. (1994) *Bibliography: Students' Alternative Frameworks and Science Education* (4th ed.), Universität Kiel, IPN.

Part I
Teaching and Learning

Introduction
Phil Scott

The papers on teaching and learning in science classrooms point to the ways in which researchers in Europe are working to build on insights to children's thinking about natural phenomena, in developing science teaching approaches which can be (and have been) used in 'real' classrooms. Three of the papers presented here focus on the use of analogies in teaching science concepts; two others consider how particular symbolic systems can be used to introduce children to some of the basic concepts of science. Opportunities for children to *talk* about their ideas concerning particular concepts or issues are prominent in all of the teaching sequences; two papers explicitly investigate the nature of talk, around computer-based problem solving exercises on the one hand and group concept mapping on the other. The content domains addressed through the teaching sequences include areas of fundamental importance such as circuit electricity, energy and nature of matter; this pure science focus is extended with two papers which consider teaching and learning about science-related social issues. Methodologically the approach taken in virtually all of these papers is to systematically and carefully lay out for the reader the approach taken in the teaching sequence and to make explicit the assumptions and rationale guiding its development. Koos Kortland goes one step further in this process in setting out 'teaching hypotheses' for his sequence; these are hypotheses which attempts to anticipate the ways in which children will interact with, and respond to, the developed teaching activities. Finally it is interesting to note that the analytical approaches to curriculum development, which characterize the planning and teaching described in these papers, are being researched across the age range, in primary as well as high school science teaching.

1 Developing New Teaching Sequences in Science: The Example of 'Gases and Their Properties'

Björn Andersson and Frank Bach

Abstract

This paper is about developing new teaching sequences in science. It is concerned with integrating knowledge from various sources into a series of lessons, which are then tested. Of particular interest to us is the students' long term conceptual retention. The 'pedagogical content knowledge' (Shulman, 1987) created during this research into what happens in the classroom and what lasting learning takes place may contribute to strengthening the research base for teacher training, at present almost exclusively based on 'well-tried experience'. We regard such knowledge, reported in writing, as important for the cumulative development of science teaching. In the paper we first describe how we created and tested a teaching sequence about gases and their properties for 13-year-old students at compulsory school (grade 7), before discussing how our work can be improved.

Point of Departure and Problem

In 1992 a national evaluation of Swedish compulsory school students' knowledge and understanding of science was carried out in grade 9 (15 years of age). One of the areas investigated was gases and their properties, which was probed by five tasks. Two of these were as follows. (Percentage of students for each alternative in parenthesis, n = 3103, random national sample. Gases and their properties are dealt with in grade 7.)

> John draws air into a plastic syringe and seals the opening with a rubber stopper as shown in the picture. No air can now get into or out of the syringe. The distance from the bottom of the syringe to the piston is 10 cm (see picture). John holds the rubber stopper against a wall (see picture) and tries to push the piston into the syringe. What will happen? Mark one alternative!

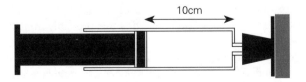

- It is not possible to push the piston [42%]
- It is possible to push the piston a millimetre or so [33%]
- It is possible to push the piston a centimetre or so [15%]
- It is possible to push the piston several centimetres [4%]
- It is possible to push the piston all the way to the bottom of the syringe [4%]

A can of paint is standing on a shelf. If the lid is taken off, you can smell the paint after a while. Which of the following alternatives is the best description of what happens?

- Molecules from the paint spread in all directions from the can. When they get into your nose you can smell paint. [16%]
- A smell spreads in all directions from the paint, but no molecules leave the can. Your nose can notice the smell. [12%]
- Fumes spread in all directions from the paint, but no molecules leave the can. Your nose can notice the smell. [61%]
- Molecules from the paint spread in all directions from the can. A smell flows out of the molecules. When the molecules are near your nose, you can notice this smell. [9%]

The performance on these tasks (and of the other three) were judged by us to be unsatisfactory. This naturally led to the formulation of a problem: How might we improve the teaching about gases in compulsory school?

Why Teach about Gases?

The question 'Why?' is important in the Swedish didactic tradition. Teaching time is expensive and one must be able to justify why this time should be spent on particular content. Also, answers to the question may influence the teaching process in various ways.

A knowledge of gases is necessary if we are to understand important processes in the world around us. An example is combustion in more than 400 million cars and innumerable fireplaces in the world, implying that oxygen is taken up from, and carbon dioxide and water given off to, the atmosphere. The amount of carbon dioxide generated and spread out in the atmosphere can influence the climate on Earth.

The Swedish hydrologist Malin Falkenmark deplores the fact that man is 'water blind', — cannot see water in its gaseous state. She means that if we could, we might economize better on water in areas of shortage. You can generalize this and observe that we are also 'gas blind', and that this probably influences our world view and our actions. If, for instance, we could see sulphur dioxide and nitrogen oxides being generated by point sources and spread out in the environment, we might more easily understand that these substances can influence groundwater and organisms and therefore act more cautiously. If we could see gaseous products of combustion, we might not so readily believe that it is possible to get rid of unwanted matter by burning it. Our gas blindness certainly has advantages. But the disadvantages have to be compensated by school helping students to develop science concepts which are sufficiently adequate to address such issues.

These answers to the question 'Why?' have led us to deal with the spreading of gases in our experimental sequence as well as the conservation of matter in various transformations (changes in volume and temperature of a gas in closed systems and emission of gas from open ones).

The gas concept is also of importance within the area of science, including the development of an appreciation of the nature of science. It offers, among other things, an opportunity to let the pupils have a go at reasoning with models, i.e., the interplay between a multiparticle model and the macroscopic properties of gases. The model can be developed to include the liquid and solid state, as well as changes of state. It can also be seen as a preparation for the more complicated atomic theory of chemistry and molecular interpretations of biological phenomena, e.g., osmosis.

A Constructivist View of Knowing and Learning

Our general view of knowing and learning is a constructivist one. Valuable in this connection are the many studies of students' everyday conceptions of science phenomena. These demonstrate time and time again that the pupils have difficulty in understanding what the teacher considers to be easy or assumes that the pupil already understands. This type of knowledge, as well as the many details of the pupils' answers, stimulates new attempts at teaching. We have used results reported by Brook, Briggs, and Driver (1984), Nussbaum (1985), Séré (1982, 1985, 1986, 1990) and Stavy (1988). These studies have influenced the way we introduce the concept of air, how we sequence the first five lessons and how we formulate the particle model. They have also led to the construction of many new problems for the pupils to solve.

Our constructivist model of the knower-known relation gives a certain direction to our thinking. Information about the initial state of the student becomes important to the teacher, and the model leads away from transmission of knowledge and towards helping the student in her attempts to develop, from her initial state, new structures in accordance with the aims of the science curriculum. However, the model is not a theory of teaching but a way of looking upon knowing and learning. Therefore, it does not imply a certain teaching method. The individual may construct and learn under a variety of circumstances — lectures or individual discussions with the teacher, a test on homework or a brain-storming session, solving a problem or reading some text, alone in introspective thinking or participating in a group discussion, dissatisfied with existing conceptions or just curious about something new that is presented. Our approach towards teaching methods is therefore a pragmatic one. We try to be as well informed as possible about various methods that have been tried, e.g., analogical reasoning (Brown and Clement, 1989), cognitive conflict (Stavy and Berkovitz, 1980) and the sequence 'eliciting students' prior ideas', 'providing restructuring experiences', 'providing opportunities to apply new ideas' and 'reviewing any changes in ideas' (CLIS, 1987a). But we do not think one particular method has to be used all the time. While awaiting unambiguous empirical evidence, we choose methods for a given situation on the basis of experience and intuition. It must be said, though, that knowledge of students' conceptions and difficulties in understanding, described in the literature, constantly informs the teacher's decisions in class in responding to what the students express and do.

Two Teaching Programmes about Matter and Molecules

We have, with interest, studied two attempts to teach not only about gases but also about the liquid and solid states. These are 'Matter and molecules' or shorter MAM (Berkheimer, Andersson and Blakeslee, 1988; Berkheimer, Andersson, Lee and Blakeslee, 1988) and 'Approaches to teaching the particulate theory of matter' (CLIS, 1987b). The student's work according to MAM is directed by a text that is read individually. Certain experiments are also included. The text takes students' alternative conceptions into consideration in various ways. e.g., by contrasting them with the conceptions of school science, but the general tone is one of transmission of knowledge. MAM has stimulated us to write a very different students' text.

An element of the CLIS programme is a consistent attempt to build on the students' preconceptions of matter and its structure. It becomes apparent that many students aged 13–14 have some sort of particulate conception of matter. By the students presenting various conceptions of this kind in class and having them questioned by classmates and the teacher, the programme creates conditions for conceptual development. We find this method attractive but have nevertheless chosen another approach, namely to introduce a scientific particle model without eliciting students' alternatives. On the one hand, we believe that, to the extent that there are alternatives, it is not a question of robust and tenacious ideas but rather guesses, which the students are prepared to change. On the other hand, elicitation and discussion of the students' alternative conceptions takes time. We prefer to spend this on arranging opportunities to discover the explanatory power of the scientific model. The teacher is well informed about the students' conceptual difficulties, for instance, the belief that matter is both continuous and static, and gives the students opportunities to revise these by, e.g., solving problems in small groups.

Unlike both MAM and CLIS we have limited ourselves to gases only, and a particle model for those, in order to make the introduction to particle thinking somewhat less complicated.

On Lasting Conceptual Retention

A central educational question is to what extent lasting conceptual retention is achieved. Without this, the student does not have any tools which he can use, e.g., when he tries later in life to understand scientific information communicated via the media. Neither MAM nor CLIS has really tried to tackle this crucial question, although some progress has been made. The MAM group has carried out pre- and post-tests and demonstrated significant gains in the students' conceptions of matter (Lee, Eichinger, Andersson, Berkheimer and Blakeslee, 1989). There is, however, no information about whether the post-test was given immediately after the teaching or considerably later. Nor is there any information about results of individual tasks, which makes it impossible to judge if the students can use their knowledge in new contexts (most of the tasks are very close to the specific content of the programme, but a few are new to the students.)

CLIS has carried out a diagnostic pre-test, which was also given as a post-test (Johnston and Driver, 1991). Here again, there is no information about when the post-test was given. Some good progress is observed, but the results are only reported in an appendix without comment, which suggests that the project group was not particularly interested in the problem of long-term conceptual retention.

A Teaching Sequence about Gases

From the knowledge-base described above we have developed a teaching sequence in the form of a teacher's guide, including a collection of problems. Some of these are intended for work in small groups. Each group is expected to produce a common answer to a given problem. The rest of the problems are used in various ways, e.g., in homework. A student's text with the title 'Ludvig, Lisa and the air' is also part of the sequence. During the introductory lessons, air is studied macroscopically. Then a particle model is introduced making it possible to explain earlier observations and new ones. During the remaining lessons, the pupils try to solve various problems and explain experimental results by means of their new model. Also, gases other than air are studied. In summary, the aims of the teaching sequence are that the pupils should develop an understanding of macroscopic properties of gases and be prepared to use a qualitative particle model as an explanatory tool.

Lesson 1

Various research studies demonstrate that it is not self-evident to the students that air exists and occupies space. The aim of the lesson is to let the pupils experience this by practical work of their own. To begin with, each group of two students gets a transparent vial with a tight-fitting lid of soft plastic, in which a small hole has been punched. They also get a small funnel, a plastic tube which can be squeezed through the hole, a glass tank with water and some beakers. The task is to fill the vial with water without taking off the lid. (Just immersing it in water will not work.) If this succeeds, the next task is to empty the vial with the lid still on. When this has been done, the pupils are asked to find out what happens if you hold an upside down beaker above a floating cork and press the beaker under the water surface. Where in the beaker is the cork? Further, the students are asked to fill a beaker with water, immerse it in the tank and turn it upside down. A second beaker is turned upside down, immersed in the water, held right below the first beaker and, finally, turned upright. What happens? One more experiment is to put a pea in the mouth of a bottle and try to blow it into the bottle. For all experiments, the pupils are asked to write down what they have done and what they have observed and to try to explain the results obtained.

Lesson 2

The experiments from lesson one are discussed and the concept of air is introduced. What is in the vial and has to get out in order for the water to get in, what makes the cork remain in the opening of the beaker, what is transferred between the two beakers under water, and what prevents the pea from going into the flask is *air*. The lesson continues with the question 'Where is there air?' Is there air in an empty bottle with or without a cap, in a flat tyre, furthest down in the far corner of the room, at the South Pole? The concept of *atmosphere* is introduced. During the remaining part of the lesson, the students are asked to predict what will happen if a flask with a balloon stretched over the neck is heated, firstly, from below with the flask upright, and then from above with the flask upside down. The experiments are carried out and discussed. The students get a section of the text 'Ludvig, Lisa and the air' as homework. In this section the two young people are thinking about how to take home some air from the seaside. In other words, it is a question of how to solve the problem of delimiting a sample of air, which is a precondition for studying its properties. Preliminary research findings demonstrate that some students do not consider this as possible, because air is 'all one thing, all one mass'.

Lesson 3

The students ponder individually over whether it is possible to push the piston into and pull it out of a sealed syringe and, if so, how far (compare the national evaluation task in the introduction). They also think about what will happen if the syringe is heated in boiling water and cooled in ethyl alcohol at -50°C. They write down their thoughts, carry out the experiments in pairs, and discuss the results in groups of four.

Lesson 4

Weighing of air. Among other things, the students discuss whether an inflated football weighs more, the same or less than an uninflated one. A flask with a volume of half a litre is weighed before and after evacuation.

Lesson 5

What the class knows about air is now summarized. The number of observations to think about is increasing, and many of the pupils want to understand e.g., why is it possible to compress air? In order to manage all this, a theory is introduced. The teacher presents the following one for the class to test:

- Air consists of many, many small particles called molecules. Between the molecules there is nothing.

- The molecules are matter. They have mass and weight although they are very small.
- One litre of air consists of millions of millions of molecules.
- Each molecule of air moves at a high speed in a straight line until it collides with a molecule of another substance (e.g., in the walls of a flask) or with another molecule of air. Then it changes direction and speed. Thus the speed varies but is on average high (500m/s).
- The molecules which together make up an amount of air (e.g., the air in a flask) move in all directions and independently of each other.
- On average, the molecules are quite far from one another.
- If air is heated, the speed of the molecules increases. If it is cooled, the speed of the molecules decreases.

The theory is used in a few examples and is then given as homework.

Lessons 6–11
The theory is used. The students spend two lessons solving ten given problems in groups of four, producing a common set of written answers to each problem. Five of these deal with the spreading of gases. Air pressure is treated, amongst other ways, by demonstrating and discussing the classical experiment with the Magdeburg spheres. Physical properties of other gases than air are studied (are e.g., hydrogen, nitrogen, carbon dioxide compressible?) Technical applications are taken up, e.g., the suction pad. One lesson is devoted to a written test.

The Collection of Problems

Two examples of the problems worked on in groups of four are:

- Beaker A contains a solvent that is easily transformed into a gaseous state. You remove the lid of the beaker. After a while Karl notices the smell of the solvent (the air is draught-free). How is it possible for his nose to notice the smell of the solvent? There's some distance between the beaker and his nose, after all. (Picture of the beaker and Karl.)
- Trees and other plants give off oxygen. But this oxygen does not remain close to the plants but is mixed uniformly in the atmosphere. Explain this.

Students' Text

Since the contents of the sequence described above are, to a great extent, not dealt with in the teaching material traditionally used (e.g., the use of a model to explain and predict), we have written a text for the students. This provides, amongst other things, an opportunity to revise and to encounter the contents of the lesson in a new way. We have tried to make the text personal by letting two young people, Ludvig and Lisa, be the main characters. They discuss air and its properties as part of their daily lives and also carry out experiments now and then. Sometimes their teacher,

Carolina, takes part in the action by providing new decisive ideas to develop further. The text is 'constructivist' in the sense that everyday conceptions are brought out and discussed among the young people. Some non-scientific elements arise, including the relationship between Ludvig and Lisa. We have tried to avoid being 'educational' in a narrow sense. For instance, the reader is not asked any questions but simply introduced to what is going on and, hopefully, stimulated to think and reflect.

The text is linked up with current international debate in two ways. Firstly, Fensham (1994) has proposed that one main task for research and development in the future is to write and test constructivist textbooks. Our text is a contribution, if small, to this work. Furthermore, Lemke (1990) and Sutton (1992) have pointed out that science texts are written in a very impersonal and spartan way. Insofar as this is also true for school science textbooks, it may have a repellent effect on teenagers encountering science for the first time. As mentioned above, we made our text personal.

Results of Pre- and Post-tests

The teaching sequence has been tried out in one class. One week before the start a pre-test with eight written problems, mostly of the 'open answer' type, was given. The same test was given six months after teaching was completed. This design does not control possible learning effects of the pre-test. You can also expect some pre-test-teaching interaction. The long time lapse between the end of teaching and the post-test should, however, ensure that any superficially learned details are forgotten. We shall return to the problem of design in the concluding discussion, but first some results will be reported. There were twenty-two students who did both the pre-test and the post-test.

Problem 1
A boy is supposed to pour juice from a large container into smaller bottles. But the funnel won't keep still. So he makes it stick with plasticine, which seals the space between the funnel and the bottle. Then the juice won't run into the bottle. Why? Before the lesson, eight students answered in an acceptable fashion, e.g., that the bottle was full of air that had to come out before the juice could come in. Afterwards twenty students gave this explanation.

Problem 2
The same problem as the one presented right at the beginning of this article, testing whether the students know that the air in a syringe can be compressed. We regard 'a couple of centimetres' and 'several centimetres' as acceptable answers. Four students answered in this way beforehand and sixteen afterwards. We have also asked for explanations. None of the students gave particle explanations in the pre-test. Of the sixteen pupils who tick 'a couple' or 'several' centimetres on the post-test there are four who simply say that air is compressible. The remaining twelve give explanations to do with particles. Of these half are unclear. The students say, e.g., that 'it's possible to press the molecules together a little'. Do they

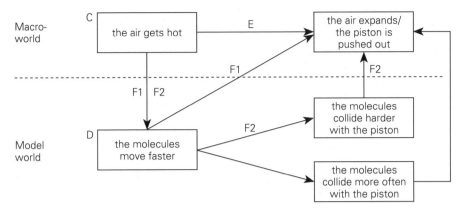

Figure 1.1: Schematic description of statements and links in the students' answers to problem 4

mean that it is the molecules themselves that can be pressed together, or are they referring to a multi-particle system with spaces between the particles? The rest of the arguments are acceptable and clear: e.g., 'There is a vacuum between the molecules so that when the piston is pushed in, you 'push' the molecules together'.

Problem 3

Here it is a matter of answering whether the piston in the previous problem can be pulled out and, if so, how much. Nine students in the pre-test and nineteen in the post-test believe that it's possible, either several centimetres or all the way out. No students give particle explanations before, but twelve do so afterwards. All the particle explanations have shortcomings. The model may lead to false conclusions, e.g., 'No, [it's not possible to pull out the piston] because the molecules press from the outside (where there are more).' Another example: 'No (you can do it a little, then the piston is pushed back again). The pressure from the molecules on the outside is as great as inside the piston. That's why it's impossible to pull out the piston.' In other cases, the students focus only on the air inside the piston: 'Yes. It's possible to pull the molecules apart a little bit.'

Problem 4

This concerns heating the air in a syringe. Does anything happen then and, if so, what? The answer has to be explained. We describe the result in somewhat more detail to demonstrate the techniques we use when attempting to treat complex answers with elements of both the macro- and the model world.

It is possible to give an answer only on the macro-level, e.g., that when the air in the syringe is heated, it expands and then the piston is pushed out. A good answer making use of a particle model is that when the air is heated, the molecules increase speed. Then they collide more often and harder with the piston, which is pushed out. Schematically, the various answers from the students can be presented as in Figure 1.1. The answers contain in varying degrees the statements (the text in the squares) and the links (arrows) shown in the figure.

Table 1.1:   *Distribution of categories before and after teaching (problem 4)*

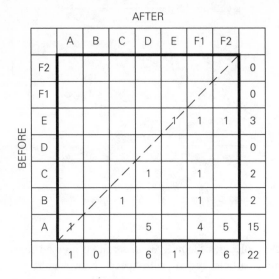

AFTER

	A	B	C	D	E	F1	F2	
F2							/	0
F1						/		0
E					1	1	1	3
D				/				0
C			/	1		1		2
B		/	1			1		2
A	✗			5		4	5	15
	1	0		6	1	7	6	22

BEFORE

Categories

A. No or irrelevant answer. B. Nothing happens. C. The air gets hot.

D. Only particle description: the molecules move faster.

— 'No. The molecules only move faster, but nothing happens.'

E. Air expands (when heated)/the piston goes out.

— 'The piston goes outwards. Air expands when it gets hot.'

F1. The movement of the molecules increases, the piston goes out.

— 'The piston goes outwards. The molecules move faster when it gets hot.'

F2. The movement of the molecules increases, more powerful molecular collisions/ increased pressure, the piston goes out.

— ... 'the heat makes the 'm' move faster and faster and collide with the piston at such a speed that it goes out.'

Thirteen students are in category F after teaching, but seven of these give vague answers. They do not specify any mechanism for the effect on the piston.

Problem 5

This is analogous with 4 but concerns cooling. Here there are nine students who explain in the post-test (none did in the pre-test) that the piston goes in because the molecules move more slowly. Three say only this, three others also point out that the molecules collide less strongly with the piston, and the remaining three add that the piston goes in because the pressure outside is now greater.

Problem 6

This is a matter of explaining why a suction cup can stay fixed to a smooth wall. There are seven students who give a particle explanation in the post-test (none in the pre-test), e.g., 'More molecules exert pressure from outside than from inside.'

Problem 7

This problem considers whether you can notice the smell of methylated spirit in a draught-free room if there is a distance of several metres to the flask. The answer has to be explained. There are twelve students who give a molecular answer in the post-test (none in the pre-test). They express in various ways that smell consists of molecules that spread.

Problem 8

This is about four students discussing helium, which is in a container. They each put forward an alternative suggestion, which is illustrated in a drawing. The students are to state which they think is best.

A. Gas is a single thing, a single mass, so that there aren't any atoms in the container. [five students before, none after]

B. There are helium atoms in the container with helium gas in between. [eight students before, six after]

C. There are helium atoms in the container. They are very close together. (the picture shows that the atoms have contact) [five students before, one after]

D. There are atoms in the container moving around each other. There is a vacuum between them. (The picture shows the atoms sparsely spread out) [No student before, fifteen after]

Generally speaking, we can establish that for problems 2 to 7, eighteen of twenty-two students use particle thinking at least once, twelve do this at least three times, and five at least five times. In view of the fact that six months have lapsed since the course, and that the post-test was entirely unprepared, we consider these results to be positive — the students venture to use particle thinking, and this in ways that in many cases are not bad at all. One student explains, for instance, that the piston in the syringe goes in on cooling by 'The piston is pressed together because when it's cold the molecules move slowly and the pressure from outside is then greater than the pressure inside the container.' Note, however, that the range in the quality of the answers may be considerable. Another answer to the question about cooling reads: 'All the molecules become stuck like stones. It gets so cold that they crumple up.' Other differences in the quality have been evident in the earlier descriptions of how each of the problems was answered.

What Did the Students Think about the New Sequence?

After this course was completed the students were given the opportunity to say what they thought of it compared to other 'normal' teaching in scientific subjects. Typical comments are that the experiments are fun, that you have to think a lot, that the text 'Ludvig, Lisa and air' is easier to read and understand than the normal physics book, and that it is more difficult than usual to know what you ought to learn. As far as having to think a lot is concerned, some students find this fun, while others do not appreciate it at all.

Discussion

Design Issues

The results obtained have to be assessed according to the experimental design we have used. The students were given the same problems to solve both in the pre- and post-test, and the contents of these problems have in one way or another been dealt within the teaching, often in situations similar to those described in the test problems. We therefore see the improvement in the answers as an effect of pre-test plus teaching. We also note that one consequence of our design is that no problems test whether the students are able to use their particle model in a situation that is new to them.

How then may we improve our design? A control group is desirable, and it is conceivable that comparable classes at the same school, which are being taught about gases in the conventional manner, could act as such. In practice, this could be difficult to achieve for social reasons. Also, the time factor is not controlled — conventional teaching of the corresponding area takes about half the time. We therefore imagine a design as follows:

G1 T1 X (6 months) T2
G2 T2 X (6 months) T1

The students in a class are randomly assigned to the groups G1 and G2. G1 are given the pre-test T1, and G2 are given T2 on the same occasion. T2 contain quite different problems from those in T1, but both concern gases. Both groups are taught together, since they make up one class (X means teaching about gases). At the post-test six months later T2 is given to G1 and T1 to G2. G1's post-test results are compared with G2's pre-test results, and vice versa. In other words, the groups are each other's controls. If the experiment is carried out in more classes in accordance with this design, larger groups for comparison can be obtained. If the contexts of a given problem are not dealt with in teaching, they can be used as a test of the students' ability to apply the particle model in new situations. The design presented has a number of advantages. It does not interrupt school activities. It allows a conclusion to be drawn that an observed improvement is a result of the teaching.

There is, however, one question that the improved design does not answer. What aspects of the teaching were particularly important, and which were less important, with reference to achieving the observed result? What effect did the sequencing, problem-solving in groups, the new text, the systematic elicitation of the students' conceptions actually have? If one wishes to tackle this problem, one needs to compare two teaching sequences that differ only in respect to one of the above factors. This type of study is technically and practically demanding, but nevertheless desirable. For example, does encouraging the students to talk about their everyday concepts have a positive effect on long-term retention? Such a study would entail treating the experimental and control groups equally in a number of respects: the same sequencing, the same students' experiments, the same problems, the same student text, teaching time of the same length, etc. But in one group the students are never asked to describe or discuss their everyday concepts, in the other this is done on a regular basis. A study of this type can provide results implying that beliefs, hopes and interesting hypotheses are replaced with knowing, obtained by scientific methods.

We regard the types of study discussed above as a desirable supplement to the process studies carried out in recent years, e.g., in which the 'learning pathways' of individual students have been described and analysed (see for example Scott, 1992).

Meta-understanding: What Is Science?

The syllabuses for science subjects in the Swedish compulsory school state that students should obtain an understanding that concepts and models are human constructions that change over time. This makes new demands on teaching, and therefore it is welcome that students' understanding of the 'nature of science' has been investigated in recent years (Carey, Evans, Honda, Jay and Unger, 1989; Driver, Leach, Millar and Scott, 1993). A descriptive instrument with three levels has been developed from both of these studies. At level 1 the student does not distinguish clearly between theory and evidence and between explanation and description. Knowledge is seen as a copy of reality and science as something that has to do with discovering things. At level 2 the student distinguishes between theory and evidence, but theories arise through doing experiments and making observations. Science is a matter of finding out how nature works, e.g., by varying the conditions in experiments. Knowledge is 'out there' waiting to be discovered. At level 3 theories are seen as human constructions, developed through interaction with observations. Levels 1 and 2 are common when students begin science at the age of 12–13. Level 3 is rare at the upper level of the compulsory school. (Those who know Piaget's work will observe that levels 1 and 2 are reminiscent of his general descriptions of the concrete operational stage and level 3 of the formal stage.)

Teaching about gases offers opportunities to discuss the nature of science. According to what was just said, most students in grade 7 are at level 1 or 2. They imagine that science is a question of discovering things and finding out how the world around them functions. The teaching offers elements in line with this e.g., the discoveries that gases can be compressed and that they expand on heating. But what status do the air molecules have? You cannot see them, not even with a microscope, so the students cannot make any discovery here. Perhaps the teacher or some scientists have seen the molecules? If no one has seen them, the gas model must from the start be a 'human construction' and not anything that has entered one's head as a result of careful observation. But the model can be used to predict and explain. Does this mean that the molecules do exist 'out there' and behave in the way the model claims? Do the *students* think that the properties of gases are more easily understood with the help of the model? Does this, in its turn, lead to *their* being more convinced of the existence of molecules? It remains to be seen whether students in grade 7 think that it is fun to philosophize about these and other questions.

References

BERKHEIMER, G.D., ANDERSSON, C.W. and BLAKESLEE, T.D. (1988) *Matter and Molecules Teacher's Guide: Activity Book* (Occasional paper No. 122), East Lansing, Michigan State University, Institute for Research on Teaching.

BERKHEIMER, G.D., ANDERSSON, C.W., LEE, O. and BLAKESLEE, T.D. (1988) *Matter and Molecules Teacher's Guide: Science Book* (Occasional paper No. 121), East Lansing, Michigan State University, Institute for Research on Teaching.

BROOK, A., BRIGGS, H. and DRIVER, R. (1984) *Aspects of Secondary Students' Understanding of the Particulate Nature of Matter*, Leeds, Children's Learning in Science project, Centre for Studies in Science and Mathematics Education, University of Leeds.

BROWN, D. and CLEMENT, J. (1989) 'Overcoming misconceptions via analogical reasoning: Abstract transfer versus explanatory model construction', *Instructional Science*, **18**, pp. 237–61.

CAREY, S., EVANS, R., HONDA, M., JAY, E. and UNGER, C. (1989) '"An experiment is when you try it and see if it works": A study of grade 7 students' understanding of scientific knowledge', *International Journal of Science Education*, **11**, pp. 514–29.

CLIS (1987a) *Approaches to Teaching Energy*, Leeds, Centre for Studies in Science and Mathematics Education, University of Leeds.

CLIS (1987b) *Approaches to Teaching the Particulate Theory of Matter*, Leeds, Centre for Studies in Science and Mathematics Education, University of Leeds.

DRIVER, R., LEACH J., MILLAR, R. and SCOTT, P. (1993) *Students' Understanding of the Nature of Science: Resumé and Summary of Findings*, University of Leeds, Children's Learning in Science Group and University of York, Science Education Group.

FENSHAM, P. (1994) 'Postscript', in FENSHAM, P., GUNSTONE, R. and WHITE, R. (Eds), *The Content of Science*, London, Falmer Press, pp. 263–4.

JOHNSTON, K. and DRIVER, R. (1991) *A Case Study of Teaching and Learning about Particle Theory*, University of Leeds, Children's Learning in Science Group.

LEE, O., EICHINGER, D., ANDERSSON, C.W., BERKHEIMER, G.D. and BLAKESLEE, T.D. (1989, April) 'Changing Middle School Students' Conceptions of Matter and Molecules', Paper presented at the annual meeting of the American Educational Research Association, San Francisco.

LEMKE, J.L. (1990) *Talking Science: Language, Learning and Values*, Norwood, NJ, Ablex Publishing Corporation.

NUSSBAUM, J. (1985) 'The particulate nature of matter in the gaseous phase', in DRIVER, R., GUESNE, E. and TIBERGHIEN, A. (Eds), *Children's Ideas in Science*, Milton Keynes, Open University Press, pp. 124–44.

SCHULMAN, L.S. (1987) 'Knowledge and teaching: Foundations of the new reform', *Harvard Educational Review*, **57**, 1, pp. 1–22.

SCOTT, P.H. (1992) 'Conceptual pathways in learning science: A case study of the development of one student's ideas relating to the structure of matter', in DUIT, R., GOLDBERG,

F. and NIEDDERER, H. (Eds), *Research in Physics Learning: Theoretical Issues and Empirical States*, Kiel, Germany.

SÉRÉ, M.G. (1982) 'A study of some frameworks used by pupils aged 11–13 years in the interpretation of pressure', *European Journal of Science Education*, **4**, pp. 299–309.

SÉRÉ, M.G. (1985) 'The gaseous state', in DRIVER, R., GUESNE, E. and TIBERGHIEN, A. (Eds), *Children's Ideas in Science*, Milton Keynes, Open University Press, pp. 105–23.

SÉRÉ, M.G. (1986) 'Children's conceptions of the gaseous state, prior to teaching', *European Journal of Science Education*, **8**, pp. 413–25.

SÉRÉ, M.G. (1990) 'Passing from one model to another: Which strategy?', in LIJNSE, P.L., LICHT, P., DE VOS, W. and WAARLO, A.J. (Eds), *Relating Macroscopic Phenomena to Microscopic Particles*, University of Utrecht, Centre for Studies in Science and Mathematics Education, pp. 50–66.

STAVY, R. (1988) 'Children's conception of gas', *International Journal of Science Education*, **10**, pp. 553–60.

STAVY, R. and BERKOVITZ, B. (1980) 'Cognitive conflict as a basis for teaching quantitative aspects of the concept of temperature', *Science Education*, **64**, pp. 679–92.

SUTTON, C. (1992) *Words, Science and Learning*, Buckingham, Open University Press.

2 Exploring the Use of Analogy in the Teaching of Heat, Temperature and Thermal Equilibrium

Michael Arnold and Robin Millar

Abstract

This paper describes the development and trial use of a water-flow analogy for a system in thermal equilibrium. The analogy is intended for use in a teaching sequence designed to help students in the early years of secondary schooling to understand the scientific model of thermal phenomena and, within it, to differentiate the concepts of heat and temperature. The study provides guidelines for the use of this particular analogy in teaching basic ideas of thermal equilibrium. It also throws light on learners' abilities to make use of analogies in general. Though the analogy itself was well understood by most students, spontaneous mapping of the analogy on to a system in thermal equilibrium was achieved by only a minority of students.

Introduction

One characteristic of scientific reasoning is the use of mental models to frame explanations of classes of phenomena. Whilst students are apt to use the surface features of their school science experiences in constructing explanations, scientists weave abstract concepts into internally consistent explanatory 'stories', using specialized language and imagery (Sutton, 1992). An important aim of science education is to enable students to appreciate, comprehend and use these scientific mental models in their own thinking, and to make sense of novel situations.

There is, however, a large body of research evidence which shows that students' existing knowledge, before teaching, differs markedly from the scientific view, and that these 'alternative frameworks' are remarkably resistant to change by teaching (Driver *et al.*, 1994). Science educators have attempted to deal with this problem by carefully grading and sequencing the conceptual demands of their lessons and by ensuring that they begin instruction 'where the child is at' by relating the experiences they provide to the child's existing knowledge. One problem with this approach is that what appears to an expert to be a logical conceptual progression may not be seen as such by the learner. Another problem is that some concepts only make sense as part of a whole 'story'. Teaching each concept in turn can, therefore, lack coherence in that the learner is not able to integrate the separate concepts into a meaningful and useful model. This is, in our view, the principal teaching and learning problem in the area of elementary ideas about thermal phenomena.

Teaching about Heat and Temperature

Research suggests that elementary thermodynamics presents problems of teaching and learning which are particularly acute. Students have great difficulty in coming to accept the scientific model (Erickson, 1979, 1985; Tiberghien, 1983, 1985). The terms 'heat' and 'temperature' are frequently not differentiated by students. Heat may be thought of as a substance which moves spontaneously into objects being heated, or out of those being cooled (Erickson, 1985, p. 60). Students are unlikely to consider the surroundings of an object as being of importance in its thermal history; nor do they believe that objects in the same thermal environment will necessarily have the same temperature. Instead, they often estimate the temperature of objects on the basis of their size, material or use (Tiberghien, 1985, p. 77). The idea of thermal equilibrium, when, as a result of heat transfer, two objects initially at different temperatures come to a common temperature, is, therefore, not well understood. When two samples of water at different temperatures are mixed, many students predict the final temperature by adding rather than averaging the initial temperatures (Strauss and Stavy, 1983), thus neglecting to treat temperature as an intensive quantity (Erickson, 1980). The Children's Learning in Science Project (Brook *et al.*, 1984) found that the proportion of 15-year-olds able to apply scientific ideas about heat transfer varied from 6 per cent to 15 per cent depending on the context, with many students using the constructs of both 'heat' and 'cold' in their explanations.

An in-depth interview study by Kesidou and Duit (1992) explored the ideas of students aged 15–16 in detail, corroborating many of the earlier findings. They describe students' difficulties in learning energy ideas, making use of a particle model and differentiating heat and temperature as 'severe', whilst noting that students' intuitive ideas lead to correct conclusions concerning the equalization of initial temperature differences and the irreversibility of thermodynamic processes. Finally, in a major study of the teaching of elementary thermodynamics, Linn and Songer (1991) evaluated a series of curricula, each a development of earlier versions, based on teaching a 'pragmatic model' of thermal phenomena centred around the idea of heat-flow. Alongside many positive learning outcomes, they again report the persistence amongst many students of a rather undifferentiated concept of heat/temperature.

The Scientific Model of Thermal Processes

The two pivotal ideas in the scientific 'story' about thermal events are heat and temperature. These, however, are inter-dependent, and their definition draws upon a third concept, that of thermal equilibrium. Thus the temperature of a body is that property which determines whether or not there will be a net flow of heat into it or out of it from the surroundings, and in which direction the heat will flow (Isaacs, Daintith and Martin, 1984, p. 688). Heat is energy being transferred from one body to another as a result of a temperature difference (Isaacs, Daintith and Martin, 1984, p. 317). The bigger the temperature difference between these two objects, the greater the rate of heat transfer. In time they both reach the same temperature and are then said to be in *thermal equilibrium* with each other. It is the teacher's task to communicate this complex mental model to students, though not, of course, in such abstract terms.

Teaching the Scientific Model

We would argue that the scientific 'story' must be accepted as a whole if it is to make sense to students. Presenting temperature as a measure of the 'degree of hotness' of an object, and heat as a form of energy which is present in hot objects and which may, under certain circumstances (usually not specified in detail), move from one place to another, as many science texts do, does not encourage understanding. (We would also note, even though it is not part of our argument here, that such accounts are not merely simplifications but include ideas which are 'wrong' from a physics perspective.) The commonest sequence is probably to begin instruction with the idea of temperature, often in primary school, linking the concept to the child's bodily sensations of hot and cold. The idea of heat is introduced later, as the 'quantity' responsible for causing high temperature, or change in temperature. Thermal equilibrium is often not explicitly discussed, though the concept may be assumed to be derived inductively by the student from experience of heat transfer situations.

It seems to us that the central ideas of heat, temperature and thermal equilibrium are so closely interdependent that any attempt to clarify one of them requires some appreciation of the others. Even understanding how a thermometer works involves the ideas of thermal equilibrium and heat transfer as well as temperature. Rather than beginning with activities designed to focus on either heat or temperature, we would argue that it is necessary to consider situations in which all the main elements of the scientific model are involved. At first sight, this approach appears doomed to failure, as there is no direct relationship between two of the ideas involved (heat and thermal equilibrium) and the student's sensory processes, and no obvious way of linking the model to the child's experience. We addressed this problem by using an accessible analogy to introduce the elements of the model and to display their inter-relationships. Then, rather than attempting simply to differentiate distinct concepts of heat and temperature from the child's composite heat/temperature conception, this differentiation is set within the progressive development of a coherent mental model of thermal processes. Through the use of this model the concepts of heat, temperature and thermal equilibrium can be applied and talked about together, to enable students to make personal sense of everyday and laboratory thermal events.

Using Analogy in Science Teaching

Analogies are widely used by scientists in explaining their work to their peers or to the general public, and by science teachers to clarify an area of knowledge for students. Recently, considerable attention has been paid by science education researchers to the ways in which analogies are, and might be, used to support teaching and learning (Duit, 1991; Glynn, 1991; Treagust *et al.*, 1992). An analogy can help understanding by abstracting the important ideas from the mass of new information, making clear the system boundaries and introducing the appropriate language in which to frame a scientific explanation. In some cases, the analogy provides a model problem solution which the learner may be able to 'look up' in future, and extend to solve other problems. For example, Gick and Holyoak (1980) investigated subject's attempts to solve Duncker's (1945) radiation problem: how to destroy a tumour with radiation when the intensity required to kill the tumour exceeded the safe dosage for surrounding healthy tissue. (The problem is also discussed in Kahney, 1986.) Subjects were given one of three analogous stories about an army attacking a fortress. The attackers could not approach the fortress in one group (because of its defences) and so split into smaller forces and re-assembled at the target. They found that very few subjects could use the analogy spontaneously to solve the radiation problem but, when given a hint, 80 per cent could then do so. Their particular solution was influenced by the story they had been given as the analogy. One difficulty students face, Gick and Holyoak report, is to decide the level of abstraction at which to match the two examples. The optimum is evidently that level at which similarities are maximized and differences minimized, but how subjects arrive at such distinctions is not clear.

Gentner and Gentner (1983) presented subjects with two analogies for electric current: the 'moving-crowd' and the 'flowing-fluid' models. As predicted, different aspects of electric circuit behaviour were best understood by the most apposite analogy. Black and Solomon (1987), in a similar investigation, concluded that taught analogies did not confer any great advantage on students in answering typical test questions, but that students were better able to interpret novel situations. In a related type of study in mechanics, Clement (1987) has used a series of 'bridging analogies' to overcome students' 'misconceptions'. His approach involves finding an 'anchoring intuition', which is a conception held by the student which approximates to the scientific view, and bridging from this base to the intended scientific end point with intermediate cases which can be shown to be equivalent to examples on either side of them in the sequence. Unfortunately for our purposes, no anchoring intuition in the domain of heat and temperature could be found (see Arnold and Millar, 1994), necessitating the teaching of a 'base system', which we term the water analogy, followed by the extension of this model into the required domain.

The Water Analogy of Thermal Equilibrium

To introduce children to a model for thinking about thermal phenomena, we provided direct first-hand experience of two practical contexts. The simple system

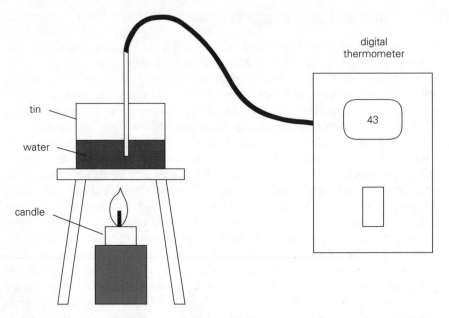

Figure 2.1: The water heating experiment

exemplifying the thermal model we wished the children to adopt and learn to use consisted of a metal tin containing a small quantity of water, which was heated by means of a candle (Figure 2.1). The water temperature was recorded every minute. In a short time the temperature stabilizes as the water in the can loses heat to its surroundings at the same rate as it gains heat from the candle flame.

The 'base system' we used to teach the inter-relationships of the three important concepts of the thermal model involved a logically equivalent analogy consisting of water running into and out of a glass container (Figure 2.2). By controlling the rates of inflowing and outflowing water, it is possible to arrange an infinite number of equilibrium positions in which the level of water in the container is constant. If the inflow exceeds the outflow, the water level in the container rises, and vice versa. Though the outcome of the teaching of the analogy is intended to be a dynamic mental model of the system (a more complex mental representation than that provided by a simple set of propositions), examination of the propositions underlying the two cases illustrate the close mapping between the water analogy and the heating activity when the temperature has equilibrated.

Water analogy propositions

1 Water is being supplied to a container by a tap.
2 The water is supplied at a constant rate by the tap.
3 The water level in the container remains constant.

Heating activity propositions

1 Heat energy is being supplied to water by a candle.
2 The heat energy is supplied at a constant rate by the candle.
3 The temperature of the water remains constant.

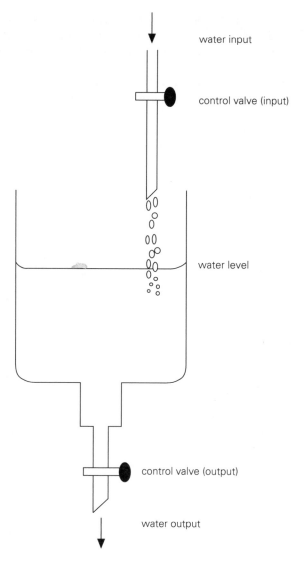

Figure 2.2: Water flow analogy

4 Therefore water must be flowing out of the container.

4 Therefore heat energy must be flowing out of the water.

5 The rate of water loss must be exactly equal to the rate of water supply.

5 The rate of heat loss must be exactly equal to the rate of heat supply.

The water analogy was presented and discussed with a group of middle school (age 13, n = 12) and upper school students (age 15, n = 10) before they performed the candle and tin heating activity. Equal sized and nominally equivalent control groups were simply given the candle and tin activity to perform. In both cases, students completed a structured worksheet requiring open-format answers to several questions about the system(s) they had investigated and (in some cases) observed.

A total of twenty-two experimental and twenty-two control students was therefore involved in the study. All forty-four were subsequently interviewed individually. Interviews were tape-recorded and transcribed. In reporting transcript evidence below, student comments are prefixed by initials (NA, ZM. . . .). The interviewer's remarks are indicated by 'I'.

Control Groups' Conceptions about Thermal Equilibrium

None of the control group students was able to explain the outcome of the water heating activity in scientific terms, as anticipated on the basis of the earlier study (Arnold and Millar, 1994). Nearly all students predicted a temperature rise (nine students aged 15, seven students aged 13) with three 13-year-olds predicting a levelling because the water boils, and one 15-year-old because 'heat escapes from the water'. When confronted with the fact that the temperature in fact stabilized, students' explanations included the inadequacy or constancy of the heat supply (nine students), the effects of the air (two students) and the insensitivity of the thermometer (one student). Language difficulties may also have contributed to some students' failure to generate an explanation.

What are the reasons for the equilibration of temperature? Written responses are quoted below.

The most popular explanation for the temperature stabilizing in this activity was the limited ability of the candle to heat the water.

> **AN** Is it because that it's ..er..mm.. the heat that's being given off is a certain amount . . . and that it's not enough to make it even more hotter.
>
> **PH** The candle's not . . . the water . . . you know the water in the tin . . . it's not . . . if you put a bit less in it might boil but the candle's not warm enough . . . if you put a Bunsen burner underneath . . . it'll boil, it gives off more gas . . . and more heat.
>
> **JA** Because it's blowing the same heat every time . . . all the time.

In attempting to explain why the candle was able to heat the water at the beginning of the activity, but was not able to raise the temperature further by the end of the activity, the initial coldness of the water was the most popular explanation. Cold water was thought to be better able to 'absorb heat'. Once the amount of heat absorbed reached a certain amount, more powerful heat sources than candles were seen to be required to 'pump' more heat into the water.

Another response appears to suggest that the temperature of the water rises until a constant temperature has been achieved throughout all the water. When asked what decides the point at which the water temperature levels off, the student (AN once more) replied:

> **AN** I think it's the water . . . all of it's at constant temperature.

Some students used the properties of the tin to explain the experimental results.

> **PH** Oh! I know! Because it's a metal tin . . . and metal conducts.
> **I** So here the water's heating up because the tin conducts . . . but doesn't the tin conduct here?
> **PS** It can't conduct because the can . . . it's . . . can give so much amount of heat . . . the candle.
> **AN** Er . . . this metal thing here . . .
> **I** Metal?
> **AN** Er . . . it's conducting the heat.

This student also stated that on blowing out the candle, the temperature of the water would remain constant for some time due to the metal of the tin.

> **I** What would happen if I blew out the candle . . . and kept on taking the temperature?
> **AN** That might have stayed at constant because . . . you know metal . . . sometimes keeps (noise) insulator . . . (noise) . . . conductor . . . of the heat . . . might be . . . might drop down later on.

The other student quoted above also believed the metal of the tin would affect the behaviour of the apparatus when the candle was extinguished.

> **PH** Then it'll drop . . . no heat to keep it warm . . . and the metal'll run out of heat . . . you know . . . what's been conducting . . . it'll run out and just go.

As well as the idea that absence of heating is sufficient to account for an object cooling, this student's concept of 'conducting' appears to include the notion that the metal stores and passes on heat to the water rather than passively allowing it to flow.

Any scientific interpretation of the experimental outcome must incorporate the idea of heat loss from the system. A small number of students did recognize that heat was lost to the surroundings from the water in the tin. For example:

> **SZ** The heat is spreading all around?
> **I** Spreading . . . mmm . . . where to?
> **SZ** The rest of it . . . and going out (waves arms).
> **SQ** Might be . . . escaping from the water first then it's starting to come out . . .

The crucial realization, that heat was being lost from the can and its contents, was entertained by very few students and the idea of a balance of heat input and heat loss was not mentioned by any students in the control groups.

Teaching the Analogy

The behaviour of the water analogy was demonstrated to the twenty-two 'experimental' group students and extensively discussed to show them that a variety of equilibrium conditions could be established, such as:

- input full on, output fully open
- input half on, output half open
- input very slow, output very slow

and even:

- input off, output off.

Students were then asked to write down the three important features of the system (input rate, output rate and water level) and to reason about the system when in equilibrium and when the container was filling or emptying.

Students' Understanding of the Water Analogy

Most students answered the questions perfectly. For example:

NA   The same amount of water that is dripping out is flowing in.

In response to questioning about a rule for controlling the behaviour of the system, most students correctly concentrated on achieving a balance between input and output to keep the water level constant. The second example cited below even includes the method of raising and lowering the water level in the explanation.

MM   The water is going in at the top and the same amount of water is going out so the level stays the same.

AK   To control the water level if you want it to balance pour the same amount of water at the top and the bottom. If you want the rise you will have to put more water from the top and less from the bottom. And if you want the water to go down make less water come from the top and more from the bottom.

The analogy was, therefore, well understood by nearly all students, who could reason about it and record their understanding with impressive fluency.

Analogy Groups' Understanding of Thermal Equilibrium in the Heating Activity

The twenty-two students who had been shown the water analogy were then asked to perform the heating activity, and thirteen were eventually able to use the analogy to help them successfully interpret the outcome at interview. Three students (all aged 15) saw the connections spontaneously, whilst a further ten students (including six 13-year-olds) made the link when reminded about the analogy and asked for perceived connections between it and the heating activity. The parallels with Gick and Holyoak's (1980) findings are striking; in particular, the lack of spontaneous mapping of the analogy on to the target problem is again noted, as is the better success when a hint is provided. The interview transcripts provide some insight into the ways in which students were able to draw on ideas from the analogy to make sense of the thermal situation.

Reasons for Equilibration of Temperature

When accounting for the stable water temperature in the heating activity, several interviewees at first used the 'it can't get any hotter' explanation, either on its own or coupled to an 'inadequacy of the candle' argument. Only one came straight to an explanation, interrupting the interviewer's question with 'Heat's going out', and adding 'It evaporates away'. The conversation continued:

I Right . . . if the temperature is neither going up or down . . . what does that tell you about the . . .

NA (interrupts) Same as . . .

I . . . amount of . . .

NA It's same . . . er . . . level of heat going up . . . and coming out.

I This is what we did last week . . . can you see a connection between what we talked about last week . . .

NA Yes sir that's . . . was . . . er . . . water going in, the same level of water's going in and coming out and the level stays the same in t'water.

I Yes . . . and this one?

NA Same level of heat . . . same amount of heat going in and coming out . . . and t'temperature stays the same.

The interviewer's use of the phrase 'going up or down' in the first line of this extract, which reflects back the student's earlier usage, and the student's later use of the words 'level of heat', both suggest a transfer of ideas and imagery from the analogy to the heating context. One other student was also quick to connect the two activities.

I Think about last week's experiment . . . what does that tell you must be happening in this week's experiment?

> **BA** Heat must be coming out.
>
> **I** So what's happening . . . here then . . . when the temperature's steady?
>
> **BA** The same amount of heat's going out . . . as is coming in.

Another student in this group knew that the room temperature was important, and that the water lost heat into the room. He used the word 'balancing' to explain why the temperature of the water stabilized. When reminded of the water flow task he saw the connection immediately, though he confused heat and temperature in his reply.

> **I** Can you see any connection between what we did last week and what we did this week?
>
> **JA** Sir like the same . . . like water coming in and water going out . . . temperature . . . er . . . heat going in and heat going out.
>
> **I** Brilliant. So what was it last week that was balanced . . . this week it's the temperature . . . what was it last week?
>
> **JA** The water level.

Of those students who took somewhat longer to see the link, a variety of cues triggered a positive response. One was asked what the consequence of blowing out the candle and continuing to record temperatures would be. When he thought about the resulting temperature drop he realized heat loss would also be occurring when the candle was lit.

> **AR** We're losing the heat at the same time as the candle's on . . . letting off heat and that's losing heat at the same time.
>
> **I** Right . . . now . . . have another think about that then . . . can you see any reason why . . . it might end up being level?
>
> **AR** Sir it's the same as that tap thing . . . er . . . that you were showing us . . . it's the same idea as that . . . that has . . .
>
> **I** Go on . . . tell me more.
>
> **AR** It loses . . . er . . . it's got the same amount of heat lost as . . . the same amount of heat's coming up . . . up at it.

No explicit mention was made by the interviewer of the analogy in this case; the interviewer's use of the word 'level' picks up on an earlier use of the word by the student in the conversation leading up to this extract. The student appears to make the link with the 'tap thing' analogy spontaneously. It is apparent, however, that language limitations are affecting the student's ability to reason about the problem.

Several restatements of the question were required before two other students were able to make the connection. By asking them to itemize the three quantities of importance in the analogy and match them to the heating activity, the 'missing' heat loss was discovered, and then the problem solved. When this girl had eventually realized that the inputs of water and heat were analogous, and that the water level and temperature were constant in the two activities, she was asked what factor from the water flow task was missing from the heating task. The exchange continued:

SN   Coming out.
I   Coming out? . . . right?
SN   Heat comes out.
I   Yes . . . go on.
SN   The equal amount of heat is coming out than . . . the same amount is coming in.
I   That's very good . . . where's it going?
SN   Into the air.

All of the transcript extracts presented above come from interviews with the 15-year-old group. In general the 13-year-olds were slower to appreciate the relationship between the water analogy and the heating activity. All students in this group needed to discuss and consider the two cases before they perceived the similarities. Four appeared to understand the analogy after some discussion, two others had some understanding and six could not see the relevance of the water analogy even after a structured discussion of the features of the two activities.

Conclusion

The aim of this study was to allow an informed decision to be taken about the choice of the heating activity, along with the water analogy, as introductory activities for a course in basic thermodynamics. We took the outcome — that thirteen out of twenty-two pupils shown the water analogy were able, with prompting, to make use of it to provide explanations of the heating activity — as indicating that the approach was worth pursuing. As the target age for this course was thirteen, we were encouraged to note that six of the twelve 13-year-olds were able to make use of the analogy, with help. This evaluation study also suggested that, in the teaching programme which we developed and taught as a result of this development study, the mappings between the analogy and the corresponding thermal system should be made explicit. Students cannot be expected to make the connections spontaneously and many will require considerable assistance and prompting.

The analogy is clearly not a panacea. Nine of the twenty-two children were unable to make use of it in explaining the thermal situation. It would be unreasonable, however, to expect any teaching intervention to work for all learners. Following the evaluation reported above, we used the water analogy in this way as the basis of a short course in elementary thermodynamics for 13-year-old students and monitored in detail the learning outcomes (see Arnold, 1993; Arnold and Millar, forthcoming). Here there is space only to state that the approach was a significant improvement, in the school setting where it was used, on previous, more 'traditional' approaches.

References

ARNOLD, M.S. (1993) 'Teaching a scientific mental model. A case study: Using analogy to construct a model of thermal processes', Unpublished DPhil thesis, University of York.

ARNOLD, M.S. and MILLAR, R. (1994) 'Children's and lay adults' views about thermal equilibrium', *International Journal of Science Education*, **16**, 4, pp. 405–19.

ARNOLD, M.S. and MILLAR, R. (forthcoming) 'Learning the scientific "story": A case-study in the teaching and learning of elementary thermodynamics', *Science Education* (in press).

BLACK, D. and SOLOMON, J. (1987) 'Can pupils use taught analogies for electric current?', *School Science Review*, **69**, 247, pp. 249–54.

BROOK, A., BRIGGS, H., BELL, B. and DRIVER, R. (1984) *Aspects of Secondary Students' Understanding of Heat: Full Report*, Leeds, Centre for Studies in Science and Mathematics Education, University of Leeds.

CLEMENT, J. (1987) 'Overcoming students' misconceptions in physics: The role of anchoring intuitions and analogical validity', in NOVAK, J.D. (Ed) *Proceedings of Second International Seminar on Misconceptions and Educational Strategies in Science and Mathematics*, Ithaca, NY, Cornell University.

DRIVER, R., SQUIRES, A., RUSHWORTH, P. and WOOD-ROBINSON, V. (1994) *Making Sense of Secondary Science: Research into Children's Ideas*, London, Routledge.

DUIT, R. (1991) 'On the role of analogies and metaphors in learning science', *Science Education*, **75**, 6, pp. 649–72.

Duncker, K. (1945) 'On problem solving', *Psychological Monographs*, **58**, whole number 270.

ERICKSON, G.L. (1979) 'Children's conceptions of heat and temperature', *Science Education* **63**, 2, pp. 221–30.

ERICKSON, G.L. (1980) 'Children's viewpoints of heat: A second look', *Science Education*, **64**, 3, pp. 323–36.

ERICKSON, G.L. (1985) 'Heat and Temperature: Part A', in DRIVER, R., GUESNE, E. and TIBERGHIEN, A. (Eds) *Children's Ideas in Science*, Milton Keynes, Open University Press, pp. 55–66.

GENTNER, D. and GENTNER, D.R. (1983) 'Flowing waters or teeming crowds: Mental models of electricity', in GENTNER, D. and STEVENS, A.L. (Eds) *Mental Models*, Hillsdale, NJ, Lawrence Erlbaum, pp. 99–129.

GICK, M.L. and HOLYOAK, K.J. (1980) 'Analogical problem solving', *Cognitive Psychology*, **12**, pp. 306–55.

GLYNN, S.M. (1991) 'Explaining science concepts: A teaching-with-analogies model', in GLYNN, S.M., YEANY, R.H. and BRITTON, B.K. (Eds) *The Psychology of Learning Science*, Hillsdale, NJ, Lawrence Erlbaum, pp. 219–40.

ISAACS, A., DAINTITH, J. and MARTIN, E. (Eds) (1984) *Concise Science Dictionary*, Oxford, Oxford University Press.

KAHNEY, H. (1986) *Problem Solving: A Cognitive Approach*, Milton Keynes, Open University Press.

KESIDOU, S. and DUIT, R. (1992) 'Students' conceptions of the second law of thermodynamics: An interpretive study', *Journal of Research in Science Teaching*, **30**, 1, pp. 85–106.

LINN, M.C. and SONGER, N.B. (1991) 'Teaching thermodynamics to middle school students: What are the appropriate cognitive demands?', *Journal of Research in Science Teaching*, **28**, 10, pp. 885–918.

STRAUSS, S. and STAVY, R. (1983) 'Educational-development psychology and curriculum development: The case of heat and temperature', in HELM, H. and NOVAK, J. (Eds) *Proceedings of the International Seminar: Misconceptions in Science and Mathematics*, Ithaca, NY, Cornell University, pp. 292–303.

SUTTON, C. (1992) *Words, Science and Learning*, Buckingham, Open University Press.

TIBERGHIEN, A. (1983) 'Critical review of the research aimed at elucidating the sense that the notions of heat and temperature have for students aged 10–16 years', in *Research on Physics Education: Proceedings of the First International Workshop*, La Londe les Maures, 26 June–13 July 1983, Paris, Editions du Centre National de la Recherche Scientifique, pp. 75–90.

TIBERGHIEN, A. (1985) 'Heat and temperature: Part B', in DRIVER, R., GUESNE, E. and TIBERGHIEN, A. (Eds) *Children's Ideas in Science*, Milton Keynes, Open University Press, pp. 67–84.

TREAGUST, D.F., DUIT, R., JOSLIN, P., and LINDAUER, I. (1992) 'Science teachers' use of analogies: Observations from classroom practice', *International Journal of Science Education*, **14**, 4, pp. 413–22.

3 Developing Scientific Concepts in the Primary Classroom: Teaching about Electric Circuits

Hilary Asoko

Abstract

Science teaching in the primary school is frequently characterized by the provision of activities designed to broaden children's experience and promote exploration and investigation of natural phenomena. Whilst not denying the importance of this, if science learning is viewed as a process of enculturation into the ideas and explanations of the scientific community and if science teaching at the primary level aims to develop children's conceptual understanding, then it is not sufficient. Teachers also need to find ways to make appropriate ideas of science accessible to young children and support pupils in relating these ideas to experiences. Case studies provide one way of alerting teachers to some possible strategies to achieve this and to stimulate discussion, reflection and development.

This study reports on a primary classroom in which an analogy was used to support the development of children's understanding of energy transfer and current flow in simple electrical circuits. Aspects of children's learning are discussed and related to teacher action.

Introduction

Science has, over recent years, become an accepted part of the primary school curriculum in many countries. In England and Wales the introduction of the National Curriculum in 1989 identified science as a core subject, along with English and mathematics and required that it be taught to children from the age of 5. Curriculum requirements, however, must be translated into classroom practice. How this occurs depends on teachers and is influenced by, amongst other things, their views of teaching and learning, their knowledge of, and attitudes to, science and their feelings of confidence and competence about what is to be taught.

The majority of primary teachers in England and Wales do not have a strong personal background in science. Surveys of teacher knowledge of science reveal that 'the majority of primary teachers in the sample. . . . hold views of science concepts that are not in accord with those accepted by scientists' (Summers, 1992, p. 26). Their ideas are more likely to correspond to those of the children they will teach (Carré, 1993; Webb, 1992). Teachers do, however, frequently have firm beliefs in the value of 'hands-on' experience for learning and of the role of the teacher as provider of opportunities for children to explore and investigate. Coupled with ideas that science knowledge is revealed by careful observation and investigation this has led to classroom practice which is frequently based on 'discovery learning'. Teachers have developed a repertoire of activities which appear to be science based, and from which they expect children to learn science, though they often have difficulty in articulating the purpose of such activities in terms of precise conceptual aims. In countries other than Britain, primary school teachers may deal with their lack of subject knowledge by relying heavily on other strategies such as lecture and text-based teaching approaches (Stofflett and Stoddart, 1994).

However, if science learning is viewed as a process of enculturation into the ideas and explanations of the scientific community (Driver *et al.*, 1994), and if science teaching at the primary level aims to develop children's conceptual understanding, then simply providing opportunities for exploration and investigation is not sufficient. Neither will simply presenting children with the science ideas to be learnt be effective. Teachers need to find ways to make appropriate ideas of science accessible to children and provide opportunities for exploration both of the ideas and of the phenomena to which they relate.

In-service and pre-service training courses therefore need to provide opportunities for teachers both to develop their personal understanding of science concepts and to integrate this with their knowledge of children and classrooms to develop 'pedagogical content knowledge' (Shulman, 1986).

Shulman has suggested this would include knowledge of:

(i)   what makes the topic easy or difficult to understand — including the preconceptions about the topic that students bring to their studies;

(ii)   those strategies most likely to be effective in reorganizing students' understanding to eliminate their misconceptions;

(iii) a variety of effective means of representing the ideas included in the topic — analogies, illustrations, examples, explanations and demonstrations. (Shulman, 1986 p. 9)

It is relatively easy to introduce teachers to the ideas students bring to their learning in selected topics. Research evidence abounds and comprehensive summaries have been produced (e.g., Driver *et al*, 1994). When teachers are encouraged to investigate children's ideas in their own classrooms, their interest is captured. Consideration of why students might respond as they do not only provides opportunities to consider children's learning but also prompts teachers to reflect on their own understandings and whether these are scientifically acceptable. If, as learners of science, teachers are introduced to a range of teaching strategies, and the reasons for their use are made explicit, theoretical perspectives on conceptual change teaching may make more sense. Stofflett and Stoddart argue that 'Learning through modelling (i.e., being told about and observing a demonstration of an innovative instructional approach) must be distinguished from active involvement as a learner experiencing the pedagogy . . . it is the latter experience (i.e., understanding what it means to reconstruct your personal understanding of a concept) that will enable teachers to change their instructional approach' (1994, p. 46). Studies have been reported on the development of primary teachers' knowledge and understanding of science and how this may be translated into classroom practice (Smith and Neale, 1989; Neale *et al.*, 1990; Kruger and Summers, 1993). Such studies highlight the difficulties teachers face. Even if teachers have appropriate subject knowledge, are aware of the ideas children bring to their science learning and are convinced of the value of using a range of teaching strategies including those which may be termed 'conceptual change' they are still left with the problem of how to implement all of this in their own classrooms.

Kruger and Summers reported that although the two teachers in their study elicited the ideas of their pupils when teaching science, this was seen 'as a guide to the conceptual level of the knowledge to be presented rather than to the nature and range of existing ideas which had to be changed or displaced' (1993, p. 23). They also commented that 'each teacher showed some examples of the transformation of conceptual understanding acquired from training into learning experiences for children by choosing everyday examples to illustrate a force in action . . . however neither teacher used any other analogies or metaphors as teaching tools and both chose practical activities designed to develop process skills rather than to promote conceptual change'. Kruger and Summers suggested that 'classroom-tested representations (e.g., analogies, metaphors) must be devised to translate scientific knowledge into something comprehensible to young children in ways which are meaningful and related to their interests and experiences' (1993, p. 23).

This study is part of a programme which aims to develop and explore the use of such representations in the primary classroom and thus focuses on the third element of pedagogical content knowledge. Material from such studies can be used to illustrate to teachers and student teachers how subject knowledge can be combined with knowledge of children's thinking and general pedagogy to develop teaching with clearly specified conceptual aims and strategies for achieving these. This material is not envisaged as 'the best way' to teach a topic, but rather to highlight important aspects of teacher action and decision making, to stimulate discussion and to sensitize teachers further to the sense-making of children. One case study, relating to teaching about the behaviour of light, has already been reported (Asoko, 1993). This paper will focus on aspects of teaching and learning about electricity, in particular the use of an analogy to explain the behaviour of simple circuits.

Background to the Study

In learning about simple electric circuits, practical experience is clearly important. However, experience from both pre-service and in-service courses demonstrates that teachers, like children, may achieve technical competence without developing scientific understanding. When pressed to explain what they see happening they readily confess to difficulties in accounting for the effects observed in circuits and to confusion about terms such as current, voltage and power. In order to move thinking towards a scientific understanding it is necessary to differentiate the notion of 'the electricity' into the components of energy transferred from the battery to the bulb, and current which flows around the circuit. Geddis (1993) describes pre-service secondary science teachers coming to understand the importance of this notion and developing a teaching approach as a result. These students already had a background in science which included at least one year of university physics. They were therefore familiar with the scientific view but had not necessarily considered its implications. Primary teachers, on the other hand, frequently need support in understanding the scientific view in the first place. A range of analogies have been developed to represent ideas about the behaviour of circuits (see for example Dupin and Johsua, 1989). In working with teachers to develop their own understanding, I usually introduce them to several of these, typically one based on a water circuit, one on a carrier model and the third which is a 'string circuit'. In this last analogy participants sit in a circle with one hand extended. A length of smooth string is passed around the circle and tied to make a continuous loop, loosely supported between the first finger and thumb of each person. The hands represent the wires in the circuit; the string represents the electric current. One person, acting as the battery, makes the string (current) circulate. The string moves simultaneously in all parts of the circuit and the same amount of string returns to the 'battery' as leaves it. The 'battery' transfers energy to the movement of the string. If a person in the circuit grips the string gently, acting as a resistance in the circuit, the string moves less easily and that person feels energy transferred to them through the heating of their hand. The analogy can thus be used to help students to focus on the need for a continuous circuit and to differentiate between current flow and energy transfer.

This analogy enthuses the majority of primary teachers, partly because they find it helpful for their own understanding but also because they see it as something which they could use with their own pupils. There is, however, a danger that teachers who are not science specialists, learning science in a way which they feel is effective, may transfer, not just strategies, but specific analogies, examples and activities to their own teaching without considering whether these are appropriate for their pupils. Teachers should be aware that thought needs to be given to selecting the most useful analogy, to when and how to use it and to ways to capitalize on the learning opportunities it provides.

These concerns therefore prompted an investigation into:

- whether 'the string analogy' made sense to primary school children;
- the learning outcomes which resulted from its use as part of a teaching sequence on electricity; and
- teacher action which affected the progress children made in developing their understanding.

The Classroom Context

The teacher who volunteered to conduct the study was working with a class of thirty-one 8–9-year-olds in a primary school on the edge of a city in northern England. He was familiar with some of the research on children's learning about

Session	Time	Nature and focus of activity
1	90 min	Exploration of initial ideas about electricity and its effects. Discussion of battery and mains electricity. Safety.
2	45 min	Observational drawing of a light bulb. Exploration of how to make a bulb light using 1 battery, 1 piece of wire (initially unstripped) only. Recording/discussing arrangements which do/do not make bulb light.
3	30 min	Exploration of how to make the bulb light if it does not touch the battery. Recording and discussing arrangements which do/do not make bulb light.
4	60 min	Introduction and discussion of the string analogy. Making circuits (now using battery and bulb holders) and relating them to the analogy. Discussion.
5	60 min	Exploring different combinations of components: 1 bulb, 2 batteries; 2 batteries, 1 bulb; motors.
6	60 min	Review of main ideas. Investigating conductors and insulators. Summary.

Figure 3.1: Outline of teaching sequence

electricity, in particular the work of Shipstone (1984), Tasker and Osborne (1985) and Osborne *et al.* (1991) and was especially interested in the notion of 'clashing currents' as a model which children seem to develop spontaneously. The children had had some experience of making simple circuits, in school, three years previously.

The teacher and researcher discussed the classroom work before it commenced and whilst it was in progress. During the teaching the researcher acted as a participant observer in the classroom, providing support for the teacher in the form of an extra pair of hands when children were working in small groups. The researcher provided feedback and suggestions to the teacher but made no teaching input. Data were collected in various forms, including observational field notes, copies of children's written work, audio-tapes of teacher talk to the class and discussions of both teacher and researcher with individuals and groups. A number of children were interviewed, individually, following the post-test.

The Teaching Sequence

The teaching sequence, determined by the teacher, is outlined in Figure 3.1.

At different times during the teaching children worked as a whole class, in small groups, in pairs and as individuals. This reflected their normal pattern of working. They were encouraged to record, by writing and drawing, their observations and explanations. Many of the activities with which children were involved — making devices such as bulbs, buzzers and motors work, exploring which materials would conduct electricity and investigating the effects of varying the number and arrangements of components in a circuit — are typical of those found in primary classrooms. However there was, in addition, an explicit intention to introduce, through the use of an analogy, ideas of energy transfer and current flow. These ideas were considered necessary to provide children with an explanatory mechanism for the effects they observed. The 'string model' was seen as appropriate for a variety of reasons. Unlike many other analogies, such as the water circuit, it makes few assumptions about children's existing knowledge since they are directly involved in the physical model at the time. It is a simple model, involving no complicated equipment and with little to distract children's attention. Although it includes no physical representation for energy, it was anticipated that children's ideas about their own 'energeticness' could be utilized.

During the initial exploration of circuits, children's attention was focused on the structure of the bulb, the connections which needed to be made to it and the filament as part of the circuit. Children were encouraged to find as many ways as possible to make the bulb light and to keep records of all arrangements of battery, bulb and wire, whether the bulb lit or not. Cylindrical, R20 1.5 volt batteries were used and observations of, and discussion with, children suggested that they saw the top or positive end of the battery as significant in terms of making the bulb light.

Many children when using two wires to connect the bulb to the battery, explained, usually using hand movements, that 'energy' 'electricity' or 'power' left the battery from both ends, met in the bulb and caused it to light. Other children had some notion of a circular movement and related this to the + and − signs on the battery. For example, having demonstrated how to light the bulb using one wire, a pupil, Paul (P) explained to the researcher (R):

P The positive is coming off that bit on to there.

R The positive is coming off that bit. . . .

P Yes this knob bit. (points to top of battery)

R Positive what?

P Positive power . . . it's coming up through one wire then the filament, back down and out and into this bit, then the power you take it from here and down into the negative.

R So power's coming out of the battery . . .

P From there.

R Through the bottom of the bulb, up that little spike inside, across that little thing in the middle, making it glow, down the other side and out the case, down the wire and back to the battery?

P Yes and when it's flat all the positive power that's in it has been replaced by the negative bit.

R So are you saying that it comes out as positive power and goes back as negative?

P   Yes, and when it's gone into there it fills up with negative power and
the positive's gone out — and when the negative gets up to there
(indicates top of battery) the battery goes flat.

Paul, a confident boy who spent time at home playing with electric train sets,
understood the need for a complete circuit and recognized that something, which
he termed 'power' travelled round it. This power was changed, from positive to
negative, on passing through the circuit.

Following the initial exploratory stage of sessions 2 and 3 (see Figure 3.1) the
teacher introduced children to the string analogy for the circuit in session 4. Working with half of the class at a time he began by reviewing with the children their
ideas about energy. These, as anticipated, included 'energy helps us to do things'
and 'energy makes things work'. He then set up the string circuit, involving all the
children in the group and with himself as the battery, and encouraged the pupils to
describe what was happening. With the string still circulating, the teacher explained
why a model was useful:

> one of the problems is you can't actually see inside the wires — we're
> talking about what's happening here (with the string) to help us to under-
> stand what might be happening where we can't see.

He went on to focus children's attention on the relationship between the parts
of the model and a circuit consisting of a battery and bulb connected by two wires.
At no point was there a practical demonstration that the current in both wires was
equal. Rather the emphasis was on providing a common-sense reason for the need
for two wires. The main ideas emphasized at this time were:

* the battery provides energy;
* energy travels to the bulb;
* energy makes the bulb light;
* energy is carried by current; and
* current travels round the circuit and back to the battery.

Children were then asked to construct circuits and explain them using these
ideas and were supported in doing this by the teacher.

In session 5 children were encouraged to use ideas about energy transfer when
trying to explain the relative brightness of bulbs in different series circuits. The
directional flow of current was drawn on to explain why reversing the connections
to a motor reversed the direction of rotation. Session 6 provided an opportunity for
children to construct their own circuits to investigate whether different materials,
selected by them, conducted electricity.

Evaluating Learning

One week after the end of the teaching children completed a post-test which was
repeated after six months (see Appendix 1 for details). Questions 1 and 3 focused

Table 3.1: *Responses to post-test questions 1 and 3*

Circuit	Number of children correctly identifying whether bulb will/will not light		Number of children who could indicate appropriate correction*	
	Post-test n = 26	Delayed post-test n = 31	Post-test n = 26	Delayed post-test n = 31
Alan	25(96)**	28(90)	21(84)	22(79)
Anne	24(92)	30(97)	—	—
George	21(81)	23(74)	—	—
Sue	8(31)	18(58)	6(75)	14(78)
Carol	24(92)	31(100)	15(63)	10(32)***

* Figures are numbers of children who correctly identified that the bulb would not light and unambiguously described an appropriate correction.

** Figures in brackets are percentages.

*** 55 per cent of responses were ambiguous or not specific (e.g., 'she needs another wire').

on children's ability to recognize the features of a complete circuit. Questions 2 and 5 provided opportunities for children to explain what happened in the circuits. Question 4, which will not be discussed here, related to their investigations of conductors of electricity.

Table 3.1 shows the numbers of children who could correctly identify whether the bulbs would, or would not, light. Also summarized are the numbers of children who, having correctly indicated that a bulb would not light, could suggest an appropriate alteration to the circuit.

Overall, children achieved a high degree of success on three of the circuits. The majority of children who indicated, incorrectly, that George's circuit would *not* work advised that a second wire be used to connect the top of the battery to the base of the bulb. Interestingly, George's arrangement is the one which most children had used, successfully, to light their own bulbs in the initial activities of the teaching sequence. The fault in Sue's circuit (both connections to the side casing of the bulb) was not identified by many children, despite the fact that the connections to, and circuit through, the bulb had been highlighted during the teaching.

When indicating alterations to circuits, children's responses were frequently not specific, rather than incorrect. This was particularly true for Carol's circuit where many children simply stated that 'she needs another wire' without specifying where it should be connected.

In question 2 the level of explanation given for the circuit was determined by the child. Responses ranged from simple descriptions of the arrangement of the circuit:

the batrey is making the boulb light.
the wire is on the batrey and it gos to the boulb and makes it light. (Alex explaining Anne's circuit)

Table 3.2: Use of ideas about current and energy

Ideas used	Number of responses	
	Post-test n = 26	Delayed post-test n = 31
The battery provides energy	10	13
Energy travels/is carried to the bulb	7	8
Energy makes the bulb light	11	11
Energy is carried by current	5	4
Current travels 'round'	10	6
Total responses	43	42

to explanations in terms of current or energy:

> The electric current is going round and round from the battery to the bulb then going back to the battery for more power. (Ryan explaining Anne's circuit)

> The energy is going round in the crocodile clips and the electricity is going round and as the electricity carriages the energy the bulb lights up. (Danielle explaining Anne's circuit)

It was recognized that, although children might not spontaneously use ideas about current or energy in explaining the circuits, this did not mean that they did not have some understanding of the ideas in this context. Question 5 therefore specifically asked children to try to use the words. Question 2 and question 5 were used together to gather evidence of children using the ideas to which they had been introduced during the teaching. A summary of these responses is given in Table 3.2.

In the post-test the forty-three responses were given by twenty-one children. Four other children used an undifferentiated notion of 'the electricity' only; one used the terms current and energy in ways which were uncodable.

In the delayed post-test the responses came from twenty-four children. Three others used an undifferentiated notion of 'the electricity' whilst four responses were uncodable. Although several children wrote that the current travels 'round' the circuit it should be noted that this does not necessarily imply that children mean from the battery, through the components and back to the battery.

Evidence from all five questions was used to make a judgment about the predominant model which children were using in their thinking about electricity (see Table 3.3).

Children judged to have a 'source to consumer' model described the circuit in terms of something travelling from the battery to the bulb. Energy was specifically identified as being transferred from the battery to the bulb in sixteen of the eighteen responses in this category, whilst two responses used a generalized notion of 'the electricity' only. It was not clear from these responses whether children considered the energy to be travelling through both wires or only one.

Table 3.3: Predominant models of the behaviour of electricity

Model	Number of children	
	Post-test n = 26	Delayed post-test n = 31
No model evident	1	7
Source to consumer	10	8
Both wires carry something to the bulb	2	5
Circular flow of something other than current	6	6
Circular flow of current	7	5

However, in a further seven responses there was a clear indication that children believed something to be travelling along *both* wires to the bulb. This 'something' was not electric current but 'energy', 'power' or 'electricity'. Several of these responses appeared to incorporate elements of a model of circular flow into a source to consumer view e.g.,

the energy is coming from the battery and flowing into the light bulb. It is coming from the battery and going into both the wires and meeting together to make the light bulb light. Some of the energy goes right round and some goes into the bulb. (Nicola's response to question 5, post-test)

Other responses in this category showed a partial differentiation of the notion of 'electricity' so that different elements travelled along each wire:

When the electricity and the energy come together it makes the bulb light up. (Danielle, question 2, delayed post-test)

Notions of circular flow, indicating an understanding of something moving around the circuit from the battery to the bulb and back to the battery, were evident in twenty-four responses. Twelve of these did not use the word 'current'. Instead three children talked of energy passing all the way round the circuit, eight of electricity and one of 'it'. Children had therefore grasped the idea of circular flow but apparently had not differentiated between current and energy.

However, a further twelve responses specified that current flowed around the circuit, from the battery, through the components and back to the battery. Nine of these responses stated that the current carried energy to the bulb.

Twelve of the twenty-six children who completed both the post-test and the delayed post-test used the same model when explaining circuits both times. Three of these were using a model of circular flow of current carrying energy.

Discussion

Children at primary school are frequently introduced to simple electric circuits. It is possible to focus on developing technological competence through such activities, so that the aim becomes 'making things work', leaving the children to make what sense they can of their experiences. If, however, teachers wish to introduce a scientific way of thinking about the behaviour of circuits, then they need to find ways to help children to begin to conceptualize electricity in terms of two components, energy and current and to use these ideas in developing explanations and making predictions.

The string analogy used in this study is one which appeals to teachers. These children appeared able to understand the analogy and its relationship to the circuits they had constructed at the time it was presented to them. Not unexpectedly, when making circuits in subsequent sessions, they needed prompting from the teacher to use the analogy to help them to develop explanations for their observations.

In their thinking about the behaviour of electricity most children developed elements of a scientific understanding. For some children this was largely in terms of energy transfer from the battery to components in the circuit, whilst for others a notion of circular flow was evident. In a minority of children this notion was differentiated into an appreciation of the circular flow of current as a carrier of energy from the battery to the bulb. Later teaching would need to take account of this diversity to build on and develop the ideas children have.

Evidence from this study and others (Jabin and Smith, 1994) suggests that such analogies can be used to make scientific ideas about electricity accessible to young children. However, it is important for teachers to recognize that simply presenting an analogy as another activity in a teaching sequence will not, of itself, produce understanding. The analogy selected must be appropriate to the experience of the pupils and to the teaching points to be made. In order to capitalize on the opportunities which the introduction of an analogy provides, teachers need to be aware of its strengths and weaknesses, consider at what point it could most usefully be presented and focus children's attention on salient features. Children will need time, opportunity and support to use the analogy to guide and structure their thinking and the teacher needs to provide for this and monitor their developing understanding.

In the classroom described in this study, the teacher's main purpose was to provide children with a plausible reason for the second wire needed in their circuits. This took the form of a pathway for something to 'go round', as opposed to two paths for something to travel from the battery to the bulb. In order for children to reconcile the apparent contradiction of something travelling from, and returning to, the battery with their everyday knowledge that batteries go flat, the idea of energy transfer was crucial. The teacher utilized children's existing ideas about energy to good effect, both when talking about the 'string circuit' and in subsequent discussions, and children appeared able to use the idea of energy transfer both in discussion and in their written work. However, when talking about the current transporting the energy the teacher sometimes used the term 'electricity' and sometimes 'electric current' and this seems to be reflected in the children's thinking where these terms are not clearly distinguished. Once the term 'current' is introduced it needs to be used consistently and unambiguously. The analogy was introduced in session 4 of a sequence of six sessions (see Figure 3.1). Opportunities for children to use it to develop explanations of their observations arose in sessions 4 and 5 and, to a much lesser extent, in session 6. It is possible that the inclusion of a further session spent on activities designed to provide opportunities for children to discuss and use the new ideas might have helped pupils to consolidate their understanding and, perhaps, to move further towards a more scientific view.

A study of the use of analogy by seven high-school teachers in their normal practice revealed that, in the forty lessons observed, there were only six clear indications of analogy use (Treagust *et al.* 1992). The authors assert that 'effective use of analogies in regular classroom science teaching needs to be founded on a well-prepared teaching repertoire of analogies, using specific content in specific contexts . . .' (1992 p. 421). Such a repertoire does not exist amongst primary school teachers. Pedagogical content knowledge in secondary science teaching has evolved over many years. In contrast we know very little about how to teach science concepts appropriately and effectively in the primary school. There is a need for teachers and researchers to work together to identify, develop and evaluate ways of representing scientific ideas to young pupils and to share their developing expertise. In this way we can begin to provide opportunities for children to explore and investigate, not just the phenomena of interest to science, but also the ideas.

References

ASOKO, H. (1993) 'First steps in the construction of a theoretical model of light: A case study from a primary school classroom', *Proceedings of the Third International Seminar on Misconceptions and Educational Strategies in Science and Mathematics Education*, Ithaca, NY, Cornell University.

CARRÉ, C. (1993) 'Performance in subject-matter knowledge in science', in BENNETT, N. and CARRÉ, C. (Eds) *Learning to Teach*, London, Routledge.

DRIVER, R., ASOKO, H., LEACH, J., MORTIMER, E. and SCOTT, P. (1994) 'Constructing scientific knowledge in the classroom', *Educational Researcher*, **23**, 7, pp. 5–12.

DRIVER, R., SQUIRES, A., RUSHWORTH, P. and WOOD-ROBINSON, V. (1994) *Making Sense of Secondary Science: Research into Children's Ideas*, London, Routledge.

DUPIN, J.J. and JOHSUA, S. (1989) 'Analogies and "modelling analogies" in teaching: Some examples in basic electricity', *Science Education*, **73**, 2, pp. 207–24.

GEDDIS, A.N. (1993) 'Transforming subject-matter knowledge: The role of pedagogical content knowledge in learning to reflect on teaching', *International Journal of Science Education*, **15**, 6, pp. 673–83.

JABIN, Z. and SMITH, R. (1994) 'Using analogies of electricity flow in circuits to improve understanding', *Primary Science Review*, **35**, pp. 23–6.

KRUGER, C. and SUMMERS, M. (1993) 'The teaching and learning of science concepts in primary classrooms: Two case studies', *Primary School Teachers and Science Project, Working Paper 19*, Oxford University Department of Educational Studies and Westminster College, Oxford.

NEALE, D.C., SMITH, D. and JOHNSON, V.G. (1990) 'Implementing conceptual change teaching in primary science', *The Elementary School Journal*, **91**, 2, pp. 109–31.

OSBORNE, J., BLACK, P., SMITH, M. and MEADOWS, J. (1991) *Primary Space Project Research Report: Electricity*, Liverpool University Press.

SHIPSTONE, D.M. (1984) 'A study of children's understanding of electricity in simple d.c. circuits', *European Journal of Science Education*, **6**, pp. 185–98.

SHULMAN, L.S. (1986) 'Those who understand: Knowledge growth in teaching', *Educational Researcher*, **15**, 2, pp. 4–14.

SMITH, D.C. and NEALE, D.C. (1989) 'The construction of subject matter knowledge in primary science teaching', *Teaching and Teacher Education*, **5**, 1, pp. 1–20.

STOFFLET, R.T. and STODDART, T. (1994) 'The ability to understand and use conceptual change pedagogy as a function of prior content learning experience', *Journal of Research in Science Teaching*, **31**, 1, pp. 31–51.

SUMMERS, M. (1992) 'Improving primary school teachers' understanding of science concepts: Theory into practice', *International Journal of Science Education*, **14**, 1, pp. 25–40.

TASKER, R. and OSBORNE, R. (1985) 'Science teaching and science learning', in OSBORNE, R. and FREYBERG, P. (Eds) *Learning in Science: The Implications of Children's Science*, Heinemann.

TREAGUST, D.F., DUIT, R., JOSLIN, P. and LINDAUER, I. (1992) 'Science teachers' use of analogies: Observations from classroom practice', *International Journal of Science Education*, **14**, 4, pp. 413–22.

WEBB, P. (1992) 'Primary science teachers' understanding of electric current', *International Journal of Science Education*, **14**, 4, pp. 423–9.

Appendix 1: Electricity

Name: _____
Age: _____

1 Some children are using batteries to try to make bulbs light.

Alan did this: Anne did this: George did this: Sue did this: Carol did this:

Will his bulb light?		Will her bulb light?		Will his bulb light?		Will her bulb light?		Will her bulb light?	
YES		YES		YES		YES		YES	
NO		NO		NO		NO		NO	

2 Choose one of the arrangements which makes the bulb light.

Whose have you chosen? _____

Explain carefully what is happening to make the bulb light.

3 Look at each of the arrangements which will *not* make the bulb light. What would you tell each child to do to make their bulb light?

4 Your friend asks you whether electricity will go through their plastic ruler.

Do you think it will? _____

Why do you think that? _____

Draw and write about how you would show your friend that you were right.

You can use any equipment you like.

5 Here are some words that we used when we talked about electricity:

battery	bulb	wire	switch	filament
current	conduct	energy	circuit	

Write two or three sentences about electricity. Try to use these words at least once:

CURRENT   ENERGY

4 Teaching Electricity by Help of a Water Analogy (How to Cope with the Need for Conceptual Change)

Hannelore Schwedes and Wolff-Gerhard Dudeck

Abstract

A teaching strategy aimed at an understanding of electric circuits on a system's level, is described and evaluated. The teaching device is based on the use of a specific water model that allows, by analogical inferences from the water to the electric domain, to reach this aim. The main difficulties of this approach lie in the development of adequate concepts for water circuits. The essential steps of the instruction unit together with some measures to overcome known learning difficulties are described for the water domain as well as for the electric circuits. The main findings in testing the teaching device can be summarized as follows: Students could reach a systemic view for water circuits and by analogical reasoning also for electric circuits. Students could easily draw inferences by analogical reasoning but the level of understanding electric circuits was limited by the extent of understanding water circuits. Much time has to be spent on learning the adequate concepts for the water model which means the hydrodynamics of the water circuit will constitute a curriculum element in its own right.

Introduction

After about ten years of research in science education on students' views on electric circuits it became clear in 1988 with the publishing of the 'Study of students' understanding of electricity in five European countries' (Shipstone *et al.*, 1988) that students' everyday life views on electricity were very much the same in all European countries, and it seemed to be a hard job to guide students to accept and use the physicists' view of electric circuits. As the study of Shipstone *et al.* shows the responses given to the questions of the 'Europe Test' were very similar for all countries involved, and the proportion of acceptable physics answers was equally disappointingly low. Whatever teaching approaches were used students' everyday views on electricity prevailed and could not be shaken. A consistent physics understanding of electric circuits had not been developed at all or only in elementary steps.

　　Various approaches to handling the problem have been developed: explicitly indicating the everyday life ideas, or let students — in cases of inadequate views on electric circuits — find out what everyday life concept has guided the problem solving process (Rhöneck and Grob, 1993); or making explicit the various ideas in students' minds and in the course of discussion of such ideas, inducing cognitive conflicts which should lead to the accepted physics view. Another possibility was seen in the use of analogies, partly in order to support a strategy of conceptual change, partly — as it is proposed by the authors here — in order to explain by means of analogical reasoning the function of electric circuits (Black and Solomon, 1987; Dupin and Johsua, 1994). Along with testing new teaching approaches many studies have been made to try to elucidate how students approach physical explanations of electric circuits; learning pathways have been described (Niedderer and Goldberg, 1994), but also doubts have been stated with respect to the aims of the teaching. If an adequate physics understanding of electric circuits is not achievable, one should perhaps limit teaching to the most important application, i.e., electricity at home, to parallel connections, and stress the aspect of energy saving as has been proposed for the national physics curriculum in the Netherlands (van den Berg and Grosheide, 1993). Aiming at an understanding of electric circuits as systems we present a teaching strategy based on an elaborated water model which overcomes several of the difficulties experienced in the teaching approaches mentioned.

Dealing with Students' Ideas on Electricity

The findings about students' ideas in the field of electricity collected in science education research in recent years show that students' alternative conceptions do not exist separately and independently from each other but normally are part of an inner conceptual relationship and are interrelated to each other. We therefore consider the combination of: battery is a constant source of X, bulb is consumer of X and X is sequentially transported from the source to the consumer as one complex concept which we call the consumption concept. These interrelationships constitute part of the difficulty encountered with changing students' concepts as always several concepts have to be changed at a time. Thus, in cases where students realize that one concept will have to be revised for an understanding of the physics explanation, this often does not mean a great success as the remaining concepts still support the old everyday life concept in its unquestioned form, and before revising these other remaining assumptions students rather tend to reintegrate the questioned concept possibly even in its original unchanged shape. Only a series of experiences questioning several concepts together will be able to initiate a search for a fundamental re-orientation and re-structuring of concepts.

Because of the interrelation of students' ideas and their incorporation in a specific everyday life view of things we consider the cognitive conflict methodology to offer little promise as the panel of concepts to be questioned by each individual student varies significantly as well as the perception of the cognitive conflict intended in each case — so that these requirements cannot be fulfilled for several students at the same time during instruction. This has been proved by others in our studies on conceptual changes in electricity (Schmidt, 1989).

Through our analogy-oriented approach we chose a strategy which aims at the construction of concepts and acquiring laws and principles in a rather new domain (water circuits) much less charged with everyday life ideas obstructing the learning of adequate circuit rules. These laws and principles can later be used as tools to handle questions and problems in the domain of electricity by means of analogical reasoning.

It has to be noted that, of course, the domain of water circuits is not free of alternative conceptions. With respect to such hydrodynamic systems students, for example, make a number of assumptions comparable to the consumption concept for electric circuits, e.g., the idea that the water flow is reduced by every flow-meter and continues to flow more slowly behind it. Also the idea of a constant current source is found in many cases. The strategy of sequential reasoning implied in these assumptions must be repeatedly made explicit to students, and must be restricted to areas where it is adequate or be replaced by a systemic approach. In earlier studies lack of understanding water circuits could be traced back to a lack of a continuity concept, an incompletely elaborated concept of current intensity as well as to a neglect of feedback and a lack of systemic reasoning (Schwedes, 1983; Schwedes and Schilling, 1983; Schwedes, 1985).

The Teaching Strategy

The Idea of Analogy

The main idea of our teaching strategy is to use a water analogy to prepare a suitable arrangement of concepts and a system's approach that can be a proper basis for understanding electric circuits. So the main intention of our teaching strategy is not to confront students with their misconceptions and inconsistencies of argument but to prepare a source-domain from which a well structured knowledge of electric circuits, the target domain, can be built up and established.

The idea of using water models for teaching electricity is not new, but the appreciation of such models as being helpful for understanding is not at all shared by teachers and is not unquestioned in the scientific community (Dupin and Johsua, 1987). Indeed, the water models normally used explain only one aspect of electric circuits, so that you have to change models to clarify other aspects of the circuits. For the students then the problem arises of integrating these different aspects for understanding electric circuits as a whole, as a system. For discussion of the pros and cons of various analogies for electric circuits see Schwedes, 1995.

The basis of our special approach to the use of the water analogy is the very sophisticated device used in building up water circuits and in showing full structural identity to electric circuits (see Figure 4.1). This device can be handled and explored by students themselves in experimental group work.

Starting from a drive-impediment scheme for electric circuits the equivalent drive in the water model is the constant pressure difference Δp produced by the

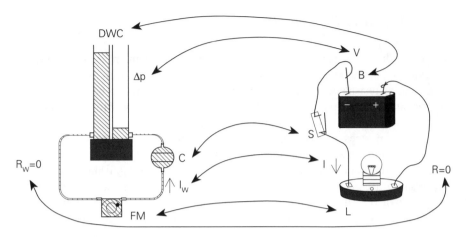

Figure 4.1: Water analogy for electric circuits
Notes

DWC	= Double Water Column	V	= voltage
FM	= flow-meter	Δp	= pressure difference
L	= bulb	I	= current intensity
B	= battery	S	= switch
R	= resistance	C	= cock

double water column (DWC), visually controllable by its constant water-level difference. The drive in the electric circuit is the constant 'electrical pressure difference' (voltage). Later on, when talking of charge or electron flow as the electric current it is possible to develop a conception of the electric pressure difference as a difference of charge, more precisely as a difference of charge density (Dudeck, 1987), or still later perhaps as a potential difference.

Resistance to flow in water circuits (electric circuits) is produced by narrowing elements, e.g., a flow-meter (resp. bulb) having a resistance R_w (R). Resistance in the tubes (wires) is supposed to be zero which is an idealization you may discuss and which can be considered as approximately fulfilled using a sufficient cross section of tubes (wires) and a sufficiently slow flow of the water (electrons). Varying water resistances (electric resistances) can be recognized by the varying pressure differences (voltages) required in order to produce a determined intensity of current. $(R_w = \Delta p/I_w)$ resp. $(R = U/I)$. The size of the current intensity results from the ratio of drive and impediment. The intensity of water current (electric current) is the quantity of water V (electron charge) passing a point in the circuit in a determined time (t), i.e. $I_w = V/t$ $(I = Q/t)$. It is measured by the rotation speed of the flow-meters.

Circuits can be built in series and in parallel — in any complexity desired — the pressure distribution in series circuits can be visualized by pressure difference meters (two thin vertical tubes connected to the water circuit at the locations requested). Variable resistances are at hand in the form of tube clamps which can be closed to a greater or lesser extent.

Building up the Source Domain

Of course, the water circuits proposed by us as source domain are by no means self-evident but the rules and conceptions such as pressure difference, intensity of current, or resistance and their relations to each other have to be elaborated by the students. This is no easy job for them and takes much time. However, through this way of learning very high aims can be achieved.

Important steps on the way to a systemic understanding of water circuits include: building up and establishing *firstly*, an idea of continuity and linked to that the elaboration of a concept of current intensity; *secondly*, establishing the concept of pressure difference as drive and cause of the flow of water current as well as the idea of a pressure difference distribution along the resistances (flow-meters) present in the water circuit; *thirdly*, the experience that if I change anything at one point of the circuit this will have an effect, at the same time, everywhere in the whole water circuit. This experience — in contradiction to the strategy of sequential reasoning which is very successful in other contexts — is the core of systemic thinking and requires repeated experience and exercising as well as constant assurance for students over a long period of time.

An overview of the sequence of instruction is given in Table 4.1.

A crucial element of the teaching method is the play-oriented approach developed in our Institute in Bremen (Aufschnaiter and Schwedes, 1989), which means that — within a determined setting — students work on self-elaborated questions, independently plan and carry out experiments and test their own predictions and hypotheses. Short teacher-oriented phases alternate with long action-oriented phases. Students are supported in their investigations by a number of questions, tasks and ideas for experiments. This procedure enables students to work on problems relevant to them and to argue on their specific problems of understanding setting their own pace. The tasks set by the teacher make sure that students are confronted with the significant phenomena to be elucidated. Moreover, students are repeatedly encouraged not to give up in their efforts to understand and also to ask for the teacher's help: 'Ask questions until you have understood!'

The Sequence

Instruction starts with elaborating the function of the DWC in team work, accompanied by open questions to investigate. The most essential outcome of this phase is the perception of the DWC as a producer of pressure difference which maintains and keeps *constant* a water level difference determinable by the experimenter, and the understanding that it is the pressure difference which makes a connected flow-meter rotate and determines the speed of its rotation. In addition students have to learn that the return pump of the DWC is not responsible for the magnitude of water current but rather it transports back to column A the water arriving at column B of the DWC and thus keeps constant the pressure difference.

Table 4.1: The electric circuit as a system elaborated by means of the water analogy

Lessons of 90 minutes:

1.–3. **The double water column in a simple circuit**
functioning of the DWC, flow-watcher and clamp as water resistance, qualitative-visual demonstration of the system

3. **Intensity of water current $I_w = V/t$**
condition of continuity

4.–6. **Free work constructing various water circuits**
parallel and series connections, water circuit rules

7.–9. **Pressure and current distribution in water circuits (qualitative and quantitative)**
pressure, pressure in liquids, areas of equal pressure

10. **Pressure (difference) distribution in the resistance tube**

11. **Varying resistances in parallel and series connections**
definition of water resistance $R_w = \Delta p/I_w$

12. **Series and parallel connection of the DWC**
exercises

13. **Test**

14. **Analogies of water and electric circuits**

15. **Conducting mechanism and how it works in the simple electric circuit**
battery as a producer of charge difference
electrons — atomic residue — model

16. **Construction of electric circuits (free experimenting)**
measurement of voltage and intensity of current, rules of electric circuits

17. **Rules of electric circuits**
elaboration of conceptions *I, R, V* (as 'electric pressure difference')

18./19. **Electric circuit system (analogy-oriented)**
determination of resistance, variable voltage source, short circuit, types of battery connection, ammeter resistances (ideal), potentiometer, condenser, linear or non-linear resistances

20. **Test**

The return pump supplies the energy required for keeping the water circuit going. The pump has a variable output and can be heard because of its working noise. It is physically perceptable as a hand-operated pump which can replace the electric pump. It is very impressive to students if several flow-meters, one after the other, are connected to the DWC in parallel and students are able only with great difficulties to maintain the constant pressure difference (difference of water levels) of the DWC, and if more flow-meters are added, they cannot maintain it at all.

Awareness of the need for the condition of continuity in water circuits and the

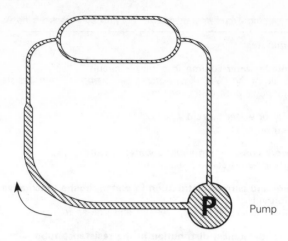

Figure 4.2: Water model with little red floating particles, showing the difference in speed in relation to the cross section of the tubes

formation of the concept of current intensity is supported by a model using floating particles, see Figure 4.2. Small red plastic particles floating in a transparent glycerine liquid are pumped in a circuit consisting of tubes of differing cross section and a branching point. Thus the higher speed in the thinner tube is clearly visible as well as the lower speed compared to the tube of equal cross section in the two parallel tubes. This initiates conceptualization of the current intensity being the amount of water passing a determined point in a period of time and makes clear the necessary distinction between water speed and current intensity.

When experimenting with a tube clamp (R = 0 to ∞) students can investigate the dependence of current intensity on the extent of impediment so that afterwards first qualitative comments on the circuit as a system can be made, seeing the connection of pressure difference, resistance, water current intensity and energy. In particular, if I modify the intensity of the water current by the clamp the intensity is changed in the whole circuit. The water flows more slowly through flow-meters and tubes, although it now has to flow faster at the point of the closed clamp than before when it was open.

In the course of the free construction work of water circuits — each group has five flow-meters at their disposal — the students elaborate qualitative water circuit rules, for example: 'in a series circuit all flow-meters rotate at the same speed, the more flow-meters are inserted in series, the slower they rotate', or 'a bridged flow-meter does not rotate at all, (almost) the whole water flows via the bridge'.

Ideas about rules are also used to make predictions on given circuits relating to the rotation speed of flow-meters, in a game we called *water toto*. Incorrect or uncertain predictions can be tested by experiment.

In the following lessons the water circuit rules are expressed by the notions pressure difference, intensity of the water current, and water resistance, for example the rule 'If I connect in parallel a second flow-meter the first one does

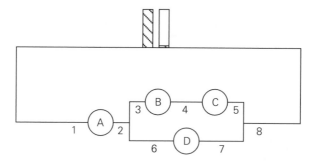

Figure 4.3:   Example of circuit diagram (for water circuits or analogous electric circuits) given for various calculations of I, R, V current intensities, resistances or (part) voltages

not change its speed. The second flow-meter behaves in the same way as the first one,' i.e., 'flow-meters connected in parallel behave as if each of them were single' is changed to 'At flow-meters connected in parallel there is the same pressure difference'.

At the end of the teaching unit on water circuits students should be able to solve tasks of the following type (see Figure 4.3).

Given the intensity of water current I_C through flow-meter C and the pressure difference of the DWC. Calculate the size of I_B, I_D, I_A. What is the size of the resistance of the parallel element (B, C, D), the total resistance (using the resistance of the flow-meter as unit), and the pressure differences $\Delta p_{1\,2}$ $\Delta p_{3\,4}$ $\Delta p_{6\,7}$ $\Delta p_{5\,8}$. State the rules you have applied for your calculation.

Equally important as the exact calculations we appraise the ability for making rough estimates, for instance, I_C is greater, equal or smaller than $I_{A/B/D}$ or $\Delta p_{2\,8}$ is greater, equal or smaller than $\Delta p_{1\,2}$, or R_{total} is greater, smaller or equal than R_A, or R_{BCD} is greater, smaller or equal than R_{BC}.

A further paradigmatic example for seeing circuits as systems is the analysis of pressure distribution in the series circuit if at one or more points resistances are modified, see Figure 4.4.

With the clamp open all resistances are equal, at each resistance there is the same (part)-pressure difference, all part pressure differences add up to the total pressure difference, while the intensity of the water current is equal in the whole circuit. If the clamp is closed now the pressure distribution — visibly — changes in the whole circuit — not only at the clamp — and, of course, also the intensity of the water current changes in the whole circuit, i.e., local changes have an influence on the whole system; the resistance at the clamp increases as well as the total resistance, i.e., the (total) intensity of the water current decreases, i.e., the (part) pressure differences at the other, unchanged resistances must decrease as well, the pressure difference at the clamp, however, must increase so that the resulting water current intensities become equal at any local point of the circuit. Because of the condition of continuity the water must flow faster through the clamp than through the other resistances.

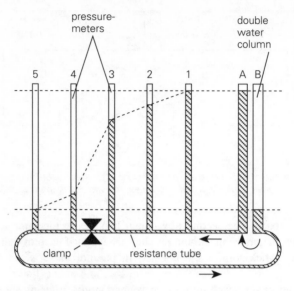

Figure 4.4: Experimental device for the demonstration of pressure distribution in a series circuit in the case of one locally varying resistor

Structuring the Target Domain

Students know the simple electric circuit with battery and bulb. So, as an introduction, the analogy relations, as shown in Figure 4.1 are summarized and discussed. To make full profit of analogical reasoning the course and the task structures very much resemble those of the water circuit unit. In the beginning students construct electric circuits following their own ideas. They are told that they are expected to apply what they have learned about water circuits. The students construct series and parallel connections as well as mixed connections including up to five bulbs. Their task is to formulate rules explaining or predicting the varying brightness of the bulbs, such as 'At a branching point the current is divided. Therefore the bulb in the common branch is brighter than the bulbs in the subsequent parallel branches.' In following problems students shall use those rules and predict the brightness of bulbs in given circuits, for example in the electric circuit analogous to Figure 4.3.

To enable students to perceive the bulb filaments as narrows in the electric circuit an experiment is carried out in which only wires of different thickness are connected to a circuit (analogous to Figure 4.2). The glowing of the thinnest wire and melting of some wax beads indicates the differing temperature of the wires during current passage. Temperature is an indicator of the different speeds of the electric charge carriers through the wire. Students' analogical reasoning is supported by the water flow model (Figure 4.2) built up again, to aid memory, next to the electric wire circuit. The electric current must pass the narrow thin wires particularly quickly (condition of continuity), and there the interaction of the current and the wire (friction) is so powerful that the wire begins to glow.

In the next stage students build electric circuits, partly according to their own ideas, partly guided by the teacher, and measure the current intensity and voltages at various locations of the circuits. Subsequently, they formulate — looking at the results of their measurements but, of course, also based on the water circuit rules developed before — adequate rules for the current intensity and voltage. For example: 'The current intensities of the parallel branches add up to the total current intensity', or 'In a series connection the current intensity is the same at any point', or 'In a series connection the part voltages add up to the total voltage', or 'In a parallel connection the voltage at each branch is equal'.

In a last step the students are asked to calculate the current intensity, part voltages and resistances in manifold branched circuits. The total voltage and the current intensity at one point of the circuit are given, the resistance is to be stated as a multiple of the resistance of one bulb.

They can use voltmeters and ammeters from the very beginning if they think these meters might be useful. The only instruction students get for these meters is that the ammeter measures current intensity and the voltmeter electric pressure difference or voltage. The rules are those formulated in Kirchhoff's Law, and students calculate resistances, current intensities and potential differences for various circuits. If they are not sure of their results, they can measure the voltages and current intensities or discuss with their neighbours and the teacher.

Results

Instruction has been carried out with one half of a grade 10 high school course, with thirteen students of which there were five girls and eight boys. Two groups of three and four students have been video-taped, all results of the group work have been collected, and at the end of the water circuit unit and the end of the whole unit a test has been written. Pre- and post-interviews have been carried out with the video-taped students. They served as the knowledge control and included more general questions on physics instruction. The post-interviews, three months after the end of instruction, served as well as a test of what had been remembered; two students have been interviewed twelve months after the end of the electricity instruction.

The test on electricity included the tasks of the Europe-Test, the test on water circuits included several tasks corresponding to those of the Europe-Test.

How the Teaching Strategy Worked

Students learned perfectly well the water model as seen from group work on video tapes and from the water test, and they had no serious difficulties in using the analogy for making inferences in the electric domain. In the phase of free experimenting at the beginning of the electricity sequences the students built-up circuits systematically. They seemed to merely verify rules they already had in mind.

Contrary to the developing phase for the water circuits all qualitative rules for electric circuits were formulated by the students on their own. Even if the water circuit rules may be considered to represent the pattern it does not seem justified to assume that the students just replaced the known rules by a few new wordings. This has become evident as the students were able to make correct predictions on the brightness of bulbs and to substantiate them using the rules they had formulated.

Similar experiences have been made in the phase of developing the quantitative electric circuit rules. The students had no difficulties using ammeters and voltmeters and to connect them correctly. Because in normal electricity courses many students repeatedly fail to solve such problems, this is another visible effect of the water analogy. The idea that the current must pass through the ammeter and that voltage may be understood as the electric pressure difference between two points of the circuit proved to be sufficient to allow students to find the correct connections, as can be seen from the classroom video-tape recording.

When explaining the demonstration experiment in which a circuit was constructed only with wires of different thickness, students' difficulties in applying the analogy with respect to resistances became evident. About half the students thought the thick wire must have the higher resistance. As the thick wire offers more 'inertia' to the current it does not become so warm. At this point it has been necessary to refer explicitly to the water analogy showing that thick tubes correspond to thick wires.

No severe difficulties could be detected in calculating for varying circuit diagrams, voltages, current intensities and resistances from some given quantities. The test on electricity in the end showed between 70 to 90 per cent correct solutions for most (for all) of the tasks taken from the Europe-test.[1]

Students reached, in their construction of meanings, the system level of complexity and could produce the arguments characteristic of an understanding of circuits as systems for the water domain as well as for the electric domain.

Even after three to twelve months students could remember the instruction unit well and describe electric circuits as a system using the concepts current intensity, resistance and electric pressure difference. Also they were able to adequately analyse or calculate parallel and series connections and mixed connections. No alternative conceptions could be detected though we tried to challenge students with bulbs of different brightness due to different resistances.

The greatest difficulty encountered was developing adequate concepts for the water circuits. Once an understanding for certain water circuit connections had been achieved they could easily be transferred to electric circuits. The reverse case that water circuit problems were handled or solved by means of electric connections has not been observed.

Several students find it difficult to regard parallel flow-meters as a total resistance which is half the size of the resistance of one flow-meter, although they know very well that the current intensity passing through each of the two flow-meters is the same as that passing through a single one, and that the total current intensity thus has doubled (if the pressure difference stays the same).

Another difficulty is to understand that in a parallel connection there is the same pressure difference at each branch. The total cross section becomes greater; if I have two or three parallel tubes instead of one, the students argue, pressure is force per area, thus doubling the cross section (two tubes) means the pressure at each tube is only half the size. With pressure difference meters this argument can be met, however, the instruction on pressure distribution in liquids and in interconnected containers of different shape is more efficient for realizing that the pressure depends only on the water height in the container.

Learning and understanding the pressure distribution in series circuits has been particularly difficult for students, especially the modification of the pressure distribution when one resistance is changed. If I increase the resistance at one point, at the clamp, students consider that the current should have a reduced speed at this point as the resistance impedes the current. The water therefore should not suddenly flow faster through the increased resistance. This is contrary to the intuitive thinking of all students. For the circuit as a whole however it holds that the total resistance increases, and the current intensity is reduced. The part pressure differences now adjust themselves so that in each section of the water circuit an equal current intensity results. This in turn leads to a faster water flow at the narrow sections than at the wider ones so that the part pressure differences adjust themselves in proportion to the resistances consistent with the knowledge that the greater the pressure difference is the greater is the speed of the flowing water. The sequence of arguments required here has — even if already understood — to be repeated and practised several times.

As we know from a preliminary study, where students didn't reach an understanding of water circuits on a system's level and thus their knowledge of circuit rules was not consistently connected and integrated, the analogical reasoning was restricted to the single rules students had in mind. To arrive at a proper combination of the several rules, which is implicated in a system's view, seemed not possible for these students though it was repeatedly tried by the teacher in the electric domain.

All students stated: 'Dealing with the water circuits was a help for understanding electric circuits.' The less students knew before about electric circuits, or the greater their uncertainty was in this domain the more they considered dealing with the water circuits to be helpful or even to make understanding possible at all. Even students not interested in electricity admitted the water circuits had opened up chances of understanding. 'With water it has been easier, there you can see it', was the unanimous judgment.

A preliminary summary of the testing of our instruction strategy can be stated as follows:

- The water analogy contributes to a better understanding of electric circuits.
- The limits of what can be learned in electricity are given by the extent of understanding the water circuits.
- Students seem to understand water circuits more easily than electric circuits.

This is due on the one hand to the closeness of many terms used for description and calculation to the sensual perception of students, and on the other hand to the hydromechanic connections which permit easier use of causal arguments which are not combined with mysterious ideas such as the invisible electric current or the dangerous electric voltage.

- Thus, the use of the water analogy leads to a certain reduction of problems of understanding electric circuits, for many students apparently the only viable way.
- The analogical reasoning produces no problems but the construction of a consistent network of knowledge about water circuits proves to be more difficult than anticipated.

Note

1 It has to be kept in mind that the results of the Europe-test represent an average of all ability levels while normally high school courses are considered to be in the upper ability range as far as performance in physics is concerned.

Messrs. MSW-Winterthur, Zeughausstr. 56, CH-8400 Winterthur, Switzerland plans to produce a set of DWC's for water circuits for demonstration purposes to be ordered via: Messrs. Müller Lehrtechnik, Pfalzgrafenweiler Str. 14, D-72285 Pfalzgrafenweiler, Germany. For do-it-yourselfers the flow-meters can be acquired from Messrs. H. Jürgens and Co., Langenstr. 76, D-28195 Bremen, and construction instructions in German language can be ordered from H. Schwedes, Universität Bremen, FB 1, P.O. Box 33 04 40, D-28334 Bremen, Germany.

References

AUFSCHNAITER, ST.V. and SCHWEDES, H. (1989) 'Play orientation in Physics Education', *Science Education*, **73**, 4, pp. 467–79.

BLACK, D. and SOLOMON, J. (1987) 'Can pupils use taught analogies for electric current?', *School Science Review*, pp. 249–54.

DUDECK, W.-G. (1987) *Entwicklung eines Spannungsbegriffes über den Einstieg mit der Elektrostatik*, Unveröffentlichtes Manuskript, Bremen.

DUPIN, J.J. and JOHSUA, S. (1987) 'Analogies and "modelling analogies" in teaching: Some examples in basic electricity', *Science Education*, **73**, 2, pp. 207–24.

DUPIN, J.J. and JOHSUA, S. (1994) 'Analogies et enseignement des sciences: Une analogie thermique pour electricité', *Didaskalia*, **3**, pp. 9–26.

NIEDDERER, H. and GOLDBERG, F. (1994) 'An individual student's learning process in electric circuits', Paper presented for the NARST Annual Meeting 1994 in Anaheim (USA).

RHÖNECK, CHR.V. and GROB, K. (1993) 'Representation and problem solving in basic electricity, predictors for successful learning', in NOVAK, J. (Ed) *Proceedings of the Second International Seminar on Misconceptions and Educational Strategies in Science and Mathematics*, Vol III, Ithaca, NY, Cornell University, pp. 564–77.

SCHMIDT, D. (1989) *Zum Konzeptwechsel*, Frankfurt, Peter Lang.

SCHWEDES, H. (1983) 'Zur Kontinuitätsvorstellung bei Wasserstromkreisen und elektrischen Schaltungen', in Kuhn (Hrsg.) *DPG-Frühjahrstagung Gießen 1983*, pp. 264–9.

SCHWEDES, H. (1985) 'The importance of watercircuits in teaching electric circuits', in DUIT, R. JUNG, W. and V. RHÖNECK, CH. (Eds) *Aspects of Understanding Electricity*, IPN, Arbeitsberichte, Kiel.

SCHWEDES, H. (1995) 'Die Leistungsfähigkeit verschiedener Analogien für elektrische Stromkreise', in BEHRENDT, H. (Hrsg.) *Zur Didaktik der Physik und Chemie*, Bd. 15, Leuchtturmverlag, Alsbach/Bergstr., pp. 283–6.

SCHWEDES, H. and SCHILLING, P. (1983) 'Schülervorstellungen zu Wasserstromkreisen', *Physica Didactica*, Heft 10, 1983, pp. 159–70.

SHIPSTONE, D.M., RHÖNECK, CHR.V., JUNG, W., KÄRRQUIST, C., DUPIN, J.J., JOHSUA, S. and LICHT, P. (1988) 'A study of students' understanding of electricity in five European countries', *International Journal of Science Education*, **10**, 3, pp. 303–16.

VAN DEN BERG, ED. and GROSHEIDE, W. (1993) 'Electricity at home: Remediating alternative conceptions through redefining goals and concept sequences and using auxiliary concepts and analogies in 9th grade electricity education', in NOVAK, J. (Ed) *Proceedings of the Third International Seminar on Misconceptions and Educational Strategies in Science and Mathematics*, Ithaca, NY, Cornell University.

5   Students' Computer-based Problem-solving in Electricity: Strategies and Collaborative Talk

René Amigues

Abstract

The purpose of this research was to study what role is played by interstudent cooperation in the use of a computer system to solve an electrical problem. Students had to search for defective resistors in two electric circuits displayed on the computer screen. In the two experimental situations, the electrical circuits were isomorphic, only the spatial organization of the components differed, according to 'canonical rules' or 'non-canonical rules' circuit diagrams. Forty-eight French tenth-grade students (age 15–16) were randomly divided into twenty-four pairs. Twelve pairs were side by side facing a computer, allowing natural collaborative talk. However the other twelve pairs were communicating through computers. In each group six pairs were working on a 'canonical circuit' diagram and six pairs on a 'non-canonical circuit' diagram. The results showed (a) in natural collaborative talk the spatial layout of information in the problem statement (the circuit diagram) orients students in the choice of a problem-solving strategy; (b) computer communication was the instigator of joint problem-solving for the pairs, whatever the spatial organization of the information. This experiment stresses the specific role of semiotic tools in supporting reflective action and shared meanings, in contrast with the normal teaching–learning situation.

Introduction

The purpose of this research was to investigate the role played by interstudent cooperation in the use of a computer system to solve an electrical problem. This study took place in an interdisciplinary project in which the aim was to study the cognitive processes involved in problem-solving strategies used by students after teaching on electricity.

Theoretical Framework

This experiment was based on the previous findings of two series of studies:

1   In France, among the various semiotic tools used in the field of electricity, electric circuit diagrams (ECDs) play a specific role. For the physicist, an electrical diagram is not an illustration of actual electric wiring. It is a graphic object which

spatialy represents physical properties: concepts and laws of electricity. Thus, it does not act as an analogical diagram, but typically, as an 'explanatory' diagram. For the student, understanding how a circuit works consists of extracting the relevant information embedded in the graphical code in order to construct a mental representation of spatial and non-spatial properties of the phenomenon considered. The major difficulty for students is to convert the temporal and spatial information into non-temporal and non-linear relations between concepts (Amigues, 1988).

Despite this specificity, the teaching of electricity does not imply specific teaching about encoding and decoding of electric circuits, in class. Moreover, the process of teaching–learning electrokinetics is based on the use of ECD as a quite 'transparent' means of facilitating the correct passage from the conceptual structure to the students' own structure and to help students understand how an electric circuit works.

These ECDs used in class as well in manuals are based on specific rules: the shape of this diagram is rectangular, the wires are drawn as horizontal and perpendicular lines. The main diagrams are further developed with detailed subdiagrams called 'canonical circuit diagrams' (Caillot, 1988): elements in series are drawn aligned and elements in parallel are drawn on neighbouring parallel lines (Amigues and Caillot, 1990). These are prototypical units for representing devices in series and devices in parallel. A second kind of rule — social rules — legitimates the conventional layout used in class.

In fact, current pedagogical practice products a socio-technical frame which induces a prototypical representation. This set of tacit rules reinforces the 'sequential reasoning' (Closset, 1983) or 'familiar procedure' which consists of giving step-by-step description of the successive elements of the circuit (like water flowing through a pipe, from the positive terminal of the cell to the negative one), without dealing with functional relationships between these elements (Amigues, 1988). This interpretation leads us to hypothesize that students' difficulties in learning electrical concepts stem from these 'tacit rules' used in class.

2 The second series of studies (Amigues, 1988, 1989; Amigues and Agostinelli, 1992) was aimed at studying the role of peer interaction on change in students' conceptions about electricity. These studies focused on the strategies used by individuals and pairs (14–15-years-old) to understand how an electrical circuit works. The students' spontaneous descriptions, even after teaching, remained based on 'naive theories' about current. These naive conceptions, which govern the 'familiar procedure' used (i.e., sequential information processing) were found to be an obstacle to describing this technical situation. The findings showed (a) that destabilization of the familiar procedure enhances the description of the technical situation; (b) that this destabilization is more frequent when students work in pairs than individually, and (c) that destabilization only occurs when students are able to elaborate a 'system of mutual meanings'. The findings showed that pairs gave more functional responses than individuals; this occurred when the pairs were able to reorganize their initial representation of the problem and regulate their own action. The results of a recent experiment (Amigues and Agostinelli, 1992) show that students transfer the learning attitude induced by the tacit rules used in traditional

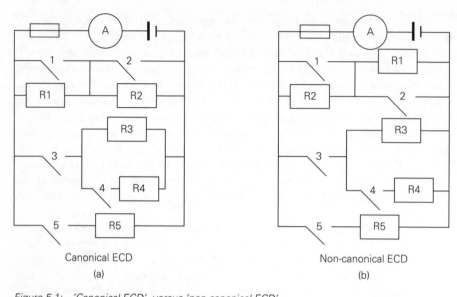

Figure 5.1: 'Canonical ECD' versus 'non-canonical ECD'
Note: The differences remain only in the spatial organization of the resistors:
(a) R1 and R2 are aligned, R3 and R4 are in parallel and this device is 'closed'.
(b) R1 and R2 in series are not aligned, R3 and R4 in parallel and this device is 'opened'.

instructional setting to the computer learning environment. In this experiment two versions of the same electric circuit were presented: one presents a pattern of information in a 'familiar manner', learned in class (called canonical); the second version presents a modified pattern of information in an 'unfamiliar manner' (called non-canonical). The results show that, in canonical version, students (working alone or by pairs) could not improve their information processing strategy to solve the problem. This is the reason why pairs only performed better than individuals in the non-canonical version of the problem. These results are very interesting because the differences between canonical and non-canonical versions are very slight (see Figure 5.1). Thus, in the coming text, canonical will mean 'familiar layout used in class' and non-canonical will mean 'unfamiliar layout'.

Based on these findings, the aim of this research was to study what role is played by interstudent cooperation in the use of a computer system to solve an electrical problem. Students had to search for defective resistors in an electric circuit displayed on the computer screen (see Figure 5.2) but they induce significant differences in the strategies used by students in each situation.

Method and Hypothesis

The study presented here is based on the results of the above experiments and in analysing the task, we noted various information processing strategies used by students to troubleshoot electric circuits. They can be classified into two main types: sequential and non-sequential. For each strategy used, we were also able to determine, among the possible troubleshooting tests run by the students, which ones were 'essential' and which ones were 'superfluous' (for more details, see Amigues, 1989; Amigues and Agostinelli, 1992).

Material

In this task students had to search for defective resistors in an electric circuit. In the two experimental situations, the electrical circuits were isomorphic, i.e., from the electrical standpoint the structural and functional aspects of the circuit were the same. Only the spatial organization of the components differed (see Figure 5.1). However in one of the situations, the circuit pattern was based on 'canonical circuit diagrams' (familiar layout used in class), whereas in the second, the information was based on 'non-canonical circuit diagrams' (unfamiliar layout).

Hypothesis

These two presentation versions were expected to induce different information processing strategies. The canonical version should induce sequential analysis based on the 'surface features' of the circuit which consists of opening and closing switches in their order of occurrence (from 1–5), without further reflection. The non-canonical version should promote non-sequential analysis based on functional aspects, which could be obtained by the combined closing of switches which were not necessarily adjacent to each other, and which can be assessed by tests run during the troubleshooting.

Sample

Forty-eight French tenth-grade students (age 15–16) were randomly divided into twenty-four pairs. Twelve pairs were side by side facing a machine, in natural collaborative talk. However the other twelve pairs were communicating via a network. In each group six pairs were working on the 'canonical circuit' diagram and six pairs on the 'non-canonical circuit' diagram. In the pairs/network the pair of students could only communicate via the machines (they could not talk to or see each other). Each pair worked on one circuit (canonical or non-canonical version) and could communicate verbally (by writing on the keyboard), or schematically (by sending part of a diagram, for instance), or via a combination of these two communication modes. The instructions asked them to cooperate and the game consisted of helping each other to find the answer without directly giving it.

Method

The screen was divided into three zones (see Figure 5.2):
In the experiment zone, the electric circuit was presented in the centre of the

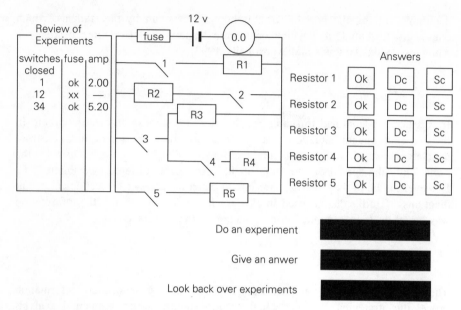

Figure 5.2:   *The screen as it appeared to the students*
Note: For each resistor the possible responses were:
non-defective   (OK)
disconnected   (DC)
short-circuited   (SC)

screen and permanently visible. The students pointed with the mouse to 'Do an experiment'. Then they could open and close the switches. The value of the current was always displayed (ammeter). Whenever there was a short circuit the fuse blew.

In the zone used to review previous experiments (on the left side of the screen) the students could point to 'Look back over experiments'. The previous experiments were displayed with the corresponding values of the current.

In the response zone (right side of the screen) the students could respond at any time by pointing to 'Give an answer'. For each resistor the possible responses were non-defective (labelled OK), disconnected (DC) and short circuited (SC). The possible computer messages were 'correct answer' (when all defective resistors had been correctly identified), 'partially correct answer' (when some of the defects proposed were incorrect), and 'incorrect answer' (when all proposed defects were incorrect).

The procedures used by the students were recorded automatically (order and timing of operations). This gave us three dependent, quantitative variables to characterize performance: the number of test experiments, the number of responses, and the number of reviews of previous experiments. The qualitative dependent variables were the choice of test experiment (essential versus superfluous) and the sequential or non-sequential nature of the strategy employed. The dialogue of each pair was also recorded on tape, or by computer.

Table 5.1: Distribution of pairs by type of strategy employed in the canonical situation and in the non-canonical situation

	Canonical		Non-canonical	
	Sequential	**Non-sequential**	**Sequential**	**Non-sequential**
D/Computer	5	7	6	6
D/Network	3	9	1	11

Table 5.2: Mean number of Experiments, Answers and Reviews for pairs (Computer and Network), in the canonical situation and in the non-canonical situation

		Experiments	**Answers**	**Review**
	Pairs/Computer	14.67	13.17	2.17
Canonical	Pairs/Network	25.50	5.17	5.00
	Pairs/Computer	18.83	5.83	1.17
Non-canonical	Pairs/Network	26.67	8.50	8.50

Results

Analysis of Strategies

A sequential strategy in this study consisted of processing information step by step by opening and closing switches in their order of occurrence (from 1–5), without further reflection. But to solve this problem, the necessary information could only be obtained by the combined closing of switches which were not necessarily adjacent to each other. This latter strategy was called non-sequential.

Table 5.1 gives the distribution of subjects according to the strategy used. In general, one half of the pairs working side by side at a computer used a sequential strategy and half a non-sequential strategy, whereas the pairs cooperating via the network more frequently used a non-sequential strategy. Results show that there is more frequent destabilization of sequential information processing in the case of the pair/network group and this effect is stronger in the case of the non-canonical diagram.

Interactive Actions with Computer

Table 5.2 indicates the mean number of (a) experiments conducted by the students, (b) answers, (c) times the subjects reviewed the results of previous experiments.

- Experiments: The pairs/network used the computer more frequently to conduct experiments than the pairs/computer did.
- Answers: In the problem-solving process the fewer the answers, the better the performance. The pairs/computer gave more answers than pairs/network. This tends to indicate that, in the canonical situation, these pairs had

Table 5.3: Mean number of essential tests and superflous tests out of the total number of experiments by pairs in the canonical situation and the non-canonical situation

	Tests	Essential	Superflous
Canonical	Pairs/Computer	8.50	6.17
	Pairs/Network	18.83	8.17
Non-canonical	Pairs/Computer	10.87	9
	Pairs/Network	18.75	7.58

more difficulty finding correct answers than the pairs/network, in the same situation, and more difficulty than their counterparts in the non-canonical presentation.

- Reviews: As a whole, and regardless of working mode, the subjects referred infrequently to past experiments before responding. However the pairs/network reviewed their experiments more than the pairs/computer.

These results suggest that interactive actions with the computer are different according to the working context. They suggest also that the activity of pairs/network was more reflective than those of pairs/computer.

The 'Quality' of the Tests Run by Students

The following results concern the 'quality' of the tests run by the students. In reference to our task analysis, tests which provided relevant information for solving the problem were labelled 'essential', and those which did not were labelled 'superfluous'. Table 5.3 gives the distribution of the tests for both tasks.

For both versions and both working modes, the students made more essential tests than superfluous ones. However, these results show that pairs did not differ in the number of superfluous tests run, but rather in the number of essential ones.

On the whole, these results show that:

- the pairs/computer were more sensitive to the sequential processing information inducted by familiar layout than the pairs/network, and their performances were poorer. In addition, the pairs/network adopt a better information processing strategy to solve the problem.
- the pairs/network adopted easily a functional strategy and these results suggest that such a strategy is based on a reflexive analysis.

Describing Paired Dialogues Using Computers

The following dialogue excerpts clearly illustrate the kind of collaborative talk used by pairs/computer, at the beginning of the task. In the example given below, two students (A and B) are simultaneously looking at the screen in the canonical situation.

Student A	**Student B**
The first resistor is OK.	
	The second, too!
The cell is . . . Where does the current come from? . . .	
	The third resistor must be disconnected
It's flowing through R1 so this wire . . .	
	I'll start to give the first answer . . .

In this excerpt, the students talk one after the other about a different element of the circuit. This type of exchange is called the 'alternation mode', because it is based on conventional turn-taking (Amigues, 1988, 1989). In this kind of conversation the implicit rules associated with the canonical diagrams emerge as rules for applying 'ready-made knowledge'. Furthermore, it is very difficult for the students to integrate a new piece of knowledge; they strive to interpret the represented properties, and each student's remarks are directed towards himself/herself, regardless of the presence of a peer.

The next excerpt is called the 'interactive mode' because it is based on mutual meanings, with the students focusing on the same object at the same time. In the example given below, two students (A and B) are simultaneously looking at the screen in the non-canonical situation.

Student A	**Student B**
Woah! This circuit is bizarre.	
	Yeah, there are five resistors.
Where is the cell?	
	Here, at the top of the circuit.
OK! That's right. And the current is leaving the cell . . .	
	It's leaving the cell from the positive connection.
Right, since the current is leaving the cell from here, and it's flowing through this wire and . . . this one	
	I wonder if they are really in parallel?

In this situation, the students could not directly use tacit rules as their counterparts did. They had to mentally reconstruct the structure of the electrical circuit, which involved considering the spatial arrangements of the diagram. Their exchanges dealt with meanings attributed to the information represented and to the goal that guided their actions.

It is important to note (a) that the two types of dialogue coexisted within each pair, in all situations, and (b) that the tacit rules learned during teaching of electricity helped students to solve problems rather than to develop an understanding of how an electric circuit works.

Describing Paired Dialogues Using a Network

First of all, note that communication via a network makes the dynamics of the exchanges between the students more explicit. For instance, the action steps to be implemented can be clearly identified for both versions of the circuit. An initial analysis of the verbal exchanges gives us the following description of the dialogue:

The problem statements serve as instructions or execution orders addressed to the partner ('You close the switches one by one, and then you compare the current'; 'Compare the current you read on the ammeter when the switches are opened, and when you close switch 5'). The student then executes an operation, and either gives an account of the result ('You're right, it works!!! It was resistor R4 that was disconnected'), or an evaluation of the process taking place ('When you try R3 then R3 plus R4, you get the same result. Why?').

Despite the physical separation and the remote communication via the machines, this sequence of statements is nothing like the alternating regulation observed before. Here, regardless of which version of the circuit is presented (canonical or non-canonical), the sequence of statements is taken into account by each pair. On this point, it should be noted that communication by means of computers points out the 'moves' mode in the solving process, which were 'masked' to a greater extent when the students were side by side facing a machine. For example, it is clear here that the students regard the 'verbal statements' as 'actions' to be carried out in order to test the state of a resistor, check a result, initiate a search, test a hypothesis, etc. These statements indicate what should be verified by focusing the students' attention on different functional aspects of the situation.

The moves or episodes in the solving process are based on the breakdown of the circuit into relevant objects — such as a series of switches, a subcircuit, or the distinction between the stable or structural objects (ammeter, intensity, subcircuits) and effect objects — which enable the students to get involved in a chain consisting of a causal agent, an action, and an effect, as in the following example: 'If you close switch 3, the ammeter says 5.20, and if you close switches 3 and 4, the ammeter still says 5.20, even though it should indicate another value.'

If these actions are viewed as 'cognitive traces' which one attempts to get the partner to share in order to construct a common representation of the problem, then we dispose of indicators which allow us to follow the on-line operations through which the problem representation is built, evolves, is transformed, by cooperative actions.

Conclusion

This study shows the potential merits of examining student dialogue in order to understand the strategies used by students in their own understanding and processes of learning. Furthermore, this study emphasizes the role of ECDs as specific semiotic tools. These semiotic tools are considered both as a 'technical tool' and a 'symbolic code'. By contrast with current teaching of electricity, or the 'natural' dialogue situation, the 'artificial' dialogue via a machine increases the 'instrumental' control exerted by the students, and the functional meanings they construct, as now their action depends directly upon how they use these tools. In this condition, computer communication was the instigator of joint problem-solving for the pairs, whatever the spatial organization of the information may have been. This is most likely the reason why, whenever the mediation of a peer alone prevails, the meanings are attributed to the activity in reference to the normative framework of the classroom and not in reference to the technical and functional aspects of the electrical circuit.

The modes of communication via the network modify the direct action of the students. These modes transform the productive and immediate actions towards a representative and regulative activity. The reflective actions about the functional aspects of the diagram emphasizes the social interaction and shape the progress of verbal interaction. This functional relationship between the technical content of didactical dialogue and communicative modes usually failed to appear in teaching–learning situations typical of current practice.

References

AMIGUES, R. (1988) 'Peer interaction in solving physics problems: Sociocognitive confrontation and metacognitive aspects', *Journal of Experimental Child Psychology*, **45**, pp. 141–58.

AMIGUES, R. (1989) 'Peer interaction and conceptual change', in MANDL, H., DE CORTE, E., BENNETT, N. and FRIEDRICH, H.F. (Eds) *Learning and Instruction: European Research in an International Context*, (Vol. 2.1) Oxford, Pergamon Press, pp. 27–43.

AMIGUES, R. and AGOSTINELLI, S. (1992) 'Collaborative problem solving with computer: How can an interactive learning be designed?', *European Journal of Psychology of Education*, **7**, pp. 325–37.

AMIGUES, R. and CAILLOT, M. (1990) 'Les représentations graphiques dans l'enseignement et l'apprentissage de l'électricité', *European Journal of Psychology of Education*, **5**, pp. 477–88.

CAILLOT, M. (1988) 'Circuits électriques: Schématisation et résolution de problèmes', in AMIGUES, R. and JOHSUA, S. (Eds) *L'enseignement des Circuits Électriques: Conceptions des Élèves et Aides Didactiques*, Technologies, Idéologies, Pratiques, 7, pp. 59–83.

CLOSSET, J.L. (1983) 'Le raisonnement séquentiel en électrocinétique', Unpublished doctoral dissertation, Université de Paris VII, France.

6   Group Concept Mapping, Language and Children's Learning in Primary Science

Steve Sizmur

Abstract

Pupils studying science are faced with the problem of learning meaningfully the language of science. Meanings arise through successful acts of communication, and this depends on knowing the relationships between concepts. Concept mapping is a way of exploring those relationships. This research investigated the discussion that took place when groups of 9–11-year-old children made concept maps together, using terms from three science domains. The talk was recorded and analysed, drawing on sociolinguistic perspectives. Patterns were identified in the discourse which could be related to the nature of the relationships the children incorporated in their maps. The findings suggested that the way the concept mapping task was structured tended to encourage the children to propose and evaluate potential relationships between the terms they were given. This was most effective when the terms were from varied ontological categories. At times, the discussion was characterized by a notable convergence on shared meanings due to the contributions of more than one group member. When pupils shared in the elaboration of a relationship between terms, this was particularly likely to result in a scientifically acceptable meaning being incorporated in the map. It was concluded that the activity was beneficial in allowing children to engage in sustained discussion about scientific meanings.

Introduction

This investigation is set in the context of a substantial body of research findings showing that children continue to use scientifically inappropriate ideas in spite of the teaching they experience in school. A burgeoning range of literature in this area over the past twenty years testifies to the seriousness with which the problem is regarded (see, for example, Carmichael *et al.*, 1990), and it hardly seems necessary to rehearse the findings again. Here, then, is the problem to be addressed in this study.

Much of the literature on this topic is driven by what has become known as 'constructivism'. This is a broad position encompassing many versions. One version focuses on the individual's internal functioning, and on how incoming data are matched to representations in memory (see, for example, Osborne and Wittrock, 1983). This present research takes a different perspective. On this view, the problem in science education amounts to a loss of meaning for the pupils (Eger, 1992). Though children can sometimes appear to apply scientific ideas successfully, they are unable to use them meaningfully.

Meaning is not primarily an individual phenomenon. In language, socially shared meaning is logically fundamental, and it is in the context of this social meaning that personal meaning develops (Wittgenstein, 1967). In Wittgenstein's view, meaning is determined within a 'language-game', an interdependent set of activities and language uses serving specific human purposes. (It might be added that this is a view that entails a kind of realism, since it is dependent on successful reference.) The same term can mean very different things in different contexts, depending on the 'game' that is being played: we have to know at which 'post' the word is stationed before we can interpret it (*ibid.*, p. 29). Part of the job of science educators, seen from this perspective, is to induct children into the language of science. This cannot mean simply adding new words to an existing 'everyday' way of talking. It is necessary to develop a scientific way of talking, through which those words acquire their meaning.

A useful development of the Wittgensteinian position has been provided by Sainsbury (1992). She locates his notion of a word's 'post' within a concept, and shows the meaning of a concept to consist in its relationship to others.

> I have described the theory-system as a network of possible connections of various kinds. Each concept is, as it were, anchored in place by these connections stretching out in different directions, its set of links with others. (Sainsbury, op cit., p. 45)

This view corresponds in many ways to the more familiar Ausubelian one (Ausubel *et al.*, 1978), but from a social rather than an individual perspective:

> For me, meaning is knowing the place of a word in its theory, not relating a word to a set of private associations. (Sainsbury, op cit., p. 99)

This brings us to the concept map: a diagram of the relationships between concepts. Using such a tool might help in understanding those relationships, whilst at the same time providing a model of the coherence of scientific theories.

Concept Mapping

Concept mapping (Novak and Gowin, 1984) has also been the subject of a vast range of literature and research evidence. The conclusion that concept mapping can have a positive effect on learning in science is hard to avoid (Horton *et al.*, 1993). However, the research that has been conducted has been carried out almost exclusively within an experimental paradigm. Hence we know quite a lot about the effect of concept mapping on the average learner in a treatment condition. Despite much speculation, we know little about how concept mapping might produce its effects, as experimental studies characteristically hold the treatment to be a 'black box'. Again, it has often been suggested that concept mapping is most beneficially carried out as a collaborative task. With the notable exception of Roth and Roychoudhury

(1992), few have ventured to investigate this claim or sought to understand the processes involved.

With teachers now encouraged to adopt approaches that best suit the learning needs of the children and the subject-matter to be learned, it would seem particularly important for them to understand just how those approaches function in practice. Group concept mapping is one strategy that could be deployed to meet a range of classroom needs. In particular, it would seem to offer a means of structuring group discussion of scientific ideas. Hence the attempt reported here to gain understanding of concept mapping in action that could help teachers use the approach to best effect.

The Present Research

The study has the aim of investigating what processes are at work when children collaborate to construct a concept map. It is assumed that, in constructing such a map, the relationships between the concepts labelled by the various terms in the map must be considered. But to what extent is the discussion that leads to the finished map constructive, a tool of present learning, and to what extent merely reconstructive, a record of past learning?

The guiding questions were:

- What processes characterize the production of a concept map by a group of children?
- Does the emerging concept map help to structure the children's activity in a way that encourages the critical sharing of meanings and the emergence of new understandings?

These questions demanded an investigation that was naturalistic rather than clinical, interpretive rather than experimental.

Sample

Groups were drawn from three classes of 9–11-year-olds, in two contrasting schools: one suburban and one an inner-city school. Each class studied a different science topic: habitats (biology); Earth in space (astronomy); sound and hearing (physics/biology). The results reported here are from across the three topics. Separate analyses are to be made for the individual topics. Altogether, fourteen groups were involved, consisting of a total of fifty-seven children.

Data

The children worked in groups to construct one concept map immediately before the topic was taught, and another immediately after. This fitted the normal classroom routine. The terms to be included in the map were chosen by the teacher (or in one case by the researcher in consultation with the teacher) to suit the teacher's plans for teaching the topic. The children were familiar with constructing concept maps, and knew that they could add terms of their own if they wished. Early trials had shown this to be an effective approach with the age group.

The resulting 'inscriptions' formed part of the data, and relationships in the maps were coded as to whether they were:

- scientifically acceptable;
- a correct, but 'everyday' meaning, or a vague relationship; and
- a definite misconception.

Generalizability of this scheme is to be evaluated, using a second coder.

Also collected were audio recordings of the groups' discussions. These were transcribed, and the utterances classified according to a category scheme that was developed through a process of iteration between prior theoretical assumptions and encounters with the data. The theoretical orientation was derived from philosophy and sociolinguistics as well as existing category systems (Searle, 1969; Halliday, 1973; Sinclair and Coulthard, 1975; Barnes and Todd, 1977; Stubbs, 1983).

The major prior assumption from this orientation was that speech is a form of situated action. To classify an utterance, it is necessary to ask what it is doing in the discourse. The category scheme featured several different kinds of action (termed 'moves'), the most important being:

- introducing a new term or relationship;
- supporting a previous move;
- elaborating on a previously introduced term or relationship;
- eliciting elaboration; and
- challenging a previously introduced idea.

The next assumption was that moves in discourse are thematically linked to form an exchange. Through some exchanges, the children developed (or attempted to develop) ideas for potential incorporation in the map. These were termed 'ideational' exchanges. Other kinds of exchange were present in the data (for example, to do with the procedures to be followed, discipline or casual chatter), but the main purpose was to discover the properties and function of ideational exchanges of different kinds. It is possible, in the course of this brief article, only to give something of the flavour of these different types of ideational exchange, together with an indication of some interim conclusions.

Results

Three distinct kinds of ideational exchange were identified as significant. The distinctive features were, firstly, whether an idea initially introduced received further elaboration during the exchange, and, secondly, whether any such elaboration

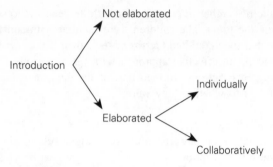

Figure 6.1: Types of ideational exchange

involved the other members of the group, who were not responsible for introducing the idea. The different trajectories may be illustrated schematically, as in Figure 6.1.

To illustrate these contrasting possibilities, some examples will now be presented. These were generated by the same group of four children, in the process of making the map shown in Figure 6.2. The children had been studying 'habitats', and most of the terms shown in the map were ones they had been given by the teacher.

In the first extract, P1 proposes a possible next link in the concept map. The relationship is expressed in fully developed form.

P1 Light is needed for survival
P2 Hmmm
P3 Hmmm ... no

The following moves by P2 and P3 show that they are unsure whether this fits. Ultimately, the idea is rejected, and they move on to a new focus of discussion. This is an example of a type (i) exchange: Introduction ⇒ Not elaborated. In the second example, P1 is trying to formulate a relationship based on photosynthesis. This is an instantiation of exchange type (ii); Introduction ⇒ Elaborated ⇒ Individually.

P1 Oh there's another one for trees
P1 Here's another one for trees, trees what-sha-ma-call-it
P2 Trees what-sha-ma-call-it?!!
P3 Trees what-sha-ma-call-it trees-
P1 No, trees, um
P3 Rain
P1 What's the word?
P3 Rain
P1 No listen
P3 Sunlight?
P1 Could you listen please? ... trees ... em ... take in carbon dioxide

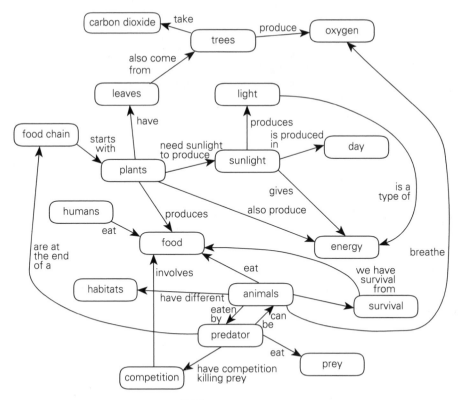

Figure 6.2: Pupils' concept map on the topic 'habitats'

P1 cannot think of the correct terms to describe what she is thinking, so uses a place holder ('what-sha-ma-call-it') to keep this open as a topic of discussion, despite the mocking remarks of the others. Eventually, P1 finds a satisfactory solution, which is incorporated into the concept map (Figure 6.2). This link is therefore mainly the contribution of one of the group.

In the various recordings, there were many instances of type (iii) exchanges; Introduction ⇒ Elaborated ⇒ Collaboratively. In the following example, the group are considering several different, but connected, strands of knowledge, all pertinent to the theory about plants' source of food. P1 has chosen to connect 'plants' and 'food', while P2 and P3 focus initially on the role of sunlight.

P1 Now plants, we put food?
P2 Nnn
P3 No
P4 Produce food
P3 Need sunlight
P4 Yeah
P1 But where's produce come then?

P1 We don't have to . . . have exact
P2 Sunlight then
P1 No put food
P3 Yeah plants need sunlight
P1 No put food and then put sunlight coming off it as well . . . put sunlight
    coming off it there
P2 Plants provide food
P2 Yeah
P3 Yeah OK

Eventually, a suitable formulation that combines the elements offered by the different group members is approved and then adopted (see Figure 6.2).

These distinct types of exchange were found to be fulfilling different functions in the discourse. Across all the groups, it was possible to draw some general conclusions about what these functions were.

Type (i) Exchanges

Exchanges of the first type were relatively common. They could range in size from a single introductory move up to ten moves or more. Figure 6.3 shows the length of different types of exchange across all the groups in the study. Most were fairly short, consisting of about two or three moves. These were often, but by no means always, evidence of exploratory talk. The children were searching the set of words they had been given for possible connections, and introducing these for consideration, often in a vague or tentative way. By leaving the nature of any connection unstated, it was possible for group members to introduce an idea as it occurred to them, without the obligation to think it through further at that stage. This kind of introduction may therefore be read as 'I think there is a connection here. Shall we pursue it?'. In some cases, the inquiring nature of the introductory move was quite plain. In still other instances, a cluster of short exchanges occurred, as the children searched for the next link.

As the first example above shows, ideas were not always introduced in an indeterminate way. However, it was not simply the degree of explicitness in the relationship that determined what happened subsequently. Decisions about whether an idea should be regarded as final or provisional depended on the status granted it by the group, as control over the discussion did not rest with the originator alone. Hence, in part, the distribution across the group members of power over the discourse created an exploratory ambience.

The extent to which the children were encouraged to explore provisional links also depended on the nature of the terms they were given, and how these were related in the domain in question. Exploratory talk was associated with sets of terms that were ontologically varied, and therefore related to each other in different ways. In the topic 'habitats', for example, there were classes of observable natural kinds ('animals', 'plants') together with more abstract theoretical notions, such as

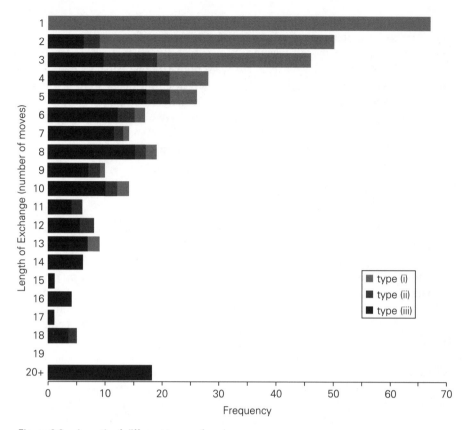

Figure 6.3: Length of different types of exchange

'survival' and 'energy'. In contrast, for the topic 'Earth in space' the terms being mapped were mainly celestial objects ('moon', 'sun', 'planet'), and the possible relationships between them were primarily class inclusion and spatial configuration. Finding links between these did not seem to challenge the pupils much, and they often expressed their ideas fully formed, allowing little scope for others to elaborate on them.

Type (ii) Exchanges

Individually elaborated exchanges were relatively rare. They could provide an extension of the function fulfilled by type (i) exchanges, allowing exploration of potential connections without the immediate need to specify the exact relationship involved, but then going on to make that relationship explicit. Such was the case in the second example above. Type (ii) exchanges could make a significant contribution to the emerging concept map. But to what extent was that contribution negotiated between the members in the group?

Examining instances across the data set provided evidence that the initial introduction was intended as provisional in many cases. One sign of this was an overt tentativeness to the introductory move, such as when it was intoned as a question. Hence, whilst these exchanges were primarily one person's contribution to the discussion, this does not imply that other group members were involved at only a minimal level. As with non-elaborated exchanges, the support of the rest of the children determined whether the ideas presented were retained, rejected, or indeed modified. A contribution presented as unproblematic could have its status redefined by somebody asking for further information or for justification.

Type (iii) Exchanges

Exchanges of type (iii) (introduction ⇒ elaborated ⇒ collaboratively) occurred very frequently, and therefore are an important feature of the data. They varied widely in length, from two moves up to a maximum of 90 (Figure 6.3). With these exchanges, not only was there the possibility of group approval or disapproval, but also of a direct contribution to the content of the idea under discussion. As with the other two types of exchange, type (iii) exchanges could be introduced with varying degrees of explicitness, and there were indications that they too were being used to introduce or develop ideas in a provisional way.

Type (iii) exchanges had two distinctive features. First, there was joint construction of the proposition written onto the map. Then there were elaborating moves that seemed to say, in effect, 'yes, we are on the same wavelength'. By extending what they assumed to be the originator's intended meaning, others in the group were able, not only to settle on a link to write in the map, but also to check that their assumptions about what the originator meant had been correct.

In many cases, there was a sense of convergence on a shared understanding, a negotiation of meaning, that was due to the contributions of more than one participant. This was not normally convergence from opposing perspectives, and there was rarely any indication that the understandings that emerged were substantively new. There was, nevertheless, a testing out of what was taken to be common understanding. This was distinct from the way the discourse progressed in the other types of exchange. In type (i) and (ii) exchanges, the extent to which members of the group other than the initiator actively checked on whether understanding could be taken as shared was very limited. Essentially, they could agree or disagree, or press for more information. In type (iii) exchanges, they could make inferences about what was being meant, about which particular way of using the terms was implied. They could then deduce consequences of that meaning in terms of propositions that would follow from it, and put those propositions before the group for confirmation.

As explained above, the links made in the concept maps had been coded as to whether or not they were scientifically acceptable. This enabled a crosstabulation of outcome by exchange type across all the groups. Figure 6.4 shows the result of this analysis. Type (iii) exchanges were, overall, more likely to result in a

All topics, pre-topic session

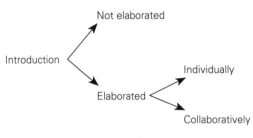

	Non-scientific outcome	Scientific outcome
	50 (77%)	15 (23%)
	11 (58%)	8 (42%)
	21 (33%)	42 (67%)

Based on data for 147 ideational exchanges across all groups

Chi-square: 24.68***
Cramérs V: 0.41

All topics, post-topic session

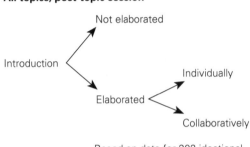

	Non-scientific outcome	Scientific outcome
	52 (57%)	40 (44%)
	5 (24%)	16 (76%)
	20 (23%)	69 (78%)

Based on data for 202 ideational exchanges across all groups

Chi-square: 24.27***
Cramérs V: 0.35

Figure 6.4: Concept map outcomes for different types of ideational exchange
Note: Tables show number and row percentage of exchanges of each type within each outcome category
*** = significant @ p < .0005

scientifically acceptable outcome than the other types. These differences were tested statistically, and found to be significant. It therefore seems that collaborative elaboration of ideas focused the discussion on the more scientifically appropriate ones. Through the process of hypothesis testing described above, the children had worked towards a shared view of which relationships should be included in the map. Ideas that were sketched out, often in a vague way at the outset, could be re-drafted, through discussion, until a mutually acceptable final form was achieved.

Conclusion

Indepth analysis of the discourse amongst these groups of children has revealed interesting features. Stepping back from the specific types of exchange, we can see what was happening in the discourse as a whole. Over the course of the discussion, the children were generating possible links between terms they were given. These possibilities were then held up before the group for critical examination, and some selected as being worth pursuing, and therefore as meaningful, while others were dropped. There are indications that the conditions under which these processes functioned were in part due to the task structure, and also to the rather different power relations that characterize collaborative group work as against more teacher-dominated discussion. Where this structuring succeeds, there is evidence that the discussion provides a valuable chance for children actually to establish communication between each other about scientific ideas over a sustained period, something that they may not get the opportunity to do in other kinds of tasks. That need to communicate leads them to negotiate meanings, and at times to adjust them towards those that are scientifically more acceptable. The resulting concept map therefore displays a complexity of meanings that is traceable, not to individual contributions alone, but to the interaction between individual views. These findings are both encouraging, in terms of the classroom use of concept mapping, and suggestive, in terms of future research.

References

AUSUBEL, D.P., NOVAK, J.D. and HANESIAN, H. (1978) *Educational Psychology: A Cognitive View*, 2nd Ed, New York, Holt, Rinehart, Winston.

BARNES, D. and TODD, F. (1977) *Communication and Learning in Small Groups*, London, Routledge and Kegan Paul.

CARMICHAEL, P., DRIVER, R., HOLDING, B., PHILLIPS, I., TWIGGER, D. and WATTS, M. (1990) *Research on Students' Conceptions in Science: A Bibliography*, Leeds, Children's Learning in Science Group, University of Leeds.

EGER, M. (1992) 'Hermeneutics and science education: An introduction', *Science and Education*, **1**, pp. 337–48.

HALLIDAY, M.A.K. (1973) *Explorations in the Functions of Language*, London, Edward Arnold.

HORTON, P.B., McCONNEY, A.A., GALLO, M., WOODS, A.L., SENN, G.J. and HAMELIN, D. (1993) 'An investigation of the effectiveness of concept mapping as an instructional tool', *Science Education*, **77**, 1, pp. 95–111.

NOVAK, J.D. and GOWIN, D.B. (1984) *Learning How to Learn*, Cambridge, Cambridge University Press.

OSBORNE, R.J. and WITTROCK, M.C. (1983) 'Learning science: A generative process', *Science Education*, **67**, 4, pp. 489–508.

ROTH, W.-M. and ROYCHOUDHURY, A. (1992) 'The social construction of scientific concepts or the concept map as conscription device and tool for social thinking in high school science', *Science Education*, **76**, 5, pp. 531–57.

SAINSBURY, M.J. (1992) *Meaning, Communication and Understanding in the Classroom*, Aldershot, Avebury.

SEARLE, J.R. (1969) *Speech Acts: An Essay in the Philosophy of Language*, Cambridge, Cambridge University Press.

SINCLAIR, J.McH. and COULTHARD, R.M. (1975) *Towards an Analysis of Discourse*, Oxford, Oxford University Press.

STUBBS, M. (1983) *Discourse Analysis*, Oxford, Blackwell.

WITTGENSTEIN, L. (1967) *Philosophical Investigations*, 3rd Ed, translated by ANSCOMBE, G.E.M., Oxford, Blackwell.

7 Using a Picture Language to Teach about Processes of Change

Richard Boohan

Abstract

Although the Second Law of Thermodynamics is considered one of the most fundamental of all scientific laws, it has a reputation for being obscure and difficult to understand, and receives only a little attention at the secondary school level. This paper describes an approach intended to make Second Law ideas intelligible and helpful to pupils aged 11–16 years. The central notion is that changes are caused by *differences* (for example, differences in temperature or concentration) and not by *energy* as is often suggested in school science courses. An abstract picture language has been developed to support the teaching of these ideas, and to stimulate discussion amongst pupils about the nature of physical and chemical changes. Although work with the picture language can be started with children as young as 11 years, the language can be developed as pupils progress, and can be used with older students to teach about more advanced thermodynamic concepts, such as entropy and free energy.

Introduction

Increasing emphasis in science education has been placed on making fundamental and everyday issues accessible to a wider range of pupils — for example, the maintenance of life and eco-systems, the origins and uses of fuels and foods, the weather, and so on. Essentially, pupils are expected to make sense of processes of change. In talking about such matters, it is natural to produce explanations about what causes these changes to occur. While the notion of 'change' is very common in experience, it happens to be rather hard to explain scientifically.

The direction of change is the concern of the Second Law of Thermodynamics, which, as well as being considered one of the most fundamental of all scientific laws, also has a reputation for being obscure and difficult to understand. A simple interpretation of the Second Law is that matter and energy always tend to become more spread out. Put more technically, the total entropy always increases in any irreversible change. Entropy is a measure of the number of microstates consistent with a given macroscopic state. So, perfume diffuses in a room because there are more microstates consistent with the state in which the perfume is spread throughout the room than when it is concentrated in one region. For a similar reason, entropy increases when a hot cup of coffee cools. But all this is very difficult to understand for an able 16-year-old, let alone an average 11-year-old.

It is unfortunate that 'energy' has been inappropriately recruited as a cause of change, and this has created a great deal of confusion about teaching energy. This in turn has stimulated much debate about these many difficulties (for example Duit, 1981; Schmid, 1982, Warren, 1982; Marx, 1983; Driver and Millar, 1986; Ellse, 1988; Ross, 1988; Solomon, 1992). One central problem is that energy is treated both as a cause of change, and at the same time as something which is conserved, mixing up a First Law idea (what is constant?) with a Second Law idea (what makes things happen?). While the total energy of an isolated system, being constant, puts limits on what it can do, it does not determine the direction of change (Ogborn, 1986).

By contrast, fuel and food are not conserved and do make things happen. When we burn some coal in a power station or run a race, we do not use up *energy* (which remains constant), but we do use up *free energy*. In everyday talk, when we say that we do, or do not, have the energy to do something, what we are referring to is much nearer the concept of free energy than the scientific concept of energy. Much research into children's ideas about energy indicates that this is their starting point — that energy is associated with activity and being alive, that it represents a power to act, that it is used up and that it can be refreshed (for example, Watts, 1983; Brook and Driver, 1984; Solomon, 1984). Such everyday ways of talking about changes will not go away — any approach to teaching about Second Law ideas must build on common-sense notions about why things change.

'Teaching about Why Things Change': A Curriculum Development Project

An approach to talking about energy and processes of change which takes account of 'Second Law' ideas has been developed by the project 'Teaching about why things change' (Boohan and Ogborn, in press), with materials being produced for school pupils aged 11 years and upwards. In developing a way of talking about change, important considerations are that it must be:

- intelligible to pupils;
- useful to teachers; and
- scientifically consistent.

Thus, by starting from common-sense ways of explaining, the approach can be potentially useful to *all* pupils in helping them to understand the world, but by remaining consistent with scientific principles, it avoids telling a misleading story.

The fundamental idea in this approach is that change is understandable as being caused by differences (Ogborn, 1990). Perfume spreads in a room because of a concentration difference. Hot coffee cools because of a temperature difference. The differences which drive change tend to disappear — the perfume becomes evenly spread out, and the coffee becomes the same temperature as the room.

(Scientifically, a difference disappearing means that entropy is increasing — or 'negative entropy' is being used up.) If differences only disappeared, the Universe would be a dull place, with matter and energy spread uniformly. But as differences disappear, they can create other differences. Thus, water can be made hot by a hotter flame. Here, a difference is being used to create another difference. This essentially simple idea is also powerful — it can be used to make sense of a wide range of phenomena, from simple temperature changes to the complexity of life.

Such an account will need to help pupils to see that many changes, while superficially appearing to be very different, have fundamental similarities — for example, coffee powder dissolving in hot water and the atmosphere becoming polluted are both changes in which matter is becoming more 'spread out'. Even though the focus is on the more familiar idea of difference rather than concepts such as entropy, making these abstractions is still not easy. The problem confronting the project has been how to find a way of developing these ideas with school pupils. To make the ideas intelligible, the project has developed a set of abstract pictures, which represent fundamental kinds of changes. Such pictures show only the essence of what is happening — for example, a difference in concentration vanishing, or energy flowing into or out of part of a system. Though the use of visual communication is very common in science education (Barlex and Carré, 1985), mostly it is concerned with the display of spatial and sequential relationships, rather than with conceptual ideas, though there have been some uses of abstract pictures to explore pupils' conceptual understanding (for example, Novick and Nussbaum, 1981).

This approach to teaching about 'why things change' will now be illustrated with some examples of the abstract picture language used with pupils.

Examples of Introductory Work with Pupils

A wide range of materials have been developed for pupils. They are designed to be used by teachers alongside their existing curriculum materials, and complement topics already taught — dissolving, fuels, photosynthesis, and so on — though with an emphasis on the nature and causes of change. Initially we developed activities in which pupils were encouraged to pay attention to abstract features of changes by asking them to identify similarities and differences between selected situations, putting groups of situations together or making matches to 'prototypical' situations. While these activities can be useful, one difficulty in using prototypical situations, is that there are surface features of these which may be distracting. This was the motivation behind developing abstract pictures of changes, which represented fundamental features of changes only.

Pupils are encouraged to be abstract about situations, so as to see different events as basically the same kind of process, by using the pictorial representations to match to different kinds of change. These kinds of tasks emphasize group activity and discussion work, as pupils debate the reason for making particular matches. Younger pupils begin by using a rather wider range of pictures which are more

Which picture best shows sugar dissolving?

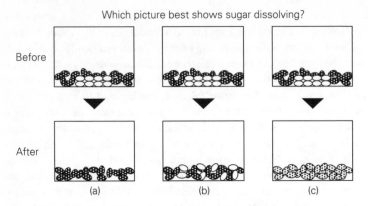

Before

After

(a)   (b)   (c)

Figure 7.1:   An example of an introductory activity: Dissolving

Which picture best shows cold lemonade becoming warmer?

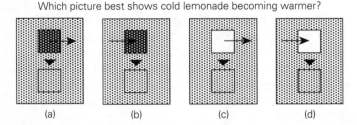

(a)   (b)   (c)   (d)

Figure 7.2:   An example of an introductory activity: Temperature difference and energy flow

related to specific situations; as they progress, they use a more restricted set of more abstract pictures. Two examples of the kind of introductory work which can be done with younger pupils are described here, before going on to look in detail at the complete 'picture language'.

Figure 7.1 shows three possible representations of what might be imagined to happen when something dissolves — the particles disappear, they mix or they change into something else (Prieto *et al.*, 1989). When matching changes to abstract representations children need to be introduced progressively to the conventions used in the picture language — here, that the two pictures represent the 'before' and 'after' states of the change with time 'running downwards' and that different substances are represented by particles of a different shading.

Figure 7.2 represents a change in which there is an energy flow due to a temperature difference. Pupils might first be introduced to Figure 7.2a as representing a hot cup of coffee cooling — higher temperature or 'concentration of energy' being represented by darker shading and the energy flow by an arrow. In making this match, pupils need to consider temperature differences and the direction and nature of the flow, for example, 'is what is flowing "hot" or "cold"?' (Engel Clough and Driver, 1985). The approach that 'energy makes things happen' finds it difficult to deal with cold objects like ice cubes (if they have less energy than water how

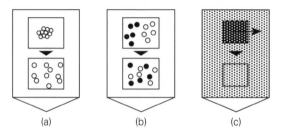

Figure 7.3: Some changes which 'just happen by themselves'

can they make things happen like cooling a drink?) or refrigerators (how can a supply of energy make it colder?). These questions are dealt with rather easily with the 'differences' approach as we shall see later.

As well as developing ideas about temperature differences, 'concentrations of energy' and energy flows, these pictures are also laying the foundations of a way of talking more generally about energy in other kinds of changes.

Differences and Change: A Brief Story in Pictures

The key idea in the approach, as outlined earlier, is to pay attention to the differences which drive change. Changes are driven by differences such as those between hot and cold, or between concentrated and dilute.

1 Differences tend to disappear

Differences tend to disappear because matter or energy or both become more spread out. Air in a balloon tends to leak out because of a pressure difference — it continues to spread out until the pressure difference disappears. Pollution spreads out and mixes with the air in the atmosphere because of a concentration difference. Eventually the concentration difference disappears. Hot coffee cools because of a temperature difference. Energy spreads out into the surroundings, as it goes from hot to cold, until eventually the temperature difference disappears. Figure 7.3 shows pictures representing (a) matter spreading out, (b) matter mixing and (c) energy spreading out as a hot object cools. The large arrow pointing down at the bottom of each box represents the direction of spontaneous change — all these are changes which 'just happen by themselves'.

If energy and matter are spread out or mixed as evenly as possible, there are no differences left. Equilibrium is where things stay the same because there is no difference left to drive a change. So, the reverse processes to the ones described above do not happen spontaneously. Figure 7.4 shows pictures representing (a) matter 'bunching together' and becoming more concentrated, (b) matter 'unmixing' and (c) an object becoming warmer than its surroundings by energy flowing up a temperature gradient. The 'up' arrow at the top of each box indicates that these are changes which do not 'just happen by themselves'.

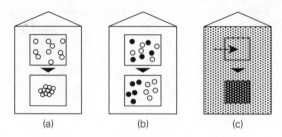

Figure 7.4: *Some changes which do not 'just happen by themselves'*

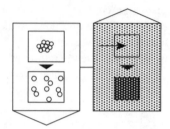

Figure 7.5: *Coupled changes (e.g., a power station and a kettle)*

2 It takes a difference to make a difference

Water in an electric kettle does not spontaneously become warmer. It happens because it is coupled to another change happening in the power station. Matter spreads out when steam in the turbine expands, and this drives the change in which energy becomes concentrated in the kettle. Figure 7.5 indicates how a spontaneous change can be coupled to a non-spontaneous change — using a change that 'just happens' to drive a change that does not 'just happen'.

Returning to a problem posed earlier, ice can make things happen even though it has less energy than water because it is (usually) different in temperature to the surroundings — it can create other differences. Plugging a refrigerator into an electricity supply makes it colder — a driving change (the power station) creates a temperature difference between refrigerator and surroundings.

The above account describes the essential features of this 'picture language' of change — representing spontaneous and non-spontaneous change and the coupling of such changes. So far, however, we have only considered a few different types of change. Figure 7.6 shows a complete set of the most fundamental elements of the 'picture language', combinations of which can be used to represent any kind of thermodynamic change.

While this is the fundamental set of pictures, further sets of pictures related to each fundamental picture allow finer discriminations to be made. Thus, Figure 7.6.4a is intended to represent any change in which energy flows down a temperature gradient; it may often be useful to refer to more specific features of such a change, for example, whether there is an energy exchange with the surroundings or

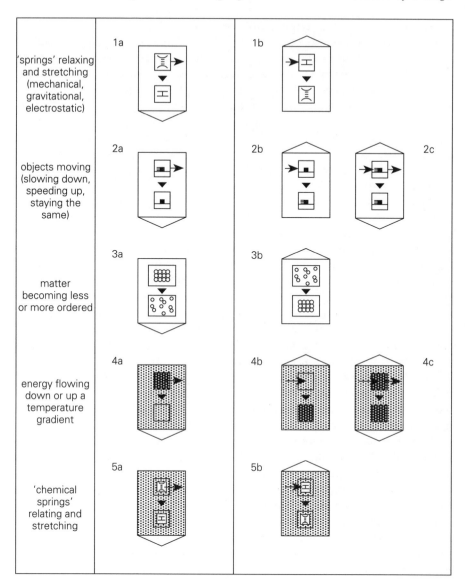

Figure 7.6: Fundamental kinds of changes

whether the system is at a higher or lower temperature than the surroundings, and a further set of pictures (not shown) allows such distinctions to be made.

Similarly, Figure 7.6.3a represents any change in which matter becomes more disordered; there are various ways in which this could happen so a further set of pictures (not shown) represent 'spreading out', 'mixing' and so on. The limited set shown in Figure 7.6, however, is more manageable when considering how changes may be coupled. The remaining pictures in this set are described below.

3 Steady states — maintaining a difference

There is more than one way of staying the same. A bottle of red wine on a dinner table and a human being may each remain at an approximately constant temperature, but for very different reasons. The wine stays at the same temperature because it is at the *same* temperature as the surroundings. A human body is maintained at a temperature which is *different* from the surroundings.

A temperature difference may be maintained by 'walling it in'. Hot coffee in a perfectly insulated flask would stay hot forever. But in the case of the human body, the difference is maintained by using a difference to keep it going. Similarly, a warm room can be kept as it is, at a higher temperature than the air outside, by a hot radiator. Figure 7.6.4c shows a steady state system maintained at a constant temperature with a continuous flow of energy — such a system, like other non-spontaneous changes, needs to be driven by a change which 'just happens'.

A similar situation arises with moving objects, which may need to be coupled to another change to keep them moving. In the real world, objects which are moving tend to slow down, and energy spreads out (Figure 7.6.2a). Objects do not start to move 'just by themselves' (Figure 7.6.2b). Objects may continue to move at approximately constant speeds — the Earth moving around the sun and a car travelling down a motorway — but for different reasons. The car maintains a constant speed because the energy input (from the engine) balances the energy output (losses due to air resistance and friction). Again, this is a steady state system (Figure 7.6.2c) which needs to be driven by a change which 'just happens'. The Earth's movement is essentially unchanging — if there is no friction, there is no energy flow and this could be represented in a picture omitting the 'up' and 'down' arrows indicating spontaneous change, but with the object continuing to move. Thermodynamically, however, nothing is happening.

There are many examples of such 'steady state systems' — a system which is maintained in the same state by a constant flow of energy or matter or both. These include simple domestic situations, biological systems, and large-scale phenomena, such as a centrally heated room, a flame, a saucepan kept boiling, a living organism, an ecosystem, a tornado, the Earth, and so on. Schrödinger (1944) noted that since living things are systems which constantly produce entropy, they need to be actively maintained (which they do by consuming 'negative entropy', for example, sunlight or food). Such systems need to get rid of entropy, and one way to do this is by being warmer than their surroundings.

4 Storing differences in potential energy differences

When you pull the two ends of a spring apart, energy is stored (Figure 7.6.1a). When you 'let go' of the spring, the energy escapes and spreads out — the spring and the air around it warm up a little. A stretched spring could stay the same forever until you release it, but once started, it is a change which 'just happens by itself' (Figure 7.6.1b). These pictures need not represent only mechanical springs — they could also represent, for example, gravitational or electrostatic attraction. (A 'perfect spring' *could* be coupled to another change without energy spreading out and with no entropy increase, for example, a perfectly elastic rubber ball

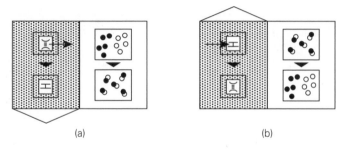

Figure 7.7:   Chemical changes (a) 'joining' and (b) 'splitting'

bouncing up and down forever — though, thermodynamically nothing would be changing.)

5   Storing differences in 'chemical springs'

Chemical reactions are perhaps the most difficult kind of change to understand, since we need to pay attention to what is happening both to matter and to energy. Combustion is an example of a change which is driven mainly by the spreading of *energy*. However, it is wrong to think of energy as being stored in the fuel itself; rather we should think of the energy being stored in the fuel–oxygen system (Ross, 1993). A good example is the hydrogen–oxygen system. When water is electrolysed, hydrogen and oxygen are pulled apart, and energy is stored — it is concentrated in the 'hydrogen–oxygen spring'. Figure 7.6.5b represents the 'pulling apart' of a 'chemical spring'. When hydrogen and oxygen are burnt together, the energy escapes and spreads out into the surrounding. Figure 7.6.5a represents the 'release' of a 'chemical spring'. Just like a stretched spring, a mixture of hydrogen and oxygen could stay the same forever, but once started, the change 'just happens by itself'. The sound of the explosion when hydrogen and oxygen react is very evocative of a large spring 'snapping shut'.

Figure 7.7 represents the change in more detail, showing both what is happening to particles and to energy. A widely held belief is that energy is stored in bonds and released when the bonds are broken. These pictures emphasize that in 'joining' energy spreads out while in 'splitting' energy is stored. (Of course, in the reaction between hydrogen and oxygen, bonds are also broken before they are formed, but this is part of the story which can be left until later.)

Hydrogen, made by electrolysing water, can be used to produce electricity in a fuel cell, but one difficulty in using it commercially as a fuel is how to store it conveniently. Nature has solved this problem neatly — in photosynthesis, after splitting hydrogen from oxygen, the hydrogen is attached to carbon from carbon dioxide to make glucose. We can look upon glucose as a 'hydrogen store' with energy still stored in the hydrogen–oxygen spring. Similarly, talk about 'high energy bonds' in ATP is confusing — when a phosphate group is attached to ADP, it is being pulled away from solvent water molecules. Energy is stored in the 'phosphate-water' spring, and escapes when the phosphate group is grabbed back by the water molecules.

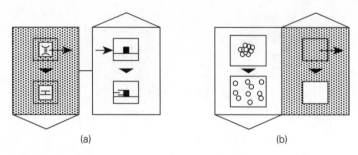

Figure 7.8: Types of coupling (a) motor car (b) water evaporating

There are different kinds of spontaneity in chemical change (Atkins, 1991). Some changes are driven by the spreading of matter, for example, dissolving. Some changes, for example, the reaction of hydrogen and oxygen, are driven by the spreading out of energy (though in most combustion reactions, matter as well as energy spreads out). In other changes, matter may become more concentrated at the expense of energy spreading out, or vice versa. When crystals form as a hot solution cools, particles become 'bunched together' — this is driven by energy spreading out. Conversely, in all spontaneous endothermic reactions particles become more disordered — this drives the 'concentrating' of energy.

6 Coupled changes — extensive and intensive

When one change drives another, there are two possibilities — a change in one system may be driving a change in another, or both changes may be part of the same system. Figure 7.8 shows an example of each, and how they are represented. In a motor car, petrol burning in the engine is a change that 'just happens'; this acts on another system to make the car move (Figure 7.8a). When water evaporates, two changes occur to the water itself — the spreading out of the water vapour can create a temperature difference if there is cooling by evaporation (Figure 7.8b).

An Example of More Advanced Work: Why Does Water Freeze?

Work with these kinds of pictures can be started with children as young as 11 years. But they can also lay the foundation for introducing more difficult concepts, such as entropy and free energy, and can complement an approach to matter and energy spreading based on probabilistic behaviour (Atkins, 1988). They can also help to illustrate quantitative work. One example, about the freezing of water, is discussed here.

When water freezes, ice crystals form, often in beautiful and elaborate patterns. But if the Universe is constantly becoming more disordered and entropy always increases, how can such patterns arise?

Molecules are more ordered in solid ice than in liquid water. We can measure this difference in order — for melting ice, $\Delta S = +22$ J mol^{-1} K^{-1}. If this was the

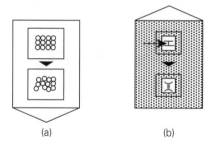

(a)   (b)

Figure 7.9:   Changes to matter and energy during melting

only factor involved, ice would always spontaneously change into water, since in this change the entropy increases (see Figure 7.9a).

However, melting ice also involves a change in the surroundings — when 1 mole of ice melts, 6 kJ of energy is taken from the surroundings. (The energy is 'stored in the chemical springs' as hydrogen bonds are pulled apart.) The size of this negative entropy change depends on the temperature of the surroundings; the higher the temperature, the less negative the entropy change (the easier it is to take energy from the surroundings and 'concentrate' it in the water) (see Figure 7.9b).

The change will take place in the direction of increasing entropy. Above the melting point of ice, the 'matter spreading' change drives the 'energy concentrating' change, and the ice melts (see Figure 7.10a). At 0°C, the changes are balanced, with the entropy change of the ice being equal and opposite to the entropy change of the surroundings, and the ice and water are in equilibrium (see Figure 7.10b). Below 0°C, the 'energy spreading' change drives the overall change (see Figure 7.10c). Matter becomes more ordered as ice crystals form, but the entropy of the Universe still increases.

Using the Approach in the Classroom

The approach is intended for pupils of a wide range of abilities and ages, and activities based on this approach have been trialled in a variety of kinds of schools. Evidence about the effectiveness of the approach have been collected from classroom observation, pupils' written work and pupils' discussions during small group work (Boohan and Ogborn, in press). In the examples below, the extracts from pupils' discussion are selected to be typical of 11–14-year-olds of modest ability.

To many teachers the abstract pictures may initially look unfamiliar and daunting, and they may feel that pupils will not understand them. When they try the activities with their pupils, they are often surprised by the ease with which pupils use the pictures. Despite their abstraction, the pictures work well in promoting useful and relevant discussion amongst pupils. The focus of their discussion is, as we would hope, about the nature of the changes and not about the meaning of the conventions. This 12-year-old pupil is explaining why 'wood burning' could be matched to the picture of a 'chemical spring':

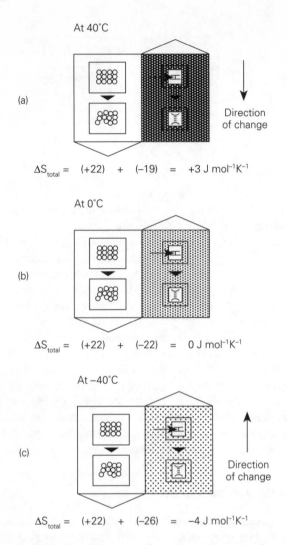

At 40°C

(a)

$\Delta S_{total} = (+22) + (-19) = +3 \text{ J mol}^{-1}\text{K}^{-1}$

At 0°C

(b)

$\Delta S_{total} = (+22) + (-22) = 0 \text{ J mol}^{-1}\text{K}^{-1}$

At −40°C

(c)

Direction of change

$\Delta S_{total} = (+22) + (-26) = -4 \text{ J mol}^{-1}\text{K}^{-1}$

Figure 7.10:   *Direction of melting/freezing is affected by temperature of surroundings*

The wood is now burning so nothing needs to happen to make it burn more. OK, I know you have to light a fire to make the wood burn but now that it's already burning you don't need anything else to make it burn more. It's burning and it's burning and it's just happening by itself and the energy is going out and the thing is getting cold.

This explanation is about the spontaneity of the change and what is happening to the energy. In contrast, these pupils are also considering the same match, but are discussing their confusion about what the pictures say. Such discussions about the meaning of conventions are rather rare.

But that's not saying that you're having energy back, that's just saying energy is escaping same as that.

No you can . . . it joins back doesn't it . . . this shows like a spring stretching coming back together again.

There are certain scientific ideas which run throughout the approach, and in general, pupils are able to use these in their discussions, for example, ideas about the distribution of particles, energy flows and spontaneity.

- sweating to stay cool
  It's because when you get rid off the water from your body it spreads out into the air. It evaporates.
- crystals forming in copper sulphate solution as it cools
  Before it was like solution right, it wasn't like apart and then it turns into copper sulphate crystals so it bunches together to make a crystal.
- similarities between 'an electric light bulb gets hot' and 'using a kettle to boil some water'
  Energy is being stored up.
  Something is driving the water, something has to drive the light bulb.
  It doesn't just happen by itself.

Most pupils are stimulated by the pictures to talk about situations in a more abstract way, with some pupils producing quite sophisticated explanations. Here, a 12-year-old is explaining what happens to a hot chocolate drink in a cup and in a vacuum flask:

The hot chocolate is hotter than the room temperature so this [the central region in the representation] is darker [i.e., hotter with energy more concentrated] and when the hot chocolate is left in the room, because the room is colder the energy goes out of it [the chocolate] and the hot chocolate starts to get colder slowly. This one in here [in the vacuum flask] — the hot chocolate is still hotter than the room so it's darker but it stays the same, as it's in the vacuum flask so no heat will come out and no heat will come in.

Of course, pupils bring to the classroom their own ideas about energy and change, and merely using a pictorial language will not move their thinking towards a more scientific point of view. However, the pictures are a useful tool to stimulate discussions and explanations in which pupils make their ideas more explicit, for example, about whether 'heat' and 'cold' are two things or just aspects of one thing.

- One choice of picture:
  It's colder on the inside, and then the energy from the heat on the

outside is going in making it warmer, and when it's in the vacuum flask the coldness just escapes making it warmer.
- A different choice of picture:
Mine's the same really but the cold doesn't escape — the insulation makes the lemonade warmer.
That's what I meant — the insulation makes it warmer.
No but that means if the insulation made it warmer the cold wouldn't go out but the heat would come in — so the heat goes in not comes out.

Conflicts about ideas can help pupils to see things in a different way, and may be used by the teacher in discussion to further the pupils' understanding.

What would count as progress? There are two ways in which we are investigating progress in pupils' performance. There is particularly striking progress in the short-term, in that the use of these pictures tends to stimulate pupils into more extended and fundamental discussion about the nature of changes. However, we are also interested in looking at the longer-term development of pupils' ideas as a result of being taught about this approach to energy and change. Learning to use the picture language is *not* an end in itself — progress must be seen in relation to pupils' conceptual development. Stylianidou (1995) has carried out a longitudinal study of 11- and 12-year-old pupils who have been taught using this approach, and reports on the development and analysis of pilot interviews intended to elicit pupils' ideas. The analysis of the main study is currently in progress.

We have developed a wide range of activities based on the approach of the 'picture language'. Topics include dissolving, pollution, separating mixtures, chemical change in everyday materials, energy transfers, insulation, weather, fuels and food, photosynthesis and respiration, synthesis and decay of large molecules, homeostasis, use and costs of domestic fuels, and eco-systems. However, such activities can only serve as exemplars. The key issue will be the extent to which the picture language can give teachers the necessary confidence and understanding about this story of 'why things change', in order to incorporate the ideas routinely into their own teaching.

References

ATKINS, P.W. (1988) *The Second Law*, Scientific American Library, W.H. Freeman and Co.

ATKINS, P.W. (1991) *Atoms, Electrons and Change*, Scientific American Library, W.H. Freeman and Co.

BARLEX, D. and CARRÉ, C. (1985) *Visual Communication in Science*, Cambridge University Press.

BOOHAN, R. and OGBORN, J. (in press) *Energy and Change*, Association for Science Education, Hatfield.

BROOK, A. and DRIVER, R. (1984) 'Aspects of secondary students understanding of energy', *CLISP Report*, Centre for Science and Mathematics Education, University of Leeds.

DRIVER, R. and MILLAR, R. (Eds) (1986) *Energy Matters*, Centre for Science and Mathematics Education, University of Leeds.

DUIT R. (1981) 'Understanding energy as a conserved quantity', *European Journal of Science Education*, **3**, 3, pp. 291–301.

ELLSE, M. (1988) 'Transferring not transforming energy', *School Science Review*, **69**, 248, pp. 427–37.

ENGEL CLOUGH, E. and DRIVER, R. (1985) 'Secondary students' conceptions of the conduction of heat: Bringing together the scientific and personal views', *Physics Education*, **20**, 4, pp. 176–82.

MARX, G. (1983) *Entropy in the School*, Roland Eötvös Physical Society, Budapest.

NOVICK, S. and NUSSBAUM, J. (1981) 'Pupils' understanding of the particulate nature of matter: A cross-age study', *Science Education*, **65**, 2, pp. 187–96.

OGBORN, J. (1986) 'Energy and fuel: The meaning of the "go of things"', *School Science Review*, **68**, 242, pp. 30–5.

OGBORN, J. (1990) 'Energy, change, difference and danger', *School Science Review*, **72**, 259, pp. 81–5.

PRIETO, T., BLANCO, A. and RODRIGUEZ, A. (1989) 'The ideas of 11 to 14-year-old students about the nature of solution', *International Journal of Science Education*, **11**, 4, pp. 451–63.

ROSS, K.A. (1988) 'Matter scatter and energy anarchy: The second law of thermo-dynamics is simply common experience', *School Science Review*, **69**, 248, pp. 438–45.

ROSS, K.A. (1993) 'There is no energy in food and fuels — but they do have fuel value', *School Science Review*, **75**, 271, pp. 39–47.

SCHMID, G.B. (1982) 'Energy and its carriers', *Physics Education*, **17**, 5, pp. 212–18.

SCHRÖDINGER, E. (1944) *What Is Life?*, republished 1967, Cambridge University Press.

SOLOMON, J. (1984) 'Prompts, cues and discrimination: The utilisation of two separate knowledge systems', *European Journal of Science Education*, **6**, 3, pp. 277–84.

SOLOMON, J. (1992) *Getting to Know about Energy in School and Society*, Falmer Press.

STYLIANIDOU, F. (1995) 'Teaching about physical, chemical and biological change', Paper presented at European Conference on Research in Science Education, University of Leeds, April 1995.

WARREN, J. (1982) 'The nature of energy', *European Journal of Science Education*, **4**, 3, pp. 295–7.

WATTS, M. (1983) 'Some alternative views of energy', *Physics Education*, **18**, 5, pp. 213–16.

8 Construction of Prototypical Situations in Teaching the Concept of Energy

Andrée Tiberghien

Abstract

In this paper we present research on the development of the design of a teaching sequence and thus on the development of a part of knowledge to be taught in the domain of energy at the high school level. In this teaching sequence, one phase has specific characteristics from the points of view of the knowledge to be taught, and of the students' activity. We term this phase a 'prototypical situation'. Our theoretical approach is developed from an epistemological and a didactical basis; it establishes relations between the knowledge to be taught, its teaching and learning. In order to validate this 'prototypical situation', the teaching sequence was trialled both in the classroom and in the laboratory. The results allow us to conclude that a valid 'prototypical situation' has been constructed.

Introduction

A current research problem in science education (or in French *Didactique des sciences*) concerns the relationship between teaching situations and pupils' learning, in which designing and evaluating teaching situations is of crucial importance. To address this problem, we introduce the notion of a 'prototypical situation', which is characterized from two points of view. On the one hand, the prototypical situation is defined by the way in which the knowledge to be taught is developed. On the other hand it is defined by how it is 'staged' (*mise en scène*) or framed in the teaching situation, in particular in terms of the respective responsibilities of the teacher and the pupils in introducing new physics knowledge. To define this notion we initially describe our position with respect to the knowledge involved in a teaching situation and its relation to learning. Then we discuss the necessary choices made in the development of the knowledge to be taught and the teaching situation, as a research result. Finally we describe the ways in which this situation was tested and validated.

Basis for the Development of a Teaching/Learning Situation

Although the affirmation 'the intention of physics teaching is that physics will be acquired by the pupils' seems obvious, it is only so in appearance since we need to make clear what we mean by physics itself.

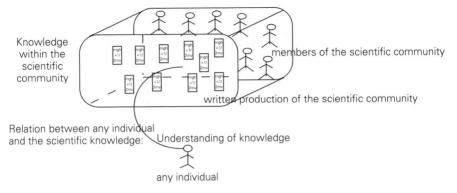

Figure 8.1:   Relationship between an individual and the knowledge within the scientific community

Scientific physics knowledge is shared within the scientific community of physicists. Nowadays no one can pretend to possess such knowledge completely; even for members of this community, at least a part of this knowledge is external to individuals but nevertheless available to them. This theoretical position, adopted within a part of the French community of research in 'didactique' (Arsac *et al.*, 1994), allows us to define the relation between an individual and scientific knowledge which is outside him/her and which, simultaneously, he or she has to deal with. We will use the term 'understanding of knowledge' to signify this relation (Figure 8.1). Other approaches, such as Chevallard (1991), use the term 'rapport with knowledge' (*rapport au savoir*): clearly, this is still an open question for debate.

The two terms 'knowledge' and 'understanding of knowledge' allow us to specify the differences between the uses of the same concepts according to the context involved. For example, the uses of the 'heat' concept by a teacher solving a problem at high school level, and an engineer calculating in a power station are not the same. The frameworks of use may be calorimetry for the teacher, and the equivalence of work and heat for the engineer. Each of these different persons might not be able to deal with the aspects of this concept involved in the other situation. In this case, the ways of understanding the knowledge about heat are different; they are not interchangeable, even if this knowledge is of the same concept.

Physics knowledge is attached to a given community. The scientific community is different from the educational system community which has the responsibility for physics teaching. The educational system community, in France for example, is composed of all the people involved in physics education, the minister of national education, all the policy makers including the inspectors, the teachers and all the other people who have a task related to the schools from the point of view of physics education. These two communities allow us to distinguish scientific knowledge from knowledge to be taught (Figure 8.2). Once this distinction between scientific knowledge and the knowledge to be taught is recognized, the question is

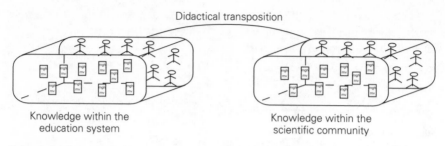

Figure 8.2: Relationship between scientific knowledge and knowledge to be taught

raised as to how the knowledge to be taught is developed, by a process that we term 'didactical transposition'. This means that scientific knowledge is modified or transposed in order to be taught at different levels. This transposition is done under several sorts of constraints. One kind of constraint follows the educational system. For example, in France policy makers and politicians decide when particular topics should be taught. From the learning point of view, knowledge to be taught must be understandable by the pupils. Theoretical studies have been carried out, particularly in mathematics, on the processes by which scientific knowledge is didactically transformed into the knowledge to be taught (Chevallard, 1991).

Our aim in designing teaching situations is that the learners should be able to construct their understanding of the knowledge to be taught, and that this construction is of value to them. We would like to avoid the situation where learners deal with knowledge mainly in order to get a good mark, to be recognized as a good pupil by their teacher or by their peers. We consider that, it is desirable that teaching situations should involve devolution from the teacher to the pupils to the extent that the pupils assume responsibility for dealing with the knowledge to be taught when this responsibility might have been taken by the teacher alone (Brousseau, 1981). In particular it means that, in situations where devolution takes place, the teacher's role is no longer to present new knowledge, even partly, but only to manage the situation so that the pupils by themselves are responsible for coming to understanding the knowledge involved. Obviously such situations can only occur at specific times during teaching.

This position implies making the following hypothesis on learning: if devolution takes place then the teaching situation offers highly favourable conditions for learning. This approach allows us to emphasize the distinction between teaching and learning. Teaching is related to development of the knowledge to be taught and to the *mise en scène* of this knowledge in a teaching situation, whereas learning is, by hypothesis, associated with the learner's construction (even if it is through social interaction) of an understanding of this knowledge.

In summary, in the proposition 'the intention of physics teaching is that physics will be acquired by the pupils', the term 'physics' can be understood in three ways: (1) scientific knowledge, (2) knowledge to be taught, (3) understanding of knowledge. This implies that research requires on the one hand development of the knowledge to be taught by taking into account as such the three types of knowledge, and on the other hand designing teaching situations. This design should specify which situations allow devolution from the teacher to the pupils in order that they can construct an understanding of the knowledge to be taught.

Principal Choices for the Development of the Knowledge to be Taught and of the Teaching Situations

From most of the research on students' conceptions (Driver *et al.*, 1985; Tiberghien, 1989), we conclude that one of the difficult aspects of physics learning is establishing relationships between the physics models which are learnt and experimental facts; in other words how to predict or interpret real events by using knowledge of theory and models of physics. We shall define the terms of 'theory' and 'model' below by giving a summary of our approach (Tiberghien, 1994).

We consider that when physicists interpret and predict experimental facts they do not directly apply a theory to the situation. Instead by using the chosen theory, they construct a model of the experimental situation. We refer here to a French epistemologist, S. Bachelard (1979) who considers that the model is always relational; it is an intermediary: 'The model is not an imitation of phenomena, ... it represents only some properties of reality'. In this perspective, in physics, interpretation and prediction imply a modelling process which puts three levels into play: theory, model, and experimental field of reference. We do not intend to give a formal definition of these levels; we specify their meanings in an operational way.

1   Scientific theory contains the explanatory system; paradigms in Kuhn's sense (Kuhn, 1972) — the set of beliefs, recognized values and techniques which are shared by the members of a given group of physicists, research questions, basic principles (conservation, symmetry, ...), and laws — are all part of the theory. A fundamental aspect of scientific theory is its hypothetical status which implies the validation process. This is a foundational aspect of modern science.

2   Models consist of qualitative and quantitative functional relations between physical quantities (implying mathematical formalisms) in order to represent selected aspects of a set of material situations.

3   The experimental field of reference consists of the experimental situations which belong to the domain of validity of the theoretical construction (theory + model) brought into play in modelling. This field includes experimental facts, experimental devices and measurements; we call this set 'objects/events'.

We assume that the three different levels are necessary in the processing of physics knowledge, and that they constantly interact.

Therefore, from our analyses of pupils' difficulties and of physics processing, we chose to develop the knowledge to be taught and a corresponding teaching sequence with its main aim as helping pupils to construct an understanding of this relation between theory/model and experiments in the case of physics teaching of energy at the high school level (16–17-year-old). This aim has two aspects, epistemological and conceptual. Knowledge to be taught can be analysed differently according to each aspect. However during the teaching sequence, which corresponds to the first introduction of the energy concept, the students are participating in a modelling task. Therefore, it is too early for them to reflect upon physics epistemology. We do not exclude some specific teaching on epistemological aspects. In this paper, although the teaching content is on energy, we do not discuss the content in itself; we only focus on our elaboration of a 'prototypical situation'.

Main Characteristics of the Teaching Sequence

The knowledge to be taught can be presented in three typical ways when we refer to our physics analysis. The introduction may start from theory/model of physics, from relations already established between theory/model and experiments, or from experiments (Johsua et Johsua, 1989). In this last case, experiments can be everyday experiences or come from problems of society such as that of the environment. Moreover the initiative of the knowledge development can be taken by the teacher or by the pupils. In our research, taking into account our aim, we chose the introduction from theory/model on the initiative of the teacher. The pupils take the initiative in constructing relations between theory/model and experiments. This choice has several consequences, in particular the theory/model should be reasonably understandable by the pupils and it should be sufficiently compatible in terms of completeness and coherence with the experimental field to allow a modelling process.

In research such as this, the knowledge to be taught and the design of the teaching situations are research products and not simply a context for collecting data (Artigues, 1988). Consequently these products have to be validated. Here we propose two types of validation, the feasibility of the teaching situations in the classroom and the effectiveness of the devolution. The first one is associated with teaching and probes how such a teaching situation may be carried out in the educational system; that it is compatible with the educational system's constraints, for example the teaching sequence duration and the possibility to incorporate the sequence into the whole physics curriculum; the second is associated with the relations between teaching and learning and tests, how in the situation where the devolution is supposed, pupils assume responsibility for a part of the knowledge to be taught.

The Development of the Knowledge to be Taught

This development was done with the following constraints: the theory/model has to be understandable by pupils and coherent with scientific physics knowledge. This development also takes into account the fact that, in France, energy is introduced for the first time in the physics teaching curriculum at this level. Let us note that

Table 8.1: Text presenting the seed of theory and model on energy

Theory (seed)	Model (seed)
Energy can be characterized by:	To build an energy chain
* its **properties**:	
— **Storage**	* the drawn symbols are to be used:
The reservoir stores energy	
— **Transformation**	[res.] for reservoir
The transformer transforms energy	
— **Transfer**	——▶—— for transfer
Between a reservoir and a transformer, or between two reservoirs, or between two transformers, there is transfer of energy. The different modes of transfer of energy from a system to another one are:	(tr.) for transformer
— *by work,*	by indicating:
There is transfer of energy under the form of mechanical work when there is movement of an object or of a part of an object during an interaction, under the form of electrical work when there is an electrical current (displacement of charges)	— in each rectangle the system corresponding to the experiment;
	— under each arrow the mode of transfer;
— *by heat,*	by putting
— *by radiation.*	— an arrow by the mode of transfer.
	* the following rules are to be used:
Energy can also be characterized by:	
	— a complete energy chain starts and ends with a reservoir;
* a fundamental principle of conservation	— the initial reservoir is different from the final reservoir.
The energy is conserved whatever the transformations, transfer and forms of storage	

the knowledge to be taught is developed in France in an institutional framework usually by inspectors and/or official commissions, and also by authors of textbooks.

We decided to introduce as such a 'seed' (that is from which further understandings can grow) of a theory and a model as complete and coherent with a limited experimental field. The customary choice is different; it consists of presenting parts of the models successively. In Table 8.1 the content of this seed is presented, and in Figure 8.3 an example is given of a symbolic representation (syntax of the model) of a simple experimental situation, a shining bulb connected to a battery. This text was developed from previous innovations (Lemeignan, 1980) and from our analysis of work on energy teaching which was the object of numerous pieces of research (Duit, 1983; Solomon, 1985; Koliopoulos *et al.*, 1986; Trumper, 1990).

Comments on the Chosen Theory

This text imposes a categorization of the world in three parts: reservoir, transformer and transfer. We consider that a categorization is a basic theoretical aspect (Levy-Strauss, 1962) which implies a way of dividing up the world. This text also presents

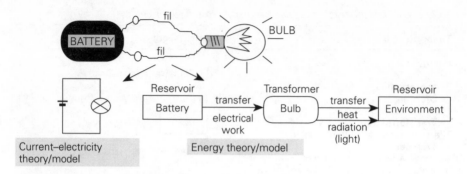

Figure 8.3: Example of an energy chain of a battery bulb setting in comparison with the current: Electricity representation

the physics principle of energy conservation which is largely recognized as belonging to theory level. Our development of this knowledge to be taught is also based on our learning hypothesis and on students' conceptions research which shows the importance of pupils' use of simple causal reasoning (Shipstone, 1985; Tiberghien, 1989; Rozier, 1988). This reasoning consists of a causal agent acting, through an intermediary (the mediator), on an object (Tiberghien, 1984). The seed of the theory is compatible with such causal reasoning even if it does not imply it (Kuhn, 1971) since we only have one physical quantity, energy and because initially energy can be associated with a causal agent.

Comments on the Chosen Model

This model is very limited in comparison with the physics model usually introduced in physics at this level. It contains a symbolic representation: two types of rectangles, an arrow and some rules (Table 8.1). The rule, 'a complete chain starts and ends with a reservoir', is the expression of the principle of conservation. Let us note that, in this model, the transformer is characterized by a constant energetic state (the quantity of internal energy remains the same) between the beginning and the end of the experiment; and the evolution of physical quantities with time can be taken into account: transitory and stationary behaviour (power constant or not) can be differentiated.

Comments on the Chosen Experimental Field of Reference

The experimental field consists of everyday situations using domestic appliances. This choice comes from two considerations. The first is related to the objectives of this teaching sequence. We would like the learners to establish relations between their everyday context and physics teaching; the second is related to our modelling approach: the experimental field should consist of situations which can be interpreted by the taught theory/model; this is the case with the situations selected from the daily environment.

Comments on the Modelling Approach

This modelling approach implies only qualitative aspects, at least initially. The relations between the two worlds of theory/model and objects/events are simpler than if both qualitative and quantitative aspects are involved. We considered that this activity allows a basic understanding of fundamental theoretical physics knowledge. For example, this seed of theory and model is sufficient to make a differentiation between current–electricity theory/model (*électrocinétique* in French) and energy theory/model which implies a drastically different partition of the world. In current–electricity theory/model, the considered system is only composed of the battery, the wires and the bulb whereas in energy theory/model, the whole universe (or the room by approximation) is seen as the isolated system, and the world is divided into three systems: the battery, the bulb and the rest of the universe (named environment) (Figure 8.3). We develop the seed of the theory and model by introducing the physical quantity 'power' as associated with transfer. The relation between energy, power and time has been selected from the body of scientific knowledge because it is an essential element of prediction for the chosen field of reference of the model. This choice is related to the priority given to modelling which requires respecting the coherence of the model with respect to its field of reference. This choice is different from that taken in the usual curriculum where power is not very important.

Design of the Teaching Sequence

We consider two main aspects in the design of the teaching sequence: the way of sequencing the knowledge (order of presentation) and the way of presenting the knowledge, particularly the design of the pupils' tasks.

As we stated, the teaching includes several phases, we only present the two first ones; the aim of the third phase is to refine the theory/model and to enlarge the experimental field.

Phase 1 (Text Construction) (Task 1 and 2)

The aim of the first phase is to present the students with tasks that lead them to recognize the need for a new energy theory/model. The students have already been taught about electricity. In task 1, the students produce a text that should describe an experimental situation (such as one with a battery, wires and a bulb). They are then asked to classify the text into two categories: statements referring to objects and events, and statements referring to physics knowledge (task 2). The latter

Andrée Tiberghien

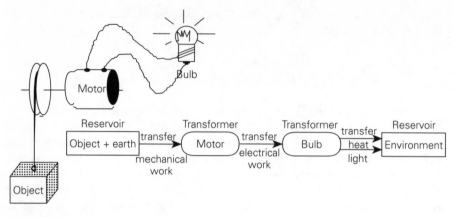

Figure 8.4: Experiment in task 4 and solution

category is divided into two subcategories referring to electricity, and 'other' statements. They are then asked to give an appropriate name to the 'other' category of statements which could be named 'energy'.

The initial 'seed' of a theory/model of energy is given to the students between phase 1 and phase 2 (Table 8.1).

Phase 2 (Constructing an Understanding of the Theory/Model) (Tasks 3 to 5)

Aim and Content

The aim of this second phase is to enable the students to construct an understanding of the energy theory/model by establishing relations between the theoretical construction (the 'seed') and the objects and events observed in the experiment.

The students' tasks are thus to produce an energy chain for three different experiments using the 'seed' of the theory/model as a source of information. The problem statement is the same for each task: 'By using the "energy model" (given before, Table 8.1), build the energy chain corresponding to the experiment'. Let us note that the choice of the first experiment, a battery — bulb, is based on findings about students' conceptions: as a matter of fact we know that most pupils consider the battery to be a reservoir for something (Tiberghien, 1984). After task 3, a correct solution is given (Figure 8.3). For the two other experiments shown in Figures 8.4 and 8.5, the students are asked to produce the corresponding energy chain. In laboratory situations no information is given between tasks 4 and 5, and in the classroom situation the correct solution of task 4 is given.

Prototypical Situation

These three tasks are designed to allow 'devolution' of responsibility for managing the taught knowledge involved. In their design, the theory/model and the

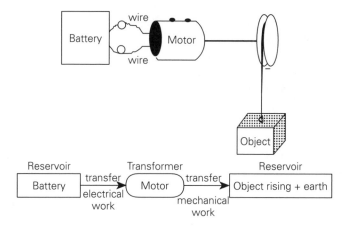

Figure 8.5:   Experiment in task 5 and solution

experimental field are radically separated in the situation. The former is provided in a written form (Table 8.1), and the latter is available in the concrete situation. The explicit goal for the students is to construct a symbolic representation, in terms of the model, of the experimental setting. In this situation, the knowledge at the theory/model level is under the entire responsibility of the teacher, as a representative of physics knowledge, whereas the description of the experimental field and the relations between the two worlds theory/model and objects/events are the entire responsibility of the pupils. Therefore, the devolution involved bears on a specific aspect of the knowledge to be taught — the relations between the two worlds of theory/model and objects/events. In this perspective we consider that, if it is validated, we have a prototypical situation of physics teaching on modelling.

Experimentation of the Teaching Sequence

Our research had several steps. The first was carried out in the framework of experimenting with a new teaching sequence with three teachers. It consisted of a dialectic approach involving development of the knowledge to be taught and the teaching experiments in the classroom. The duration of such work is necessarily long since an experiment can be done only once a year. The results of this step was that the teaching sequence was feasible. The second step consisted of refining the teaching sequences in order to produce a final version. This version was then to be tested as such. Let us note that this is a researcher's decision because, most of the time the teachers who participate in the experiment like to modify the presentation, even slightly, in order to improve their teaching. In the third step, we carried out experiments in laboratory situations and in eight classrooms in order to test the feasibility and the devolution in terms of pupils' understanding of knowledge. The experiments in classrooms were possible because starting with this year (94–95), our content is very close to the official curriculum.

Feasibility

The data taken to determine feasibility were the teachers' advice, the written pupils' work, the video-tapes of three classrooms out of the eight, and the video-tapes of two groups of two pupils in each of these three classrooms. The test of feasibility, for which we focused our analysis on phase 2, was globally positive from three points of view:

- the teacher's opinion after the teaching sequences. The eight classroom teachers considered the sequence interesting for them and for their pupils. The main remark was that their own pupils made progress in their successive solutions of the three tasks without their intervention.
- the implementation of teaching phase 2 (prototypical situation) in the classroom without modification on the part of the teacher. We verified this by video, taping three classrooms out of the eight and by collecting students' written work from the eight classrooms. This is rare as in the classroom, teachers almost systematically modify teaching situations.
- the improvement of the students' solutions along the three successive tasks (see Table 8.2 showing the results for seven classrooms). This table shows that the improvement between the first and the second task is more marked than that between the second and the third one, and that this last improvement varies a lot depending on the classroom. We do not discuss this last point here, because to do so we must introduce a new notion: the didactical contract.
- the comparison with the laboratory situations, both pupils' solutions and the pupils' problem-solving were very similar (in three of the classrooms we video-taped two groups of pupils).

Students' Understanding of Knowledge

As we have already presented, the aim of tasks 3 to 5 is to allow the pupils to establish relations between the theory/model and the objects and events of the experiments (experimental facts). Constructing these relations is a way of developing an understanding of the theory/model. Consequently our *a priori* analysis consisted of forming a hypothesis for the different possible ways of establishing relations. We used Giere's approach (1988) which consists of considering that between the two worlds, theory/model and objects/events, there is no identity but only similarity relations.

In this paper we do not present the different relations in detail, we only give the main categories. Several mechanisms of different complexity are involved in establishing relations between the two worlds (theory/model and objects/events). We present three of these by order of complexity:

- selecting an element of one world and trying to find the corresponding element in the other world through a common property is a simple way of

Table 8.2: Class results for the three tasks and students pairs' results

Classes	n	Battery–Bulb				Falling object				Rising objects		
		Task										
		Correct	(a)	(b)	other	Correct	(b)	(c)	other	Correct	(d)	other
(1)	16	4 25%	10 62%	0	2 12%	12 75%	4 25%	0	0	11 69%	4 25%	1
(2)	28	2 8%	21 81%	5 19%	0	10 38%	3 12%	10 38%	5 19%	22 84%	6 23%	0
(3)	29	4 14%	20 69%	4 14%	1 3%	20 69%	2 7%	6 21%	1 3%	9 31%	19 65%	1 3%
(4)	11	1 9%	7 64%	3 27%	0	2 18%	2 18%	5 45%	2 18%	2 18%	6 54%	3 27%
(5)	9	0	7 78%	2 22%	0	0	4 44%	5 56%	0	4 44%	3 33%	2 22%
(6)	28	2 7%	22 79%	0	4 14%	12 43%	7 25%	5 18%	4 14%	14 54%	7 27%	7 27%
(7)	22	2 9%	11 50%	6 27%	3 14%	9 41%	2 9%	11 50%	0	5 23%	14 64%	3 14%
Tot	143	15	98	20	10	65	24	42	12	67	59	17
%		10	69	14	7	45	17	29	8	47	41	12

(a) a battery at the two extremities, closed chain
(b) Correct structure with wrong terms
(c) Chain with 3 objects
(d) Mistake for the last object

establishing a relation. Moreover in some cases as the properties of the elements of the model are partially mastered at the beginning, the matching can be done through linguistic knowledge (Collet, 1994). For example, if a battery stores and a reservoir stores then a battery is a reservoir. Here the common property is 'storage'. We call this relation 'identity of behaviour'.

• treating elements of both worlds at the same level in order to construct an interpretation seems to be a necessary condition of understanding, and even the physicist uses it in his laboratory. For example, during the second task (Figure 8.4), a student said: 'The object is the reservoir . . . The object produces energy. Do you agree? The object falls and this produces energy which goes through the motor and arrives at the bulb. And the light bulb lights up.' Here the object which falls, the motor and the bulb are at the same level of energy. We call this relation 'intermediary interpretation'.

• establishing a similarity between a large part of the real situation and of the model is more complex; this is accomplished by introducing an explanation implying a theoretical approach. For example: 'the battery allows the bulb to light up; when the bulb lights up and gives light and heat to the environment the battery wears out' is related to 'the reservoir "battery" gives energy to the transformer "bulb" which in turn gives energy to the reservoir "environment"; this energy is transferred'. The causality is at both levels. The battery or the first reservoir is the cause (the bulb lights up or the transformer receives energy). Energy is related to reservoirs, transformers and transfers. We call this relation similarity of structure.

One of the main results is that the relations used by the pupils during the three tasks evolve, they become more complex. In the first task (battery–bulb task), the main mechanism is the simplest 'identity of behaviour'. Let us note that when some of the properties, like production, are not exactly those given in the theory/model, pupils adapt or develop them, for example, a reservoir produces or gives energy. In the following tasks we discovered that pupils develop more complex relations such as 'intermediary interpretations' or 'similarity of structure' where several elements of each world are put in relation. We can conclude that the design of the tasks is validated, because the tasks allow the students to construct new knowledge. This new knowledge consists of relations that increase in complexity with each successive task, this set of relations being necessary in physics modelling.

Conclusion

We have discussed a research approach based on two main concepts — didactical transposition and devolution — which are related to a specific point of view on the nature of the knowledge involved in didactical situations. We have distinguished scientific physics knowledge, which is shared by the scientific community, the knowledge to be taught, which is attached to the educational system, and the understanding of knowledge which indicates the relations between the learner (or more generally an individual) and a form of knowledge which has been already established outside him or her.

On these bases we have designed a teaching sequence with a given content of knowledge to be taught. One phase of this sequence incorporates a prototypical teaching situation for qualitative modelling. On one hand the situation 'stages' the knowledge to be taught by separating the aspects relating to the theory and the model (in a written form) from those relating to the experimental field (in the form of a real experiment), the combination of which permits the construction of coherent relations between these two worlds. On the other hand, in this situation the pupils have the responsibility for constructing the relations between these two worlds, which requires, in a dialectical way, reinforcing their understanding of each world. This characterization of a teaching situation allows comparison of different situations having the same or complementary goals, such as on other aspects of modelling or on the development of the model itself. This possibility of comparison offered by the construction of a prototypical situation can constitute an important element in the development of research on didactics and its links with real teaching.

References

ARSAC, G., CHEVALLARD, Y., MARTINAND, J.L. and TIBERGHIEN, A. (1994) *La Transposition Didactique à l'Épreuve*, Grenoble, La Pensée Sauvage.

ARTIGUES, M. (1988) 'Ingéniérie didactique', *Recherches en Didactique des Mathématiques*, **9**, 3, pp. 281–308.

BACHELARD, S. (1979) 'Quelques aspects historiques des notions de modèle et de justification des modèles', in DELATTRE, P. and THELLIER, M. (Eds) *Actes du Colloque Elaboration et Justification des Modèles*, Paris, Maloine, pp. 3–19.

BROUSSEAU, G. (1981) 'Problèmes de didactique des décimaux', *Recherches en Didactique des Mathématiques*, **2**, 3, pp. 37–127.

CHEVALLARD, Y. (1991) *La Transposition Didactique*, 2nd Ed, Grenoble, La Pensée Sauvage.

COLLET, G. (1994) 'Apports linguistiques à l'analyse de la modélisation en physique', in *Actes du Premier Colloque Jeunes Chercheurs en Sciences Cognitives*, La Motte d'Aveillans (Isère), pp. 231–40.

DRIVER, R., GUESNE, E. and TIBERGHIEN, A. (1985) *Children's Ideas in Science*, Milton Keynes, Open University Press.

DUIT, R. (1983) 'Is the second law of thermodynamics easier to understand than the first law?', in MARX, G. (Ed) *Entropy in the Schools: Proceedings of the 6th Danube Seminar on Physics Education*, Budapest, Roland Eöstvös Physical Society, pp. 87–97.

DUIT, R., JUNG, W. and RHÖNECK, C.V. (Eds) (1985) *Aspects of Understanding Electricity: Proceedings of an International Workshop*, Ludwisburg, IPN, Kiel.

GIERE, R.N. (1988) *Explaining Science: A Cognitive Approach*, Chicago, The University of Chicago Press.

JOHSUA, S. and JOHSUA, M.A. (1989) 'Les fonctions didactiques de l'expérimental dans l'enseignement scientifique', Part II, *Recherches en Didactique des Mathématiques*, **9**, 1, pp. 5–30.

KOLIOPOULOS, D. and TIBERGHIEN, T. (1986) 'Éléments d'une bibliographie concernant l'enseignement de l'énergie au niveau des collèges', *Aster*, **2**, pp. 167–78.

KUHN, T.S. (1971) 'Les notions de causalité dans le développement de la physique', in BUNGE, M., HALBWACHS, F., PIAGET, J. and ROSENFELD, L. (Eds) *Les Théories de la Causalité*, Paris, PUF, pp. 7–18.

KUHN, T.S. (1972) *La Structure des Révolutions Scientifiques*, Paris, Flammarion.

LEMEIGNAN, G. (1980) 'Documents et activités de l'élève: L'énergie' and 'Compléments d'information: L'énergie', in *Sciences Physiques Livre du Professeur*, Coll. Libres Parcours, Paris, Hachette.

LEMEIGNAN, G. and WEIL-BARAIS, A. (1987) *Apprentissage de la Modélisation à Propos de l'Enseignement de la Mécanique au Lycée*, Rapport de fin de contrat MIR-MEN, p. 75.

LEVY-STRAUSS, C. (1962) *La Pensée Sauvage*, Paris, Plon Collection Agora.

MÉHEUT, M. and CHOMAT, A. (1990) 'The limits of children atomism: An attempt to make children build up a particle model of matter', in LIJNSE, P.L., LICHT, P., DE VOS, W. and WAARLO, A.J. (Eds) *Relating Macroscopic Phenomena to Microscopic Particles*, Utrech, CDb Press, pp. 266–82.

PERRIN J. (1903) *Traité de Chimie Physique: Les Principes*, Paris, Gautier, Villars.

PSILLOS, D. and KOUMARAS, P. (1989) 'The use of the concept of time by pupils approaching electrical circuits and the implications for modelling teaching content', Paper presented to E.A.R.L.I. conference, Madrid.

PSILLOS, D., KOUMARAS, P. and TIBERGHIEN, A. (1988) 'Voltage presented as a primary concept in an introductory teaching on D.C. circuits', *International Journal of Science Education*, **10**, 1, pp. 29–43.

PSILLOS, D., KOUMARAS, P. and VALASSIADES, O. (1987) 'Pupils' representations of electric current before during and after instruction on D.C. circuits', *Journal of Research in Science and Technical Education*, **5**, 2, pp. 185–99.

ROZIER, S. (1988) 'Le raisonnement linéaire causal en thermodynamique élémentaire', Thèse de physique, Paris, Université Paris 7.

SHIPSTONE, D.M. (1985) 'On children's use of conceptual models in reasoning about current electricity', in DUIT, R., JUNG, W. and V. RHÖNECK, C. (Eds) *Aspects of Understanding Electricity: Proceeding of an International Workshop*, Ludwisburg, IPN, Kiel, pp. 73–93.

SOLOMON, J. (1985) 'Learning and evaluation: A study of school children's views on the social uses of energy', *Social Studies of Science*, **15**, pp. 343–71.

TIBERGHIEN, A. (1984) 'Critical review of the research aimed at elucidating the sense that notions of electric circuits have for the students aged 8 to 20 years', in *Research on Physics Education: Proceedings of the First International Workshop*, Paris, CNRS, pp. 109–23.

TIBERGHIEN, A. (1989) 'Learning and teaching at middle school level of concepts and phenomena in physics: The case of temperature', in MANDL, H., DE CORTE, E., BENNETT, N. and FRIEDRICH, H.F. (Eds) *Learning and Instruction: European Research in an International Context*, Vol. 2.1, Oxford, Pergamon Press, pp. 631–48.

TIBERGHIEN, A. (1994) 'Modelling as a basis for analysing teaching–learning situations', *Learning and Instruction*, **4**, 1, pp. 71–87.

TIBERGHIEN, A., PSILLOS, D. and KOUMARAS, P. *Physics Instruction from Epistemological and Didactical Basis* (in press).

TRUMPER, R. (1990) 'Being constructive: An alternative approach to the teaching of the energy concept — part one', *International Journal of Science Education*, **12**, 4, pp. 343–54.

9 Decision-making on Science-related Social Issues: The Case of Garbage in Physical Science — A Problem-posing Approach

Koos Kortland

Abstract

Since 1993 the Dutch curriculum for physical science at the junior secondary level (grades 7 to 9, ages 12–15) includes a number of environmental issues (water, waste, energy etc.) to be dealt with in the context of decision-making by students.

Partly based on earlier research and development activities a teaching unit on household packaging waste and related decision-making has been designed, in which the teaching/learning process:

- is largely guided by the students' existing issue perception and decision-making ability in a problem-posing way, so that they themselves frame the questions that drive their learning process;
- is structured by a sequence of interrelated knowledge and decision-making ability levels, so that teaching/learning activities carefully guide students in making transitions from one level to the next higher level of issue perception and decision-making ability.

The paper will first elaborate on these general ideas about the design of the teaching/learning process, followed by a more detailed description of the way in which these general ideas have been translated into a scenario for the teaching/learning of the specific topic of decision-making about household packaging waste. The scenario provides a description and justification of the activities the students and teacher are supposed to perform, including the expected outcomes of those activities.

Introduction

Since 1993 the Dutch curriculum for physical science at the junior secondary level (grades 7–9) includes attainment targets referring to a number of environmental issues (water, waste, energy etc.) and the ability to present an argued point of view in related decision-making situations. However, in these attainment targets any indication of how these words (an argued point of view) might be interpreted is lacking. In order to assess the feasibility of the curriculum proposals a small-scale research and development project was started in 1991 with the aim of preparing and investigating teaching about one of these environmental issues (waste, limited to household packaging waste) and related decision-making in classroom practice, with grade 8 middle-ability students.

At the start of this project the need to legitimate the teaching contents, to select student ideas to pay attention to, and to develop strategies to deal with these in order to improve the quality and credibility of teaching about science-related social issues (Eijkelhof and Lijnse, 1988) was recognized. Also recognized was the need to investigate the effects of teaching on students' decisions, on the way in which students arrive at their decisions, and on the quality of their arguments (Hofstein *et al.*, 1988). As teaching units (in Dutch) meeting these requirements did not exist at that time, the first task of the project was to develop and trial one. The results of this first phase of research and development can be summarized as follows.

- The students' initial knowledge and understanding of the waste issue improved after the series of lessons, showing some progress in the recognition of depletion of raw materials as an environmental problem and (im)possibilities for re-use/recycling as a solution — but less than might be considered desirable (Kortland, 1995a).
- The students' (slightly) increased knowledge and understanding of the waste issue can be retraced in the written or oral presentation of their decision in an open-ended decision-making situation after the series of lessons, showing some progress as far as validity and clarity of the criteria used for evaluating alternatives in a normative decision-making process (e.g., Carroll and Johnson, 1990) are concerned — but still being limited with respect to the range of criteria used (Kortland, 1995b).

A Problem-posing Approach

The teaching unit developed and trialled in the first phase of the project thus did not seem to be very effective. The main reason for this lack of effectiveness might be that the teaching unit was developed in a rather intuitive way, with most of the activities drawing heavily on the traditional teaching/learning strategy of *top–down transmission*. So, the design of the teaching unit was not guided by a clear theoretical/empirical framework for the desired teaching/learning process. Therefore, the second phase of the project had to start with thoroughly reconsidering the teaching/learning process and completely redesigning the teaching unit. The experiences in other research and development projects of the *Centre of Science and Mathematics Education* at Utrecht University seemed to point to the desirability of some kind of a *bottom–up constructivist* teaching/learning process that guides students to 'construct in freedom' the very ideas that one wants to teach (Lijnse, 1995). This could be done by carefully designing a sequence of teaching/learning activities on the basis of a profound knowledge of students' pre-knowledge and of its development, building on a proper interpretation of students' knowledge as being coherent and sensible and using their constructions productively in a social process of the teacher's and students' coming to understand each other (Klaassen, 1995).

According to this view the teaching/learning process should reflect a balance between construction with 'freedom from below' (for the students) and 'guidance from above' (by the teacher/designer). This asks for a (redesign of the) teaching unit, in which the teaching/learning process meets the following requirements:

- the process is largely guided (from below) by the students' own constructions, questions and motivations in a problem-posing way, so that they themselves frame the questions that drive their learning process;
- the process is structured (from above) by a sequence of interrelated levels of concept development, starting from a proper interpretation of students' pre-knowledge and carefully guiding them in making transitions from one level to the next higher level.

The Teaching/Learning Process

The combination of both of the above requirements results in a teaching/learning process which is outlined in the following three successive phases that build on one another (van Hiele, 1986; ten Voorde, 1990; Klaassen, 1995).

The first phase is one in which students make a transition from their everyday life level to a so-called 'ground level'. In this phase the teaching/learning activities should first give students the opportunity to develop a general motivation for the topic. After that the activities should be designed in such a way that, on the one hand, they solicit and structure the students' everyday life knowledge of the topic — thus arriving at a shared body of structured pre-knowledge. But, on the other hand, these activities will also trigger questions about the topic when students disagree about their everyday life knowledge or simply don't know what happens in reality — thus arriving at a shared number of questions for further investigation of the topic. So, in this first phase of the teaching/learning process the students' pre-knowledge is used productively for reaching a *ground level*, for explictating and structuring 'what we already know' and 'what we don't know yet' — representing a *problem-posing* approach to a trivial constructivist perspective (von Glasersfeld, 1989) on the teaching/learning process.

The questions for further investigation — formulated by the students themselves, though triggered by the design of the teaching/learning activities — should provide sufficient motivation for the students to engage in the next phase of the teaching/learning process. This second phase is one in which students make a transition from the *ground level* to a so-called *descriptive level*. In this phase the students search for answers to their own questions about the topic, either by looking into the literature (written texts and/or video-tapes) or by performing experiments. The answers found by the students themselves are reported, discussed and used productively in reaching a *descriptive level*, and extending the shared body of pre-knowledge that emerged during the preceding first phase of the teaching/

learning process into a concept network describing and relating a number of relevant concepts.

During this transition from the ground level to a descriptive level the students will start having questions about why this concept network is as it is. And these questions prepare the way for the next phase in the teaching/learning process. In this third phase students make a transition from the *descriptive level* to a *theoretical level*. This theoretical level would be concerned with an explanation for the structure of the concept network constructed during the preceding transition — and so, is concerned with scientific theory. In this way the descriptive level can be seen as another ground level, now for making the transition to a theoretical level.

Research Question

After this outline of the desired teaching/learning process, for the research and development project at this stage the research question would be: is it possible to translate these general ideas about the desired teaching/learning process into a detailed scenario for classroom practice? A question like this has been empirically addressed for teaching/learning about a straight physics topic like radioactivity (Klaassen, 1995), but what about an environmental topic like household packaging waste in the context of decision-making by students?

The Scenario

The above mentioned scenario consists of: a sequence of inter-related teaching/learning classroom activities; a justification of these activities in terms of how an activity builds on the preceding one and prepares for the next one; a description of what the students and the teacher are expected to do; and a hypothesis about the outcomes of each activity. In other words: the scenario is a *design* of the desired and expected teaching/learning process in classroom practice.

An outline of the scenario for teaching/learning about household packaging waste and related decision-making is given below. However, before going into some detail it must be stressed that the scenario *anticipates* students' reactions, and thus represents *hypotheses* about the outcomes of each activity. These hypotheses are, on the one hand, based on what reasonably or logically might be expected given the structure and sequence of the activities, and, on the other hand, based on (a reinterpretation of) the experiences during the trials in the first phase of the project (Kortland, 1992; 1995a; 1995b) and preliminary findings of other classroom research on decision-making by students (Ratcliffe, 1994).

The First Level Transition: From the Everyday Life Level to a Ground Level

In the scenario this first phase of the teaching/learning process starts with an activity, intended to generate some kind of general motivation for the topic of

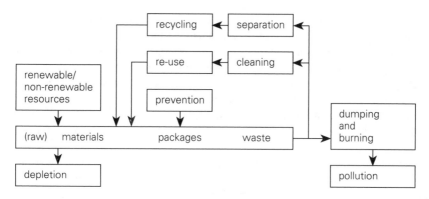

Figure 9.1: Concept network — packaging materials

household packaging waste. On the basis of a limited offer of audiovisual information about waste processing and the average composition of a typical Dutch garbage bag, students are asked to answer the question of why looking in more detail into this waste issue might be useful. It is hypothesized that students will indicate the importance of pollution prevention, in connection with the idea that they themselves do have some action perspectives with respect to the amount, the composition and the way of processing of discarded packages.

This first activity prepares the ground for the next activities in which the students' everyday life experiences with buying packaged products and throwing away empty packages act as a starting point for structuring their shared preknowledge about household packaging waste through a process of small-group work and classroom discussion. At the same time a number of questions about the waste issue will emerge, either through students' disagreeing about their individual pre-knowledge or simply not knowing. It is hypothesized that the structure of the students' collective pre-knowledge will look very much like the concept network represented in Figure 9.1.

The construction of this network by the students can be seen as part of the educational aim of the series of lessons. However, what is hypothesized to be lacking is a clear idea of the way in which the different specific packaging materials and packages fit into this concept network: which of the five most often used packaging materials (paper/cardboard, glass, tin-plated steel, aluminium and plastic) give pollution when dumped or burned, which can run out in the long-term, which are recyclable etc?

The above formulated questions for further investigation relate to knowledge about the waste issue, but what about decision-making? Near the end of this first phase students are introduced to a decision-making situation (about milk-packaging: carton or bottle), and are asked to make a decision, to reflect on the point(s), on which they compared the packaging alternatives, to list the other points on which they think the alternatives could be compared and to compare the alternatives on each point separately (the student worksheet is partly reproduced in Appendix 1). It is hypothesized that individually (each group of) students will come up with only one or two, but collectively will mention a large number of points (or criteria) for comparing (or evaluating) the packaging alternatives. And this can be transformed into another question for further investigation, now related to decision-making about the waste issue: on which points (or criteria) should packaging alternatives be compared (or evaluated) before making a decision about the 'best' alternative? And in addition, disagreements between students about the correct comparison of the alternatives on each criterion will strengthen the need for answers to the questions for further investigation of the waste issue formulated earlier.

After having formulated their existing pre-knowledge and their questions for further investigation of the waste issue and related decision-making, students are asked to construct a concept network through solving a puzzle in which they have to position and connect the different pieces, have to 'defend' their solution and have to indicate how their questions for further investigation are connected to this concept network. This structuring of existing and not-yet-existing knowledge prepares the way for the next phase of the teaching/learning process: the transition from the ground level to a descriptive level.

The Second Level Transition: From the Ground Level to a Descriptive Level

The first activity in the second phase of the teaching/learning process is fairly straightforward: small groups of students carry out an independent investigation in order to find answers to the questions (about pollution through dumping and burning, about depletion of resources, about waste separation and recycling of packaging materials etc.) formulated during the preceding phase, using a database, performing experiments, doing fieldwork etc. Each group investigates one of the questions, and in a reporting session shares the results of their work with their fellow-students. It is hypothesized that during this reporting session the concept network will expand (a bit) and will become more detailed — especially with respect to the 'paths' of different specific packaging materials and packages through the network. This constitutes the descriptive level of knowledge about the waste issue.

In the following activities this expanded and more detailed concept network is then used to formulate relevant criteria for evaluating packaging alternatives. It is hypothesized that students will mention as possibly relevant criteria something like: contribution to depletion of raw materials (or renewability); contribution to pollution through dumping and burning; necessity and re-usability of the package; recyclability of the packaging material. This development of an inventory of possibly relevant criteria is followed by a structured application of these criteria to the (already familiar) decision-making situation about milk-packaging (the student worksheet is partly reproduced in Appendix 2). By reflecting on their experiences during this activity the students themselves formulate the requirements a *well argued point of view* in a decision-making situation should meet. It is hypothesized

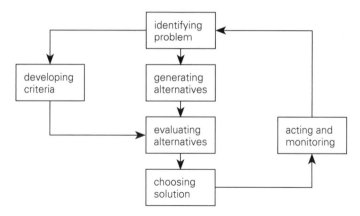

Figure 9.2: Concept network — a normative decision-making process

that the students in a classroom discussion will agree upon something like the following requirements: a correct evaluation of both alternatives on each of a complete range of relevant and clear criteria, concluded by an overall weighting of the separate evaluations on the criteria used. By reflecting on their experiences with decision-making during the preceding activities, students construct — again guided by a puzzle format of the activity, comparable to the guidance given earlier while constructing a concept network — a normative decision-making process as represented in Figure 9.2, showing a way in which such a well argued point of view could be reached.

In the final activity students are asked to formulate a well argued point of view in a self-chosen decision-making situation. It is hypothesized that students will make conscious use of the decision-making process constructed during the preceding activity. During another reporting session they are expected to use their knowledge of this process (and of the waste issue) for assessing the quality of their fellow-students' argued points of view. And with this activity students have reached a descriptive level of decision-making about the waste issue.

The Third Level Transition: From the Descriptive Level to a Theoretical Level

At the theoretical level the teaching/learning process should be dealing with an explanation of the concept network of the waste issue and related decision-making. What this theoretical level could be in environmental education is not yet very clear, let alone which kind of questions would motivate the students to start making this transition.

In any case, reaching this theoretical level is not regarded as a necessary aim for students of this age and ability level. So, for the time being this problem is 'solved'. However, in a later stage of physical science education a scientific principle (or theory) 'explaining' the waste issue and related decision-making might be

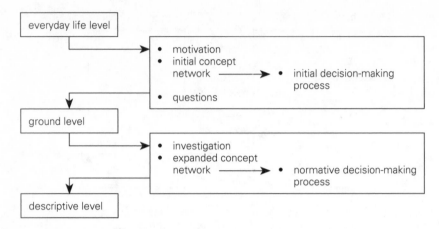

Figure 9.3: Concept network — waste and issue-related decision-making

considered useful. A possibility for such a scientific principle underlying the waste issue — in analogy with the energy issue — is the conservation and degradation of matter, with prevention of matter degradation as a general criterion for decision-making in the context of a strategy for sustainable development.

Reflection

The scenario outlined in the previous section reflects a certain interrelatedness of level transitions with respect to knowledge about the waste issue and issue-related decision-making respectively, as represented in Figure 9.3.

During both level transitions in the teaching/learning process knowledge about the waste issue is dealt with first, as a prerequisite for issue-related decision-making. Or, in other words: the development of the students' decision-making ability is being based on their existing and developing issue knowledge, so that students provide themselves with an issue-relevant conceptual input to the decision-making process.

Conclusion

The description of the scenario for a series of lessons on the topic of decision-making about the waste issue indicates that the first research question for the research and development project to a certain extent can be answered in an affirmative way. However, there is a second research question: will this scenario work in classroom practice? Or, in other words: will the expected teaching/learning process actually take place — and if not: why not, and what could be done about that in terms of a revision of the scenario?

An answer to this second research question can only be found by putting the scenario to the test after a thorough collaborative preparation by the trial teacher(s) and the researcher. These trials will take place during the school year 1995–6 in a limited number of schools.

References

CARROLL, J.S. and JOHNSON, E.J. (1990) *Decision Research: A Field Guide*, London, Sage Publications.

EIJKELHOF, H.M.C. and LIJNSE, P.L. (1988) 'The role of research and development to improve STS education: Experiences from the PLON-project', *International Journal of Science Education*, **10**, 4, pp. 464–74.

GLASERSFELD, E. VON (1989) 'Cognition, construction of knowledge and teaching', *Synthese*, **80**, pp. 121–40.

HIELE, P.M. VAN (1986) *Structure and Insight: A Theory of Mathematics Education*, Orlando, Academic Press.

HOFSTEIN, A., AIKENHEAD, G. and RIQUARTS, K. (1988) 'Discussions over STS at the fourth IOSTE Symposium', *International Journal of Science Education*, **10**, 4, pp. 357–66.

KLAASSEN, C.W.J.M. (1995) *A Problem-posing Approach to Teaching the Topic of Radioactivity*, Utrecht, CDß Press (in press).

KORTLAND, J. (1992) 'Environmental education: Sustainable development and decision-making', in YAGER, R.E. (Ed) *The Status of STS Reform Efforts around the World*, Petersfield, ICASE Yearbook 1992, pp. 32–9.

KORTLAND, J. (1995a) 'Garbage: Dumping, burning and reusing/recycling — students' perception of the waste issue', *International Journal of Science Education* (in press).

KORTLAND, J. (1995b) 'Garbage: Dumping, burning and reusing/recycling — An STS case study about students' decision-making on the waste issue', *Science Education* (submitted).

LIJNSE, P.L. (1995) '"Developmental research" as a way to an empirically based "didactical structure" of science', *Science Education*, **79**, 2, pp. 189–99.

RATCLIFFE, M. (1994) 'Decision-making about science-related social issues', in BOERSMA, K., KORTLAND, K. and van TROMMEL, J. (Eds) *Science and Technology Education in a Demanding Society*, Enschede, National Institute for Curriculum Development (SLO), vol. 3, pp. 722–32.

VOORDE, H.H. TEN (1990) 'On teaching and learning about atoms and molecules from a Van Hiele point of view', in LIJNSE, P.L., LICHT, P., de VOS, W. and WAARZO, A.J. (Eds) *Relating Macroscopic Phenomena to Microscopic Particles*, Utrecht, CDß Press, pp. 81–104.

Appendix 1: Worksheet Sample

13 Different packages
    In activity 11 you gave examples of products for sale in different packages. One example is milk. In some shops you can choose between milk in a glass bottle and milk in a carton.

a   What would you choose: a bottle or a carton? Write down your choice below. And add your reason for choosing that package.
    . . . .
    . . . .

b   In activity 13a you gave an argument for your choice. In such an argument you compare the bottle and the carton on a certain point. On which point did you compare the bottle and the carton in activity 13a?
    . . . .
    . . . .

c   You can compare the bottle and the carton on more than one point. In the table below those two packages are listed. Think about the points you could pay attention to when making a choice. Write down those points in the table's first column. And then write behind each point which package you would choose if you only pay attention to that point.

points to pay attention to when making a choice	if I only pay attention to this point, I choose the ...
. . . .	bottle/carton
. . . .	bottle/carton
. . . .	

Appendix 2:   Worksheet Sample

26   Decision-making situation: bottle/carton
In some shops you can choose between milk in a glass bottle and milk in a carton. That we call a *decision-making situation*: for the *product* (milk) there are two *packaging alternatives* (bottle and carton). And you have to *choose* between those alternatives. Therefore, you are going to *compare* the packaging alternatives on a number of points. Those points of comparison we call *criteria*.

a   In the table below the product and the packaging alternatives are already listed. In addition, write down the materials the different packages are made of.

b   Write down the *criteria* on which you are going to *compare* the packaging alternatives in the table.

c   Compare the packaging alternatives on each criterion separately: which alternative do you choose if you only pay attention to that criterion, and why? Write down your comparison in the table. Use — if necessary — the database for information about the packaging alternatives.

product	milk	
packaging alternatives	bottle	carton
packaging materials	. . . .	. . . .
	. . . .	. . . .
criteria	comparison	
. . . .	. . . .	
	. . . .	
	. . . .	
. . . .	. . . .	
	. . . .	
	. . . .	
. . . .	. . . .	
	. . . .	

d   Look carefully at the comparison of packaging alternatives in the table above. Are all criteria equally important for you? Mark the most important criteria (e.g., with a colour). Are the differences between the alternatives on all criteria equally big for you? Mark the biggest differences (e.g., with another colour).
Now what seems to be the best alternative to you, and why?

. . . .
. . . .
. . . .

10 Adolescent Decision-making, by Individuals and Groups, about Science-related Societal Issues

Mary Ratcliffe

Abstract

This study explored the skills, knowledge and values used by 15-year-old pupils in making decisions about science-related societal issues. In particular it examined the decision-making strategies of individuals, alone and in groups.

Boys, undertaking a science course in which decision-making tasks were an integral feature, were audio-taped during discussion, interviewed and had their written work analysed. Individual pupils showed stability in decision-making approach throughout the study. This has implications for intervention strategies in the classroom.

Introduction

This study sought to examine the decision-making strategies used by pupils as individuals and when working in groups. It also set out to explore the nature of scientific knowledge, personal and societal values used by pupils in deciding about socio-scientific issues.

A number of related perspectives provide a background for the study:

- There are many exhortations for encouraging 'scientific literacy' to prepare learners for an active role as 'voting' citizens (e.g., Bodmer, 1986; Krugly-Smolska, 1990). Science and technology provide a knowledge base for addressing some real issues and problems.
- The gradual introduction of STS (Science-Technology-Society) curriculum materials into British schools has been apparent over many years (Hunt, 1994). Recently, initiatives such as SATIS (Science and Technology in Society) (ASE, 1986, 1988), Suffolk Science (Dobson, 1987) and Salters Science (MEG, 1992) have sought to integrate the learning of scientific concepts more closely with real-life scenarios. Evaluations of these (e.g., Walker, 1990; Ratcliffe, 1992) have not explored in detail the interactions of learners with such curriculum materials in the classroom.
- The introduction of the National Curriculum in England and Wales includes the expectation that learners will explore the role of science in decision-making about science-related issues (DES, 1991).

Roberts (1988) gives a useful overview of seven possible emphases of science education, drawing upon views of various stakeholders in education — pupil, teacher, employer, university lecturer. One such emphasis is 'science, technology, decisions' in which he outlines a particular view of the learner:

Needs to become an intelligent, willing decision-maker who understands the scientific basis for technology and the practical basis for defensible decisions.

Implied in this view of the learner and in other researchers' views of the aims of STS education (e.g., Hofstein, Aikenhead and Riquarts, 1988) are three interrelated strands:

- developing appropriate skills, including information processing and analysis;
- understanding relevant science; and
- recognizing personal and societal values.

These strands led to three research questions:

A   How do pupils handle their decision-making?
B   What science do pupils draw upon in making decisions?
C   What values do pupils bring to bear in decision-making?

These three questions assume:

- Thoughtful decision-making not only uses an information base but also requires some systematic consideration of possible responses to the issue.
- Relevant scientific knowledge is important in coming to an informed perspective on an issue.
- Reasoning about socio-scientific issues is not based on scientific evidence alone. Value judgments are made about the worth of different responses.

This paper concentrates on decision-making strategies. Questions B and C are explored more fully elsewhere (Ratcliffe, in press).

What is Decision-making?

Decision-making implies a commitment to action. It could be more appropriate to regard this study as examining the formation of opinions rather than decisions. However, decision-making is used throughout the literature in connection with tasks of a similar nature to the ones in this study.

Part of the study examined individual decision-making strategies in the absence of a framework. The major part of the study examined group discussion for which a framework was provided. The framework built on:

- exploratory research conducted in two schools;
- research evidence of a theoretical and empirical nature regarding decision-making in general (e.g., Payne *et al.*, 1992) and in science classes in particular (e.g., Aikenhead, 1989).

Normative decision-making models give a logical structure of step-by-step processes which should be undertaken if rational decision-making is to take place (e.g., Janis and Mann, 1977). Normative models view decision-making in reality to be deficient in some way with respect to the model (e.g., Payne *et al.*, 1992). Descriptive decision-making models attempt to provide a theoretical view of decision-making as it happens (Hirokawa and Johnston, 1989). The common characteristics of most normative and descriptive models are:

- identifying the nature of the problem;
- identifying the alternative outcomes available;
- identifying the objectives to be met by the solution;
- assessing the positive and negative consequences of the outcomes; (Janis and Mann, 1977; Beyth-Marom *et al.*, 1991).

In descriptive models the process is viewed as a cycle with no fixed sequence for clarifying alternatives and objectives. The framework for group discussions incorporated these common elements and was given to pupils in the form shown in Figure 10.1. These six steps — options, criteria, information, survey, choice, review — may encourage a particular logic in group discussion. However, it was anticipated that there might be deviations from the structure according to how individuals and groups viewed and discussed the problems, in line with descriptive models.

The Research Enquiry

The study was conducted in an 11–16 city boys school, in which Salters Science was the GCSE (General Certificate of Education) course for the 14–16 age range. In 1993–4, four classes of year 10 boys (14–15-year-olds) were studying the Salters course leading to a double GCSE award, fulfilling National Curriculum requirements (DES, 1991). For these boys (ninety-three in total), a number of methods were used to examine their decision-making as individuals, and in groups, as a regular feature of their science curriculum.

Each Salters Science teaching unit of roughly three weeks duration (fifteen hours) contains a lesson by lesson outline for the teacher, accompanied by suggested pupil activities. I designed decision-making tasks for the first five units of the academic year 1993–4. The task either replaced an existing activity with the same focus or provided an additional task closely linked to the science content of the unit. The normal class teacher led the task, occupying a 50-minute lesson.

The selection of issues was framed by the units being studied and not by

Options
Make a list of all the things you could do/think of relevant to the problem.
(This statement is phrased appropriately for each different problem)
YOU HAVE MADE A LIST OF OPTIONS

Criteria
How are you going to choose between these OPTIONS?
Make a list of the IMPORTANT things to think about when you look at each option.
YOU HAVE DECIDED WHAT IS IMPORTANT IN CHOOSING AN OPTION

Information
Do you have useful information about each option?
What do you know about each option in relation to your criteria?
What information do you have about the science involved?
YOU HAVE CHECKED THE INFORMATION YOU HAVE

Survey
What are the good things about each option?
  — think about your criteria.
What are the bad things about each option?
  — think about your criteria.
YOU HAVE THOUGHT ABOUT EACH OPTION

Choice
Which option do you choose?
YOU HAVE MADE YOUR DECISION

Review
What do you think of the decision you have made?
How could you improve the way you made the decision?
YOU HAVE THOUGHT ABOUT YOUR DECISION MAKING

Figure 10.1: Decision-making framework as presented to the pupils

focusing on issues of a particular nature. Within these constraints, the following types of issue were included:

(a) personal and immediate.
  Energy: What are you prepared to do to use energy more efficiently?
(b) local and less immediate.
  Frame: Which material would you use for a replacement window frame?
  Transport: What method(s) would you choose to transport butanone from manufacturer to customer?
  Ecology: Should an 'improvement' programme for a river or a golf course go ahead?
(c) global, where judgments can be made but little action may result.
  Food: What could be done to help the world food problem?

These particular questions were chosen on the basis that:

• they had no 'right' answer;
• they linked with some identifiable scientific knowledge, from the key area chosen; and
• additional, relevant information could easily be provided.

Methodology: Following Individuals and Groups in Decision-making

The detail of the methodology was a mixture of pre-planned deliberate actions and acting on emerging data and analysis. It comprised:

1 Conducting individual interviews with thirty-one boys (from eight discussion groups — two from each class) very early in the 1993–4 academic year. In these interviews, boys were presented with four decision-making scenarios and asked *how* they would decide:

- What cups — plastic, paper or glass — they would buy for a large social gathering? (Cups)
- Whether fluoride should be added to the water supply and who should have the responsibility for the final decision? (Fluoride)
- Whether they would be vaccinated against 'flu, given some information about relative risks? (Flu)
- What advice they would give to a younger pupil about how to make decisions in general. (General)

2 Audio-taping these eight small discussion groups, on regular occasions from October to February, while they carried out the group decision-making discussions outlined above — in order to analyse the processes undertaken. All the audio-tapes were studied in some depth. Audio-tapes from four of the discussion groups (fifteen boys) were transcribed and analysed, using *Ethnograph* (Qualis Research Associates, 1988) to aid the process.

3 Collecting the written work from all the boys in four classes — in order to analyse the outcome of the tasks.

4 Collecting individual written responses from all four classes to two decision-making scenarios given two months after completion of all discussion tasks. These scenarios were linked to science content recently experienced and pupils were asked *how* they would decide:

- Whether a useful pesticide, which is considered to affect the ozone layer, should be banned? (Pesticide)
- Which treatment they would have if they were diagnosed as short-sighted — wearing glasses; wearing contact lenses; having laser treatment? (Optics)

5 Collecting responses from all the boys to a questionnaire based on one used in evaluating a North American STS course (Kelly, 1991). This sought their perceptions of the nature of science; decision-making about societal issues; involvement of different groups or individuals in societal decision-making.

6 Conducting individual interviews with fifteen boys (those in four of the eight discussion groups) two months after completion of discussion tasks. In these interviews boys were asked *how* they would decide:

- What material to use for packaging a convenience food with a long shelf-life — 'tin-cans'; aluminium containers or plastic containers? (Cans) A similar problem to the Cups problem in the earlier interview, but a different context.
- Whether fluoride should be added to the water supply and who should have the responsibility for the final decision? (Fluoride) Identical to the problem in the earlier interview.
- What advice would they give to a younger pupil about how to make decisions in general? (General) Identical to the problem in the earlier interview.

Results

The main features of decision-making strategies and group discussion are considered. In addition the features and contributions of two individuals are described in order to illustrate the extremes of decision-making strategies seen.

Decision-making Strategies: Individuals

Individual responses to decision-making scenarios were categorized according to the following scale:

0 — no response
1 — a decision is made, no reason given
2 — a decision is made, reasoning is given

Decision not made:

3i — suggests a need for further information before deciding
3e — suggests a need to carry out tests or surveys before deciding
3c — suggests one or more criteria as the basis for the decision
4 — in addition, suggests an examination of the advantages and disadvantages of the alternatives

These decision-making levels are intended to be hierarchical, as are those described by Ross (1981) and Kortland (1994), in their developments of decision-making programmes. Although there are some similarities with Ross's detailed structure, these levels emerged from the types of responses given. In each case of

1 — Flu

Simon	Don't know. I'd get vaccinated.
Researcher	Why?
Simon	So I don't get it again.

2 — Fluoride

Neil (promptly) I wouldn't put it in there, 'cos the water as it is has been going for several years like it is and nothing's happened to anyone so if they put fluoride in it, it could harm the body inside if they drank too much of it.

3i — General

Neil I'd tell him to speak to his parents about it (pause) and read books about it before he made his decision.

Researcher Why read books about it?

Neil 'Cos books tell more than your parents should really know — it tells you what can happen and what doesn't happen.

3e — Fluoride

Simon Um, do a test.

Researcher What sort of test?

Simon On an animal or something. So do an experiment and see if it comes up with any diseases or anything.

3c — Cups

Keith Um, I'd think of the recycling side of it first and then the safety side of it.

Researcher Anything else?

Keith The cost would probably come into it.

4 — General

Jim I'd um tell them to look at the information they have and weigh up the, um, what the effects of whatever the decision would be and go with the majority (effects).

Figure 10.2: Examples of responses at different levels to decision-making scenarios

interview or written response, pupils were asked *how* they would make the decision. Thus responding with a decision, even though it may be justified is given a slightly lower level than identifying strategies for making the decision. Examples of responses in these categories are given in Figure 10.2.

A number of points emerge from the analysis of the detail of individual responses.

Individual Development

In the vast majority of scenarios, the levels of response increased between pre- and post-experience interviews (Table 10.1). There was a much higher frequency of responses which discussed examining the advantages and disadvantages of each option than in the pre-experience interviews. This cannot necessarily be attributed to the experience of decision-making in the classroom.

This development shown in verbal responses was not shown in written responses to post-experience decision-making scenarios (Table 10.1).

Table 10.1: *Decision-making strategies in pre-experience interviews; post-experience interviews; post-experience written responses*
Percentage of responses at each level across *all* scenarios tackled

Level	Pre-experience interview (across 4 scenarios)		Post-experience interview (across 3 scenarios)	Post-experience written responses (across 2 scenarios)	
	SAM 15	TOT 31	SAM 15	SAM 15	TOT 78
0	1.6	(3.6)	0	0	(0)
1	3.2	(3.6)	0	0	(6.3)
2	22.2	(21.0)	1.6	28.2	(23.9)
3i	25.4	(25.3)	20.6	23.0	(18.5)
3e	19.0	(17.4)	19.0	17.9	(23.4)
3c	19.0	(22.9)	17.4	20.5	(17.6)
4	9.5	(6.0)	41.2	10.2	(10.2)

SAM 15 — the fifteen boys who were in the 4 small discussion groups and were studied in depth
TOT 31 — the total number of subjects of pre-experience interviews
TOT 78 — the total number of boys providing written responses

Similarity of Response

The similarity of the nature of the responses to identical or similar scenarios eight months apart is striking in the transcripts of many of the interviews. Although the level of response may have increased, some of the rationale behind particular criteria or strategies remained the same.

This is illustrated by Jim's and Neil's responses.

Extract 1: Responses to the Fluoride Question by Jim

Pre-experience (level 3e)

Jim I'd send questionnaires to the people affected by the change and ask them whether they thought it would be beneficial or not.
Researcher What would you do with the information?
Jim I'd then sort through it — the people who thought it was a good idea and why, and those who didn't.

Post-experience (level 3e)

Jim I'd design questionnaires and put it in newspapers, magazines, dentists surgeries all sorts of places like that and get people to decide.
Researcher What sort of questions might you ask if you're going to design the questionnaire?
Jim I'd want to know how much it would increase the price of water. How much extra they're prepared to pay. And, um, whether they want it in there.

| **Researcher** | What would you do when you'd got the questionnaire results? |
| **Jim** | Probably just average out how many people wanted it, who didn't, whether it was certain areas of the city that wanted it, and didn't. |

Extract 2: Response to Cups/Cans by Neil — A Pupil with Apparently Low Levels of Strategies Initially (There is a similarity in the starting point, although the arguments seem fuller in the post-experience interview.)

Pre-experience Cups (level 2)

Researcher	How would you choose?
Neil	Plastic cos the paper might get soggy and go everywhere and people, like, start throwing it around but plastic cups they're like soluble don't leak and they're easy to dispose.
Researcher	Anything else?
Neil	No.

Pre-experience Cans (level 3c)

Neil	Getting them home. You'd have to think of like protection for the food. If you like get a tin of sausages or something like that and you put them in a tin foil pack — they'd get all crushed up and broken or something like that. When they're in a tin they don't. There's some protection in a tin.
Researcher	Anything else?
Neil	What looks a better one. If you saw a tin one and you saw that one, which one would they pick?
Researcher	Anything else?
Neil	Um, no, um it depends if you can get more in one than the other as well, which one holds the more.

This stability in underlying strategy and rationale is also shown in the approaches of individuals across most aspects of the study. Characteristics of Jim and Neil are described in Figure 10.3. These two represent the extremes of 'information-vigilant' behaviour seen. Jim is information-vigilant from the outset. Neil shows information vigilance only in the post-experience interview.

Decision-making in Group Discussions

These extremes of information-vigilance can be set alongside more general findings from studying the group discussions and their written output.

Some of the features of discussions varied with the context of the issue and the framing by the teacher. However, there were a number of features related to the use of the decision-making framework provided, which seemed independent of

Information vigilant	**Not information vigilant**
Jim	**Neil**

in pre-experience interview:

- information seeking, identifies criteria
- considers pros and cons

- makes choices, offering reasons

in group discussion:

- guides group discussion

- evaluates information and contributions of others
- delays making decisions

- changes mind on one occasion
- understands and uses the decision-making framework in group discussion
- high achiever in science

- monitors group progress through the decision-making framework

- makes decisions early or goes along with the decisions of others

- shows little understanding of the framework but recognizes it
- below average achiever in science

in post-experience interview:

- information seeking, identifies criteria
- considers pros and cons

- suggests carrying out tests
- identifies criteria

Figure 10.3: Extremes of information vigilance

context. Some of the findings from analysing the processes of the discussions mirror those from analysing the outcomes in the form of written reports.

The common features of group decision-making, in relation to the framework were:

- Pupils were able to identify a range of possible options in examining each issue (95 per cent of written reports).
- Pupils seemed able to identify criteria for choice but spent little time on this. There was clear identification of four or more criteria in 35 per cent of written reports.
- Overt information seeking was not a dominant aspect in most cases. Information was cited in a third of reports. Only 7 per cent of reports showed any scientific evidence.
- Pupils discussed advantages and disadvantages of various options (between 15 and 48 per cent of discussion time). In the vast majority of reports (95 per cent) and discussions this was *not* undertaken systematically against identified criteria.
- Very little evaluation of decisions was undertaken (18 per cent of written reports).

Information-base Used

Very few pupils overtly drew upon scientific evidence in the discussions, even when prompted by the framework. Jim and Neil can again be used to illustrate the range of information-vigilance.

Neil did use the map provided in the Transport task to identify possible options and realized that a naked flame could set butanone alight. This was the only evidence of him using an information base in discussions.

Jim, in contrast, showed much more awareness of the information available and could discuss some of the science base. In the group discussion of measures to conserve energy (Energy) he evaluated others' contributions using information provided (Extract 3). Similarly in the Frame task in discussing the potential use of different materials for making window frames, Jim evaluated the information base and discussed some recently learnt science (Extracts 4 and 5). These types of contributions were rare.

Extract 3

Jim	The good things of loft insulation? (pause)
Steve	Good thing is it's quite cheap.
Jim	I wouldn't say that.
Scott	I wouldn't say that.
Steve	Cheapish.
Jim	Eight pounds 60 odd a roll, and you need about 2 billion rolls. It is not cheap, Steve. It is not cheap.
Steve	Cheapish, cheap and nasty.
Jim	Where does it say that?

Extract 4

Jim	Quiet, please. Aluminum costs £100, PVC costs £129. That is for the frame and glass only. It doesn't include any labour costs or installation costs. . . . Making aluminium and PVC use a lot more energy. For example, processing wood uses much, much less than processing aluminium or PVC.

Extract 5

Jim	Once aluminium is installed it corrodes on the surface but does not continue to corrode . . . Aluminium. We have down here aluminium does not rust. That's not quite correct, not quite correct.
Steve	What's that?
Scott	It oxidizes. It oxidizes.
Steve	It doesn't rust.
Jim	What do you call that on the edge, then (pointing to nearby window frame). Corrosion.
Steve	It's not rust.
Jim	Corrosion.
Steve	Yes, but it's only on the surface.
Jim	Surface corrosion.

Bob	Surface can go back can't it.
Steve	No.
Jim	We learnt that yesterday. It's only surface corrosion. It doesn't go deeper.

Values Used in Decision-making

There were a few dominant criteria for making the decisions. In about half of the discussions these were overtly identified as criteria in advance of discussing the advantages and disadvantages of options. More frequently they were justifications emerging in surveying the advantages and disadvantages of options or in the final choice.

Cost and effectiveness appeared as important considerations in all the decision-making contexts. They appeared in over 50 per cent of written reports and in all transcribed discussions. Effectiveness was, however, sometimes difficult for pupils to define. Other criteria were specific to particular contexts:

- Safety, security and environmental considerations in transporting a flammable chemical (Transport).
- Fairness to different people and animals in changing land use (Ecology).
- Altruism and selfishness in considering the world food problem (Food).
- Energy and environmental considerations and aesthetics in considering replacement window frames (Frame).

These were the most evident bases for the decisions made. Other considerations seemed not to be used.

Group Discussions and Information Vigilance

In group discussions where individuals like Jim were contributing, well-reasoned argument was evident. These groups appeared to:

- have an understanding of decision-making procedure (not necessarily following the given framework slavishly);
- make reference to available information and evaluate it;
- clarify the basis for making the decision;
- have a willingness to engage fully in group discussion (Cowie and Rudduck, 1990) (most groups fulfilled this requirement); and
- argue and justify different individual viewpoints.

In meeting the challenge to encourage this reasoning in all groups of learners there appear to be a number of difficulties.

In reaching well-reasoned defensible decisions, use of some evidence and information is expected (Kuhn, 1992). It appears that the pupils could justify decisions on the grounds of a number of criteria but this was often not clearly substantiated in terms of the information or evidence available. A contradiction was seen where boys did not draw fully on available information but, when they reviewed the decision-making process, identified the need for further information. This echoes Fleming's (1986) findings of late adolescents discussing similar socio-scientific issues.

In Kuhn's (1992) study of adolescents' and adults' justifications for opinions about social issues (crime, school failure and unemployment) she found many subjects unable to produce a coherent argument. She suggests that reasoning is to some extent predicted by educational level and experience. Kuhn argues:

> If cognitive skills exist in implicit form before they appear more explicitly, the educational challenge becomes to a large degree one of reinforcing and strengthening skills already present in implicit form, rather than instilling skills that are absent.

This strikes a chord with the underlying stability in approach found with the subjects in my study. Kuhn supports the idea of social argument in educational activities as fulfilling this challenge outlined. It seemed in my study that, for boys who had already developed thoughtful decision-making strategies, peer discussion appeared to hone these through argument. In the cases of those with less developed strategies, peer discussion did not bring about such reasoned argumentation. Although initial decision-making strategies did not seem totally dependent on educational achievement, it was the higher achieving groups who were able to sustain reasoned argumentation more fully in group discussion. Similar to Zohar's (1994) programme of teaching thinking skills, those with initially 'lower' level skills appeared to show more development, in verbal reasoning in interview, over the nine-month time span of the study than those with higher levels.

Some of the evidence from my study suggests that details of responses will vary with the context of the issue. However, within this there seems a stability of general response and underlying approach and rationale which is to do with the nature of the individual rather than anything to do with the structure presented.

General approaches may be determined by psychological and personality type and may be fairly resistant to change. Details of responses to particular contexts may vary according to the input of others, the structure of the task and the expected outcomes. The study does not show what is needed for an individual to become more information vigilant. Further and more detailed longitudinal studies would be needed to examine these issues.

Implications for Pedagogy: Developing Decision-making Skills

The inclusion of decision-making tasks into an issues-led science curriculum has encouraged Robert's (1988) 'science, technology, decisions' view of the learner.

The encouragement to use a particular approach to decision-making appears not, by itself, to develop clear reasoning skills. Evidence suggests that integrating a decision-making structure with other activities in science lessons results in no increased use of an information base. However, teacher intervention in assisting small group discussion in this study was not high. Adey (1992) has shown how in-service training for teachers in cognitive development programmes is a significant feature for improving pupils' abilities.

Intensive decision-making programmes make a number of claims for improving learners' decision-making ability (e.g., Mann *et al.*, 1988). The self-reporting mechanisms for evaluating these have been criticized (Beyth-Marom *et al.*, 1991). These programmes have not sought to examine underlying psychological traits. The development of decision-making and other higher-order skills in learners is a complex area. My study suggests that a good knowledge of pupils' abilities is needed before intervention strategies can be fully effective. Grouping pupils so that there is a mixture of skills within a group and being clear about group procedures may assist peer group interaction and individual development.

References

ADEY, P. (1992) 'The CASE results: Implications for science teaching', *International Journal of Science Education*, **14**, 2, pp. 137–46.

AIKENHEAD, G.S. (1989) 'Decision making theories as tools for interpreting student behaviour during a scientific inquiry simulation', *Journal of Research in Science Teaching*, **26**, 3, pp. 189–203.

ASE (Association for Science Education) (1986) and (1988) *Science and Technology in Society*, volumes 1–10 with a general guide for teachers, Hatfield, ASE.

BEYTH-MAROM, R., FISCHOFF, B., JACOBS QUADREL, M. and FURBY, L. (1991) 'Teaching decision-making to adolescents: A critical review', in BARON, J. and BROWN, R. (Eds) *Teaching Decision Making to Adolescents*, New Jersey, Lawrence Erlbaum Associates Inc., pp. 19–59.

BODMER, W. (1986) *The Public Understanding of Science: Seventeenth J.D. Bernal Lecture*, Birkbeck College, London.

COWIE, H. and RUDDUCK, J. (1990) 'Learning through discussion', in ENTWISTLE, N. (Ed) *Handbook of Educational Ideas and Practices*, London, Routledge.

DES (1991) *Science in the National Curriculum*, London, HMSO.

DOBSON, K. (1987) *Teaching for Active Learning: Co-ordinated Science (The Suffolk Development*, Teachers' Guide, London, Collins.

FLEMING, R. (1986) 'Adolescent reasoning in socio-scientific issues, part II: Nonsocial Cognition', *Journal of Research in Science Teaching*, **23**, 8, pp. 689–98.

HIROKAWA, R.Y. and JOHNSTON, D.D. (1989) 'Toward a general theory of group decision making', *Small Group Behaviour*, **20**, 4, pp. 500–23.

HOFSTEIN, A., AIKENHEAD, G. and RIQUARTS, K. (1988) 'Discussions over STS at the 4th IOSTE symposium', *International Journal of Science Education*, **10**, 4, pp. 357–66.

HUNT, J.A. (1994) 'STS teaching in Britain', in BOERSMA, K.Th., KORTLAND, K. and VAN TROMMEL, J. (Eds) *Proceedings of the 7th IOSTE Symposium*, Enschede, the Netherlands, National Institute for Curriculum Development (SLO), pp. 409–17.

JANIS, I.L. and MANN, L. (1977) *Decision Making: A Psychological Analysis of Conflict, Choice and Commitment*, N.Y., The Free Press.

KELLY, P.J. (1991) *Perceptions and Performance: An Impact Assessment of CEPUP (Chemical Education for the Public Understanding Programme) in Schools*, California, Lawrence Hall of Science.

KORTLAND, K. (1994) 'Decision-making on science-related social issues: A theoretical/ empirical framework for the development of students' concepts and decision-making skills', in BOERSMA, K.Th., KORTLAND, K. and VAN TROMMEL, J. (Eds) *Proceedings of the 7th IOSTE Symposium*, Enschede, the Netherlands, National Institute for Curriculum Development (SLO), pp. 448–71.

KRUGLY-SMOLSKA, E.T. (1990) 'Scientific literacy in developed and developing countries', *International Journal of Science Education*, **12**, 5, pp. 473–80.

KUHN, D. (1992) 'Thinking as argument', *Harvard Educational Review*, **62**, 2, pp. 155–78.

MANN, L., HARMONI, R., POWER, C., BESWICK, G. and ORMOND, C. (1988) 'Effectiveness of the GOFER course in decision-making for high school students', *Journal of Behavioral Decision Making*, **1**, pp. 159–68.

MEG (Midland Examining Group) (1992) *Science (Salters) GCSE Examination Syllabus for 1994*.

PAYNE, J.W., BETTMAN, J.R. and JOHNSON, E.J. (1992) 'Behavioral decision research: A constructive processing perspective', *Annual Review of Psychology*, **43**, pp. 87–131.

QUALIS Research ASSOCIATES (1988) *The Ethnograph: A Program for the Computer-assisted Analysis of Text-based Data*, Colorado, USA.

RATCLIFFE, M. (1992) 'The Implementation of criterion-referenced assessment in the teaching of Science', *Research in Science and Technological Education*, **10**, 2, pp. 171–85.

RATCLIFFE, M. (in press) 'Pupil decision-making about socio-scientific issues within the science curriculum', *International Journal of Science Education*.

ROBERTS, D.A. (1988) 'What counts as science education', in FENSHAM, P. (Ed) *Developments and Dilemmas in Science Education*, London, Falmer Press pp. 27–54.

ROSS J.A. (1981) 'Improving adolescent decision-making skills', *Curriculum Inquiry*, **11**, 3, pp. 279–95.

WALKER, D. (1990) 'The evaluation of SATIS', *School Science Review*, **72**, 159, pp. 31–9.

ZOHAR, A. (1994) 'Teaching a thinking strategy: Transfer across domains and self learning versus class-like setting', *Applied Cognitive Psychology*, **8**, pp. 549–63.

Part II
Developing and Understanding Models in Science Education

Introduction
John Gilbert

Models have a contribution to make to the provision of a more 'authentic' science education. This is because, in their various manifestations, they have roles in the conduct of science, are themselves the products of science, and are of crucial use in the teaching and learning of science. The four papers presented at the symposium discuss different aspects of this complex contribution.

Sutton reviews the way in which models are developed in science. In essence, a new way of describing an aspect of nature evolves with the adoption of a new metaphor. The new insight, as used and presented by the original user ('think of it as'), is progressively replaced by a certainty ('this is how it is') in textbooks. Sutton calls for a return to, or a recapitulation of, such tentative language in science education. Duschl's thesis is that the processes by which models and theories are produced in science are similar to those employed in learning science. To achieve a more authentic science education, it is important to have a clearer view of how models/theories are produced and changed in science. Teachers can then build these processes into learning environments for students. A series of 'generation of knowledge frameworks' are presented, showing how current insights can be integrated to these ends. Duit and Glynn, in showing how some aspects of a source are projected onto a target in order to produce a model, provide an introduction to the literature of analogy. They discuss the theory and practice of science education in this area: the introduction of students to the established models of science; the provision of teaching models to aid the understanding of these and the provision of support in the construction of that understanding. Finally, Boulter and Gilbert, adopting a contextualist view of knowledge, show how, within the broad framework of a culture, different science-related narratives are produced within diverse social situations (e.g., scientific laboratory, classroom, museum). An approach to the analysis of the texts produced by such narratives, illustrated by a primary school science classroom, is used to show how models are constructed through the processes of discussion.

These papers show the key roles played both by language and by the history of science in this field.

11   The Scientific Model as a Form of Speech

Clive Sutton

Abstract

This paper is intended as a contribution to theory about how to offer pupils a better understanding of the nature of science, and especially of the role of language in scientific insight. It begins with examples to show how a new scientific model involves a *re-description* of the phenomenon of interest — seeing it in a new way, seeing it as something else, and hence being *able to talk about it in a new way*. A change of language is thus central to the activity of scientists in building models. When other scientists, and eventually school pupils, are helped to see the phenomenon with new eyes and to join in the conversation, the change also involves using language to *persuade* (and not simply to inform) them about it. If these two points are to be brought more firmly into the awareness of teachers and pupils, we need to think of science lessons as times for attending to the storylines of science.

I end by suggesting that current ways of thinking and talking about science lessons are not wholly adequate, and that some of the expressions associated with 'construction' and 'constructivism' might be given a rest.

Modelling is central to science and should be central to pupils' understanding of what scientists do, so there is every reason to clarify just what we think it involves. To do so we might begin with physical models, which are a great support for scientists and teachers, and in the classroom there is a reassuring definiteness about having something tangible on the bench — an orrery as we talk about the solar system, a plastic 'replica' of the eye if we are discussing vision, or even just a length of Visking tubing when we discuss how intestines work. It helps to have something to handle and look at — so much better than 'mere talk'.[1] At another level, however, models *are* 'mere talk'. None of the useful toys I have mentioned could have materialized on the bench at all without their designers being guided by a mental process in which the objects helped them to explain what they had in mind. First and foremost, it is mental models which matter, and my purpose is to emphasize the role of language at two important stages in their generation and use:

- Stage 1 — when someone is developing a new model (the 're-describing' stage).
- Stage 2 — when someone else is trying to appreciate it (the 'persuading' stage).

At both stages language is integral to comprehension. It is not an extra with which something else ('the model') is explained. The language is itself part of the model.

Redescribing an Aspect of Nature

A new scientific insight involves a *re-description* of the phenomenon which is being studied. Whatever the topic of interest, there is a problem in making sense of it, and then progress is made when someone starts to visualize it in a new way, drawing on language which has not been used in that context before. Taking words from some other area of experience, they try them out in the new context, and use them as an aid to figuring out what is going on.

Thus, when William Harvey started to speak about the blood '*draining*'[2] out of the '*obscure porosities*' of the lung tissue and back to the heart, and he wrote '*I began to think whether there might not be a motion, as it were, in a circle*', he was making a crucial shift in choice of how to speak about such things. He was moving away from accounts of the heart as a spring or source of well-being from which vital spirits might emanate, and towards treating it as a point in a circuit. In his lecture notes for 1615 he also wrote of blood being carried from the lungs into the aorta '*as by two clacks of a water bellows to rayse water*', and thus he brought to the discussion of the human body a way of talking that had previously belonged in hydraulic engineering. Soon, the whole vocabulary of pipes and pumps became a possibility, and within a few years Malpighi was able to find the '*hair-thin*' (capillary) tubes which linked arteries and veins in the tail of a tadpole. By later in the seventeenth century it seemed perfectly natural to talk of the heart as a pump.

Similarly Torricelli, puzzling about the atmosphere, started to re-describe it in terms of depths of fluid. '*We live at the bottom of an ocean of air*' he wrote in 1644. This way of talking and of visualizing a material which stretches above us, but not indefinitely, was soon elaborated in highly productive ways, as investigators started moving up and down mountains with their columns of mercury, talking of '*pressure*' and '*balanced forces*' rather than about '*vacuum*' and '*abhorrence*'.

At the end of the eighteenth century, Lavoisier swung most of the scientific community towards a new way of talking about heat, which would account for melting and evaporation, and in 1789 he wrote:

> It is difficult to comprehend these phenomena without admitting them as the effects of a real and material substance, or very subtile fluid, which, insinuating itself between the particles of bodies, separates them from each other; and, even allowing the existence of this fluid to be hypothetical, we shall see . . . that it explains the phenomena of nature in a very satisfactory manner.

He gave the supposed '*subtile fluid*' a name — '*igneous fluid*' or '*matter of heat*', but in consultation with friends decided that a single word would be better and they called it '*caloric*'. Such 'fluid-talk' was helpful in many ways (for example in supporting calculations about the 'heat capacity' of objects) and it remains with us in our everyday vocabulary about the flow and conduction of heat. Other people however still 'saw' heat as an internal tremor or agitation, and that approach was eventually re-established as being a better kind of scientific insight. It was nicely summarized when John Tyndall in 1863 entitled his book 'Heat considered as a mode of motion'.

Not long before Lavoisier's efforts about heat, the investigators of electrical phenomena also began to use a language of fluids. They broke away from such expressions as '*electrical virtue*', and started to speak of a '*charge*' being built up on their friction-wheels, and then of a '*flow*' of such charge, i.e., an electrical 'current' — the beginning of a highly successful model of what may be going on in a wire, but not the only possible one.

Turning to examples from biology, in the next century Darwin re-described the interconnectedness of extinct life-forms with presently-living ones in terms of a '*great Tree of Life*' branching through the generations, which, he wrote, '*fills with its dead and broken branches the crust of the earth*'. This was an extremely powerful formulation, enough to guide investigations by palaeontologists and comparative anatomists for decades, but Darwin could not have made this leap in thought and language had it not been for the earlier efforts of geologists who had been re-describing various apparently permanent rocks as '*hardened sediments*', and speaking about a '*cycle*' of erosion and sedimentation.

In the twentieth century a breakthrough in thinking about how enzymes interact with their substrates occurred when fermentation was re-described in terms of imagined molecules which might fit '*like a lock and key*'. This example shows well that what starts as a mere figure of speech can sometimes be elaborated into a detailed model with a mental picture (and the possibility of building a physical representation) from which testable predictions can be made and experiments designed. For instance, suppose we make a small change in one component? What would we expect its effect to be? Let's try it! There has been extensive discussion of the development of metaphors into fully explicated scientific models, in the writings of Black (1962), Schon (1963), Hesse (1966) and many others. For a recent discussion of this literature and of how the new ways of talking and new ways of seeing interact see Sutton, 1992, Chapters 3 to 6.

Each new successful area of science can thus be thought of as a new way of talking about the topic of interest, a new form of conversation, closely linked with a way of visualizing in the mind's eye 'what is going on'. The visualization, its expression in words, and sometimes a mathematical elaboration as well, together constitute the model which is guiding thought. As its users spell it out and explore the implications for further investigations, the language component grows and we acquire a whole family of new expressions and technical terms. Sometimes old words take on new meanings — as in the case of the ordinary word 'rock' when it came to be used in connection with a newly shared image of cycles of 'erosion' and 'sedimentation'.

An important point is that to understand the meanings of any of these terms, you really have to 'see' the model as a whole, and allow it to re-cast your perceptions until the various expressions make sense by interaction with each other. Here are a few more examples of families of expressions which become meaningful by that kind of interaction as we take up the new way of speaking.

- wave model of light

- vibration, frequency, wavelength, interference by cancellation, refraction by skewing, etc.

- plate tectonic model of the earth's surface

- 'crust', 'plate', mid-oceanic ridge, seafloor spreading, subduction zone, etc

- template model for nucleic acid and protein synthesis

- complementary sequence, transcription, translation, code, codon etc.

Persuading Other Scientists, and then a Larger Public

The above account of the growth of science differs from the 'common-sense' view in several respects:

1 viewed in this way, progress in scientific understanding does not begin with the collection of facts; it begins with the imagining of a new model which influences what 'facts' are worth seeking, and what will count as relevant evidence.

2 language is not just incidental, an after-the-event tool for labelling what we have found, so we can tell someone else about it. Rather, language is interpretive in function, sense-making, theory-constitutive.

3 to involve someone else in your science is not just a matter of telling them what you have found; it involves *persuading* them of the usefulness and validity of the view you adopt, and the relevance of the evidence you present.

The first people who have to be persuaded are other researchers — both as individuals and as a community, for it is not much use having a new form of conversation unless others will join in. To re-describe the situation by choosing a new metaphor is the first stage, in which the innovative scientist says, in effect: 'Try looking at it like this. This is how it seems to me.' Other individuals who get the point may experience quite a sudden mental shift, as when T.H. Huxley was invited to think of evolution in Darwin's terms of population change driven by different rates of survival, and is reported to have asked himself why no-one had thought of it before. Evidently he had quickly *appreciated* the new view — or in modern American slang we could say that he 'bought' the metaphor of 'natural selection'.

Science, however, deals not just in new personal insights, but with 'public knowledge', and no new view will be generally accepted without an extended process of scrutiny that includes publication, citation, scholarly critique and many attempts at corroboration or refutation. Only if it survives all these will it become part of textbook 'science'. This process within the learned societies and research networks engages more people in using the new speech-forms as they think about the implications of the model and begin to accept or reject the evidence. It also involves a succession of different summarizing publications in which a growing acceptance of the initial suggestion or claim is signalled by increasingly definite language. Eventually we get an objectified and 'factual' kind of expression in which it is hard to hear the human voice of the speculative scientist saying 'Think of it as . . .' Instead, we have accounts which seem to say 'This is how it *is*' (for a fuller account see Sutton, 1996a). 'It has been found that . . .' replaces 'Michael Faraday suggested that . . .', and ideas which began as mere figures of speech, such as 'the field around a magnet' take on a more literal quality, as if they were simple descriptions read off directly from nature. That greater definiteness of the language is a measure of the extent to which the community has indeed been persuaded into the new view, and has come to regard it as useful and even 'true'. Strictly speaking scientists might only say 'true for the time being', or 'the best available idea that we have', but when ideas are used over and over again they get to be treated as permanent truths, and often it is in that form that they are delivered to the rest of the world.

How can people in 'the rest of the world' be helped to understand the modelling function of statements which now seem so 'simply' descriptive? Learners who are unaware of the earlier tentative stage, and who know nothing of the controversy which occurred in establishing the ideas, are almost bound to get a false impression (Sutton, 1996b), and to think they are being given a direct account of how things are. In other words, under the influence of the factualized language styles of mature science there is a constant danger of misunderstanding the status of scientific statements, unless teachers can revive some of the tentativeness of earlier times, and help the pupils to see that these statements were, and still are, human ideas. Part of the job of a science teacher is of course to show pupils what scientists are agreed about now, but another part is to help them to appreciate what scientists had in mind in talking about a topic in a certain way and to *persuade* them of the value and reasonableness of that way.

In practice, teachers oscillate uneasily between persuading and informing, especially when informing seems to be the official business of the lesson. As I shall discuss in the next section, we need a new concept of a science lesson which will banish the idea that 'tell' or 'inform' is effective as the major activity. To persuade someone into a point of view requires talking around the topic until shared meanings are developed. The teacher's personal persuasive voice is important, but learners must also have some freedom of re-expression. In order to be able to hear the scientists' language as *expression of thought* and not just 'description' of nature they must understand language as an interpretative tool, and that means having experience of using it that way themselves. I do *not* mean that they have to re-make the scientist's imaginings from scratch, but if they are to understand how scientists work they need some practice at saying 'My idea was . . .' and *even more* at saying 'I think what Darwin meant was . . .' Since models are a form of conversation, their appreciation in the classroom requires conversation; talking in an active way is indispensable.

Science Lessons As Access to New Conversation

Much of what I have written so far about models in science can also be applied to models of learning and thus to the forms of conversation which go on at conferences like this. They too are re-descriptions, and they steer our attention towards certain features of the learner's activity and away from others. The ways of talking about children's learning which have developed during the last couple of decades tend to focus interest on how those children make sense of practical experience rather than on how they gain access to existing conversations amongst adults. Some re-appraisal of that language therefore seems appropriate, and in the search for other ways of talking, I draw attention to the articles by Snow (1973) and by Solomon (1994).

1 Richard Snow was writing generally about the influence of different metaphors in educational theorizing. He put forward a technique for helping teachers to re-examine their own assumptions, using the phrase 'The teacher as . . . (x, y, z)' — e.g., the teacher as sheepdog or the teacher as gardener. In what ways might it be useful to think of ourselves as guarding the flock, steering the flock, raising tender plants, etc.? Using a variation of that method, Tobin (1990) suggested that a change of metaphor could be a 'master switch' for altering classroom practice. Perhaps we need a similar shift as a 'master switch' for researchers?

2 Joan Solomon wrote specifically about the change which the language of 'constructivism' brought about in the science education community. From about 1978 onwards there was a coming together of language as a theory in what she calls 'the rise of constructivism'. An expanding vocabulary — 'alternative conceptions', 'children's science', 'pupil as scientist', 'actively constructed understanding' — helped to mobilize the effort of researchers, and new research networks came into existence. Thus we can say that 'constructivism' has been a very productive and useful model. On the other hand, a diversity in what researchers have meant by 'constructing' is now more obvious than at first. Is it, for example, what a child does mentally on her own (learner's personal constructivism — a branch of cognitive psychology)? Or are we thinking more of a group phenomenon — either the local negotiation of shared meanings or the larger social process of building 'public knowledge'? In retrospect it looks as if some of the first enthusiasts had assumed too readily that it would be obvious what 'constructing understanding' meant, seeing it as the activity of an individual and akin to what scientists were *thought* to be doing in making sense of their experiments — hence the phrase 'pupil as scientist', which is so easily misread if one has too simplified an account of what scientists do.

With increasing awareness of science as a group activity in which the journals and scientific societies play a part in shaping consensual knowledge, the concept of 'scientist' has been changing. Studies in the history of science which recognize cultural influences on the creative imagination of scientists have also changed it, and so 'pupil as scientist' is no longer simple. Certainly any model of the 'constructivist' individual pupil working at the bench, 'finding out what happens', describing what happens and then theorizing about it, is inadequate because it draws on an inadequate view of science, and of course it is rejected by leading researchers (see Driver *et al.*, 1994). Scientific knowledge does not emerge directly from inspection of what happens at the bench. It consists of new branches of conversation built up by a community of scientists — stories of how they think the world works.

Joan Solomon's conclusion is that we need not abandon insights from the constructivist work, but we do now need other perspectives which will give a more adequate account of how children can come to understand established knowledge. She offers several possible metaphors as a basis for a new model or models — including the child as '*a stranger arriving on a foreign shore*' where unfamiliar tongues are spoken, and the child '*listening on the edge of a family circle*' where the problem is to make sense of 'what they are on about'.

My own view is that *making sense of what they are on about* is an excellent guide to the design of learning activities in science lessons. There are stories to be told and conversations to gain access to. It is important to encourage pupils to listen to those stories and conversations and to hear them with re-interpretive intent, so they are able to say 'I think these people meant . . .' The main object of attention in a science lesson, I have argued, should not be the test tube or the circuit but the scientist's 'story' of what is going on in that test tube or circuit, i.e., pupils should attend to the model, and not just to the phenomenon! (See Figure 11.1)

Before this or any other re-description of a science lesson could have an influence comparable to that of construction-talk it would have to be elaborated with examples that would capture the imagination of teachers and researchers. This might not happen easily, for within existing belief systems the idea is in some respects counter-intuitive. Are we really going to place children in the role of *listeners*, to 'overhear' the speech forms of those who are already in the know, and to gain access to scientific ways of talking by attending to the talk? It seems unlikely while the notion that understanding springs from experiment remains as strong as it is today. Teachers value 'learning by doing' a lot — and for very good reasons, but have got trapped in an unjustified extension of that idea — the belief that the practical work is the basis of the lesson, from which pupils are bound to learn. This over-confidence in experience at the bench is linked with outdated views of science and mistaken beliefs about the independence of observation from language. The whole set of beliefs holds teachers back and restricts the range of non-manual activities which they organize. If these beliefs changed, and our model of learning were to change, we would then have a much better theoretical rationale for diversifying the lesson activities, i.e., for choosing those activities which help pupils to tune in to 'foreign conversations' and to appreciate the scientific models which the 'foreigners' are using. Pupils who enter into the mental worlds that these models offer could be said to be not so much 're-constructing' anything, as

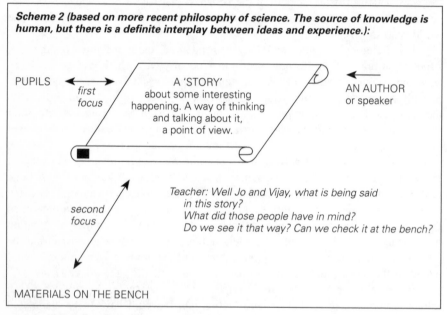

Scheme 1 (popular now, but derived from an old philosophy of science. The source of knowledge appears to be the material world, and direct investigation of it.):

PUPILS ←————→ MATERIALS ON THE BENCH

Teacher: Notice what happens. What do you see happening?
What else could you find out? How will you plan your investigation?

Scheme 2 (based on more recent philosophy of science. The source of knowledge is human, but there is a definite interplay between ideas and experience.):

PUPILS ←→ *first focus* / A 'STORY' about some interesting happening. A way of thinking and talking about it, a point of view. / AN AUTHOR or speaker

second focus

Teacher: Well Jo and Vijay, what is being said in this story?
What did those people have in mind?
Do we see it that way? Can we check it at the bench?

MATERIALS ON THE BENCH

Figure 11.1: Two conceptions of a science lesson

re-discerning what is going on, *learning to talk in new ways*, and *appreciating* that scientific ideas are an outcome of human effort.

It remains to be seen whether a range of expressions of that kind may become more important in future, but in Table 11.1 I attempt to draw some key points of contrast with the more established phrases.

The middle column in Table 11.1 is to some extent caricaturing a set of ideas that deserve that treatment mainly in order to reject them, and as I mentioned previously, experienced researchers have distanced themselves from it, but the connection between beliefs about learning and beliefs about science (shown in the second row of the table) is particularly in need of inspection and discussion amongst teachers. Caricature of earlier forms may be necessary during language shift. It was certainly present when the existing constructivist vocabulary was emerging in science education, for then it could be used to distinguish the new perspective by rejecting yet earlier models:

Table 11.1:   *Two ways of talking about children's learning*

	Constructivist model in a common (albeit simplistic) form	Changes of emphasis when learning is focused on human ideas and interpretations
1 Characteristic expressions as accounts of what the learner is thought to be doing	*constructing* a theory having an *alternative conception* keeping or changing a *framework* of thought	*appreciating* someone else's theory or story gaining a new way of *discerning* things *learning to talk in new ways*
2 Associated emphasis in beliefs about science	knowledge comes from the material world; scientists 'discover' it	ways of understanding are developed by human beings
3 Typical research situation	give a child an event in the physical world to explain or account for	ask a child to interpret 'what they're on about' and re-state it in his or her own terms
4 Focus of the researcher's attention	child making sense of a phenomenon	child making sense of a story or conversation
5 Sources of inspiration	cognitive psychology studies of perception	cultural history, and social psychology perception organized by culturally received language

- the pot-filling model — learner receives knowledge like an empty pot
- the transmission model — teacher has facts, transmits them to pupil as receiver.

All who rallied to the new ways of talking at that time were proud to think of the learner as mentally pro-active, drawing on prior mental structures to make sense of experience. We won't abandon that way of understanding; it remains a key point of awareness. We do need some others too, however, because there is at least one common interpretation of 'constructivism' which is inadequate and incomplete for the classroom. It focuses too strongly on the individual learner trying to making sense of practical experience.

There are of course many conversations in which 'constructivism' refers not to individual sense-making, but to a more general set of ideas about human understanding — i.e., a constructivist epistemology of some kind. Perhaps we might usefully distinguish 'constructivist models of learning' from 'constructivist models of science and human culture'. That's part of the problem, however — one word is meaning too many different things. This alone seems to me to be good reason for giving that word a rest and seeking some new ways of talking about what interests us.

Notes

1 A physical model to handle and to look at: Notice that these two forms of accessibility correspond to our two main metaphors of understanding, which are tactile and visual. 'I'm trying to *grasp* your meaning. I'm trying to get a grip on this idea' (Begriff in German). 'Oh, now I *see* what you mean'. The mental activity feels like looking and handling, or at least that's the way we manage to express our internal experience of it.

2 In this paper, all exemplars of language used in the context of discovery are shown in italics and delineated by quotation marks. Quotation marks elsewhere are used in the commonly accepted manner and sometimes as 'scare' marks to problematize particular words. Such usage is hopefully self-evident from the context.

References

BLACK, M. (1962) *Models and Metaphors*, Cornell University Press.

DRIVER R., ASOKO H., LEACH J., MORTIMER E. and SCOTT P. (1994) 'Constructing scientific knowledge in the classroom', *Educational Researcher*, **23**, 7, pp. 5–12.

HARVEY, W. (1615) 'The use of the water-bellows imagery' [with its 'clacks' (leather valves) comes from lecture notes which Harvey wrote in 1615, and it is quoted by Neale, E. (1975)] in *William Harvey and the Circulation of the Blood*, London, Priory Press.

HESSE, M.B. (1966) *Models and Analogies in Science*, Indiana, University of Notre Dame Press.

LAVOISIER, A. (1789) *Traité Elémentaire de Chimie*, KERR, R. (trs.), Dover Reprints (1965), p. 4.

SCHON, D. (1963) *Displacement of Concepts*, Tavistock Publications.

SNOW, R.E. (1973) 'Theory construction for research on teaching', in TRAVERS, R.M.W. (Ed) *Second Handbook of Research on Teaching*, Chicago Rand McNally, pp. 77–112.

SOLOMON, J. (1994) 'The rise and fall of Constructivism', *Studies in Science Education*, **23**, pp. 1–19.

SUTTON, C.R. (1992) *Words, Science and Learning*, Buckingham, Open University Press.

SUTTON, C.R. (1996a) 'Beliefs about science and beliefs about language', *International Journal of Science Education*, forthcoming.

SUTTON, C.R. (1996b) 'New perspectives on language in science', in FRASER, B. and TOBIN, K. (Eds) *International Handbook of Science Education*, Kluwer, forthcoming.

TOBIN, K. (1990) 'Changing metaphors and beliefs: A master switch for teaching?', *Theory into Practice*, **29**, 2, pp. 122–7.

TORRICELLI, E. (1644) 'Letter to Michelangelo Ricci', McKENZIE, A.E.E. (1960) (Ed), *The Major Achievements of Science*, vol 2, Cambridge University Press, p. 382.

TYNDALL, J. (1863) *Heat Considered as a Mode of Motion*, London, Longman, Green, Longman, Roberts and Green [First edition — The tentative 'considered as' was actually omitted from later editions!].

12 Modelling the Growth of Scientific Knowledge

Richard A. Duschl and Sibel Erduran

Abstract

The application of cognitive approaches to the design of science learning environments puts additional pressure on science teachers to coordinate science lessons. A particularly difficult task for science teachers is linking the data and evidence from laboratory and practicals to the knowledge claims contained in science texts and statements of scientific theories. We argue that teachers need models to guide them in the planning and implementation of lessons that link data to theory. Four 'Growth of Knowledge Frameworks' (GKF) derived from philosophical perspectives on the nature of scientific knowledge are presented. The GKFs facilitate the inclusion of epistemic and cognitive strategies that characterize science as a unique way of knowing.

Introduction

The prevailing opinion among educational reformers is that the actions of teaching and learning need to change away from practices and curriculum models that put an emphasis on what we know. The new 'best practices' focus is one that puts an emphasis on why, and how we know. It is hypothesized that this broader explanatory context facilitates the incorporation of higher order thinking and communication skills. Thus, we hear today requests for schools for thought (Bruer, 1993); thinking curriculum (Resnick and Klopfer, 1989); communities of practice (Brown and Campione, 1990); anchored instruction (Cognition and Technology Group at Vanderbilt University [CTGV], 1992); project-based science (Krajcik *et al.*, 1994), and alternative assessment practices (Gardner, 1991).

These new perspectives on teaching which emphasize the role of the teacher as a problem-solver and as a facilitator of learning are much more complex and challenging to implement than traditional instructional models (Cohen, McLaughlin and Talbert, 1993). Shulman (1986) argues that these 'teacher-as-facilitator' forms of teaching fundamentally rely on robust pedagogical content knowledge. In particular, he states that it is not sufficient for the teachers to report what we believe. It is necessary that they engage learners in the task of examining why it is that we believe one thing over another, adopt one method as superior compared to another, or conclude that one point of view or position is inconsistent with another. The closer examination of knowledge claims and the development of the requisite skills among learners to understand the logic, warrants, reasons, or evidence that enable one to judge one perspective or claim relative to another demands that teachers be facilitators of students' learning and harbingers of a diverse and complex pedagogical content knowledge base.

The purpose of this paper is to examine one element of this instructional challenge in the context of science education, namely, teaching about the links between scientific data, scientific theory and theory change. The paper begins with an overview of developments in the cognitive sciences and of the implications cognitive information processing and sociocultural approaches have for the design of learning environments and the adoption of instructional principles. We describe a set of goal structures that teachers of science encounter when curriculum, instruction and assessment practices are integrated and modelled upon cognitive and sociocultural approaches to knowing and learning. Then we focus on one of these goal structures: the epistemic goals. We show how teachers can use 'Growth of Knowledge Frameworks' (GKFs) to plan for instruction. Specifically, we examine how GKFs can be used as models to coordinate the complex set of knowledge structures, criteria and skills associated with understanding the link between scientific data and scientific theory.

A Model-based View of Teaching

The path to the identification of pedagogical content knowledge as a component of expertise in teaching has been cleared by the cognitive revolution. The ability to understand the arguments, evidence, reasoning, problem solving strategies, symbols and representations in a domain of scientific inquiry depends on the development and acquisition of epistemic and cognitive tools. In particular, for science teachers this ability depends on the development of models that they can employ to coordinate instructional planning, tasks and activities.

The need for models in teacher decision making is derived from research on teacher thinking (Clark and Peterson, 1986). The basic position is that teachers, in order to cope with the complex conditions classroom learning environments present, construct models of reality that guide them in making pedagogical decisions (Duschl and Wright, 1989; Shavelson and Stern, 1981). Thus, the design of instruction that affords learners the opportunities to develop target skills for linking scientific data to scientific theory must also be formatted with teachers in mind. As well, the design of instruction should afford teachers opportunities to apply functional models that guide them in instructional decision making. In this section of the paper, we will argue that the cognitive tools and skills which we seek to develop in learners must also apply to the development of decision making models for teachers.

Our image of knowing and learning has been redefined by the discoveries of cognitive scientists (Bransford and Stein, 1984; Gardner, 1991; Newell and Simon, 1972; Resnick, 1989) who have demonstrated that expertise requires the appropriation of specific strategic and procedural knowledge skills. Furthermore, they have reported the importance of individuals' prior knowledge in subsequent learning as well as the limit on the capacity to learn or process information.

The call for reforms in educational theory and practice have in large measure been prompted by this cognitive understanding of knowing and learning. Specifically, the results from cognitive science are helping to inform us about how we need to set out to make changes with respect to the design of learning environments (Bruer, 1993; Glaser, 1994). These changes while seemingly fundamental for the improvement of knowing and learning in students, have significant implications for what we demand of teachers as well. The coordination and execution of cognitively based instruction and assessment practices are difficult. However, our understanding of teacher cognition and decision making is important and it occupies a crucial area for educational research.

The challenge for teachers is made more apparent when we consider the specific ways that cognitive science learning theory apply to pedagogical practices. Glaser (1994) maintains that seven related emerging principles of instruction derived from learning theory offer insights about how we need to shape learning environments. The seven principles are (Glaser, 1994; pp. 17–20):

1 **Structured knowledge** Instruction should foster increasingly articulated conceptual structures that enable inference and reasoning in various domains of knowledge and skill. (p. 17)

2 **Use of prior knowledge and cognitive ability** ... relevant prior knowledge and intuition of the learner is ... an important source of cognitive ability that can support and scaffold new learning ... (p. 18)

3 **Metacognition: Generative cognitive skill** ... the use of generative self-regulatory cognitive strategies that enable individuals to reflect on, construct meaning from, and control their own activities ... enhance the acquisition of knowledge by overseeing its use and by facilitating the transfer of knowledge to new situations ... these skills provide learners with a sense of agency. (p. 18)

4 **Active and procedural use of knowledge in meaningful contexts** Learning activities must emphasize the acquisition of knowledge, *but* this information must be connected with the conditions of its use and procedures for its applicability ... (p. 19, emphasis in original)

5 **Social participation and social cognition** The social display and social modelling of cognitive competence through group participations is a pervasive mechanism for the internalizations and acquisition of knowledge and skill in individuals. (p. 19)

6 **Holistic situations for learning** ... competence is best developed through learning that takes place in the course of supported cognitive apprenticeship abilities within larger task contexts. (pp. 19–20)

7 **Making thinking overt** ... a significant mechanism in environments for learning is to design situations in which the thinking of the learner is made apparent and overt to the teacher and to students. In this way, student

thinking and reasoning can be examined, questioned, and shaped as an active object of constructive learning. (p. 20)

Asking a teacher to organize a classroom and execute instruction and assessment practices according to the principles advanced by Glaser (1994) is a complex problem-solving task that will require teachers to develop equally complex models of reality to coordinate. Bruer (1993) captures the complexity of the process and the challenge for teacher education when he writes:

But the new cognitive learning theory and its many potential applications don't automatically translate into better teaching practices. Knowing how to teach from a cognitive perspective . . . is not a trivial accomplishment. Although most of us don't think of teaching as a problem-solving task, . . . skilled teachers solve complex, ill-structured problems each time they teach a lesson. Research on teacher cognition is discovering the representations and the strategies teachers use to solve these problems. We are beginning to understand the differences between expert and novice teachers, between effective and less effective teaching performances. We are beginning to understand what teachers have to know and how skilled teachers use this knowledge. If teacher training programs can effectively convey this knowledge, we can have many more expert teachers who teach from the cognitive perspective. (Bruer, 1993, p. 258)

Investigations by our research group (Duschl, 1995; Duschl and Petasis, 1995; Erduran and Duschl, 1995; Gitomer and Duschl, 1995) on the development of learning environments based on cognitive science principles suggest that teachers need to make decisions about five goal states. As presented in Table 12.1, these include conceptual goals, process goals, motivational goals, epistemological goals and social goals.

In order to apply Glaser's (1994) seven principles to the classroom let us consider them as the problem space that teachers will encounter. Let us further consider how the design of curriculum, instruction and assessment models consistent with these principles can aid teachers in their roles as facilitators of student learning. We want to argue that success with the cognitively based instruction by science teachers will require the same kind of generic assistance as the students they teach. Specifically, we argue for the use of Growth of Knowledge Frameworks in classrooms to facilitate the teaching and the learning of the complex belief structures and cognitive skills which are representative of reasoning about the links between scientific data and scientific explanations or theories.

Growth of Knowledge Frameworks As Models for Instruction

Since the publication of Kuhn's *The Structure of Scientific Revolutions* (1962) the historical view of science has altered the received view of science. The view of

Table 12.1:   Goal states of cognitive psychological science learning environments

Goals	Task Environment — Research Objectives
Conceptual	Examination of learners' prior conceptual understandings and the instructional conditions that affect a change in conceptual understanding — conceptual change teaching
Process	Task analyses of the cognitive processes that function during reasoning about models, experiments, hypotheses and arguments in science domains
Motivational	Designing learning environments and instructional sequences to provide, support and sustain learners' scientific inquiry as well as development of domain specific and general knowledge
Epistemological	Examination of learners' and scientists' views of the nature of science and scientific information — what counts as evidence, theory, etc. related to process goals research
Communication and representation	Study of language, graphic organizers, drawings and representations that students and teachers generate about scientific information, knowledge claims and processes

science as an enterprise in which the growth of knowledge is an accumulation of core theories has been replaced by one that suggests that the growth of science is more accurately reflected by processes of modification, adaptation, and at times abandonment of core scientific ideas. This evolutionary image of science has come to be called the conceptual change view of science and it has raised many questions about the criteria and the practices that define the objectivity and rationality of science.

The examination of belief revision processes has led some philosophers of science (e.g., Giere, 1988; Kitcher, 1993; Thagard, 1992) to argue that cognitive processes can be used to account for the mechanisms of conceptual change. The position is then advanced that these cognitive processes can provide grounds for establishing the objectivity and rationality of science. Several authors, including one of us, have suggested that it is possible to have students examine the belief revision activities which characterize the evolutionary or conceptual change view of science (Duschl, 1990; Hodson, 1988; Matthews, 1994; Ziman, 1984). The proposals range from historical case studies of science and examination of science studies, to participation in project or problem based science that replicates discovery and justification of knowledge claims or explores authentic contexts, problems, or questions advanced by students. Setting aside the issue of which approach is best let us rather focus on the nature of the academic work (Doyle, 1983) teachers encounter when implementing instruction that seeks to capture and reveal elements of the growth of scientific knowledge. The teacher problem space is one of coordinating the diverse sets of knowledge claims — procedural and declarative — that make up a domain of scientific inquiry. One part of the problem concerns the cognitive and manipulative procedures of scientific exploration and investigation that generate data and evidence. The other part of the problem concerns learning

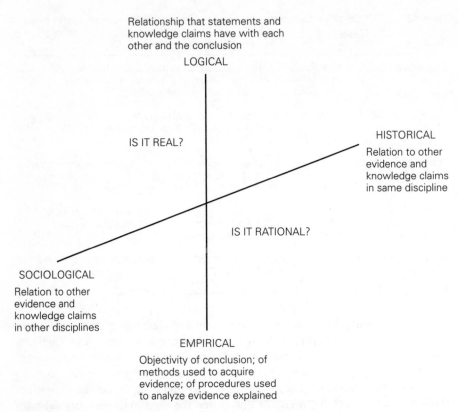

Figure 12.1: *Criteria categories for evaluating scientific theories*
Source: Root-Bernstein, 1984

the skills of argumentation and of theory development and evaluation that forge evidence into scientific explanations. Together these two parts outline the information management problem the teachers encounter when relating the evidence from practicals and experiments to the knowledge claims in texts and statements of scientific theory.

A useful model for coordinating the kind of information science teachers need to think about when linking data to theory is one proposed by Root-Bernstein (1984). According to Root-Bernstein there are four categories of criteria that are used to distinguish scientific theories from other kinds of explanations: empirical criteria, logical criteria, historical criteria and sociological criteria (Figure 12.1). The empirical criteria appeal to experiences with nature and the evidence that bears on a domain of inquiry. The logical criteria are those that apply to the form and validation of arguments that establish the merit of scientific knowledge claims. Taken together empirical and logical criteria set out conditions for the objectivity of scientific knowledge claims.

The relationship that a scientific knowledge claim has with other established claims within a domain is what characterizes historical criteria. We can ask, for example, if a theory (or the entities described by a theory) is or is not consistent with the best reasoned beliefs that comprise our understandings of this domain. In a similar manner, the relation of a claim to fundamental principles, laws, or theories in other related domains (e.g., relating claims of geology and biology to chemistry or physics) constitute sociological criteria. Together the historical and sociological criteria define grounds for the rationality of scientific knowledge claims.

Investigating and applying the criteria to actual tasks in science classrooms is what we will define as the epistemic goals of science. On what grounds do we claim that an explanation is objective or rational? How do we know if an explanation is a novel one and if it is consistent with other evidence or explanations we believe? Is there an experiment we could conduct to show, test the relationship that we think exists? Is the test a fair objective test? Is there something else we have studied that could help us with this problem?

The application of philosophical models and frameworks to curricular and pedagogical practices and analyses is an established programme of research exemplified by the work of Doug Roberts, Hugh Munby, Tom Russell, and Brent Kilborn (Munby, Orpwood and Russell, 1986). A review of their work can be found in Duschl (1994). The adoption of the Growth in Knowledge Frameworks (GKFs) by teachers as models to guide planning and instruction is, as stated above, intended to help with the management of linking evidence to theory. Two aspects of this management are the planning of instruction and the coordination of students' cognitive activities. It is our position that the GKFs can be used as models for both purposes. As models for planning, the GKFs set out for teachers the ways in which the various knowledge claims of science need to be examined. The GKFs each provide contexts for organizing the logical, empirical, historical, and sociological evidence that go into developing our best reasoned beliefs about the structure of scientific knowledge.

As models that guide learning, the GKFs provide a heuristic tool students can use to coordinate diverse sets of scientific evidence and knowledge claims. The GKFs facilitate the development of the strategic reasoning skills that are characteristic of science as a way of knowing. When used as a graphic organizer, they can capture students' representations of the evidential bases for scientific knowledge claims and thereby make possible a probe of student understanding. The completed models can be used to coordinate subsequent student activities and investigations. For example, the GKFs can provide a representational context through which students' varying perspectives can be made public, compared and discussed. Once these compare-and-contrast discussions begin to take place, an advanced application of the GKFs would be to discuss and debate the criteria for scientific objectivity and empirical accuracy as set out by Root-Bernstein (1984).

We want to advance the position that GKFs can be used as models by teachers and students to organize and achieve the epistemic and cognitive goals of science programmes. Specifically, GKFs can be used:

1   for the attainment of metacognitive general skills associated with theory development and justification (Glaser's Principle 3);

Table 12.2: Growth of knowledge frameworks as models for planning and instruction

Growth of Knowledge Framework	Characteristics of the Framework	Applications of the Framework
Toulmin's argument pattern	Models the generic link between data and conclusions	Can be used to coordinate the reporting of science investigations and to examine the guiding conceptions of scientific inquiries
Gowin's Vee heuristic	Models links between the investigative procedures of science with conceptual and theoretical foundations	Elaborates on Toulmin by providing a more diverse set of steps that link data to theory
Giere's theory arguments	Models the structure of scientific theories by coordinating empirical evidence with conceptual backings	Can be used to compare and contrast competing explanations or theories; exposes the guiding conceptions of scientific theories
Duschl's science hierarchy	Models the process of conceptual change	Three transformations facilitate the consensus making steps used in science

2 as contexts in which the active and procedural knowledge that character-ize science as a way of knowing (e.g., the Root-Bernstein criteria) can be examined in science lessons (Glaser's Principle 4);

3 to format and organize the complex set of evidence and information asso-ciated with linking data to theory and thus address the problem of working memory being limited in capacity (Glaser's Principle 1);

4 as representations for making thinking about theory claims and evidence overt (Glaser's Principle 5);

5 to facilitate the social participation and social cognition of the classroom community (Glaser's Principle 7).

Four GKFs are presented in Table 12.2. They are Toulmin's Argument Pattern (Toulmin, 1958), Gowin's Vee (Novak and Gowin, 1984), Giere's Test of Scient-ific Theory Frameworks (Giere, 1988), and Duschl's Goal of Science Hierarchy (Duschl, 1990). The Toulmin and Giere GKFs are derived from philosophical positions of reasoning and decision making carried out by scientists. The Gowin and Duschl GKFs represent applications of synthesized philosophical positions of knowledge development and revision.

Each of the four GKFs (Figures 12.2, 12.3, 12.4, 12.5) has a particular

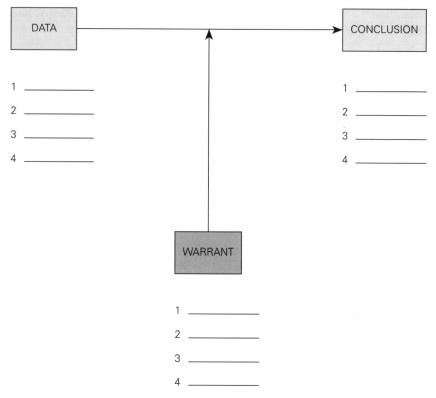

Figure 12.2: Toulmin's argument pattern
Source: Toulmin, 1958

advantage for coordinating information that links data and evidence to theory and explanation. The GKFs in combination with Root-Bernstein's (1984) epistemic criteria make it possible to discuss and evaluate the status of scientific knowledge claims. Such evaluative discussions are not a common practice in science but they have been long recognized as a critical element of teaching science as inquiry. In the *Patterns of Enquiry Project* (Connelley *et al.*, 1977), for example, an inquiry discussion strategy was developed that included a question category labelled 'evaluation'. The purpose of these questions, and a goal of the *Project*, was to entice students to assess the degree of legitimate doubt that can be attached to scientific knowledge claims. The foundational information needed to carry out this evaluation of knowledge claims is the set of guiding conceptions used by scientific inquirers. The Toulmin, Gowin, Giere, and Duschl GKFs can help teachers to make the guiding conceptions apparent and explicit. Alone or in combination, they help to coordinate complex information about the growth and nature of scientific knowledge. The GKFs provide a framework to have conversations about the status of scientific knowledge claims and beliefs.

Richard A. Duschl and Sibel Erduran

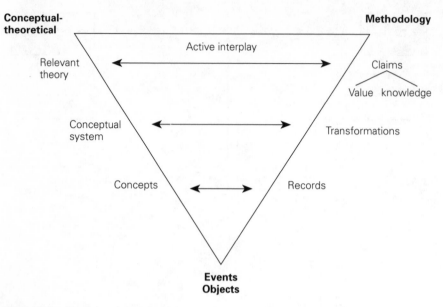

Figure 12.3:   Gowin's epistemological vee
Source: Novak and Gowin, 1984

THo   Theoretical Hypothesis — The theory being tested
IC   Initial Conditions — The states of the system and evidence (or 'known facts')
      before considering the THo
BK   Background Knowledge — The existing knowledge claims of science that should
      not conflict with THo
P   Premise — The predicted occurrence of a possible state of some real system
      described by the THo

Figure 12.4:   Schemata for testing theories with arguments
Source: Giere, 1984

Conclusion

We need to keep alive the lessons learned from previous history and philosophy
of science (HPS) and science education research. The work carried out at the
Ontario Institute for Studies in Education (Munby, Orpwood and Russell, 1986),
for example, reminds us of the importance of having models for teachers and students to employ in meaning making and reasoning. HPS frameworks thus become
quite significant. They become models to guide and judge the development of individuals' knowledge growth and sense making in science. Models are as important
for teachers as they are for learners. Thus, there are strong implications for the use
of these GKFs in teacher education programmes. They become the guidelines for
engaging, doing and reviewing science. The new philosophies of science have
come out of an analysis of actual practice and then in turn, they can and should
inform practice. The challenge for science education researchers interested in

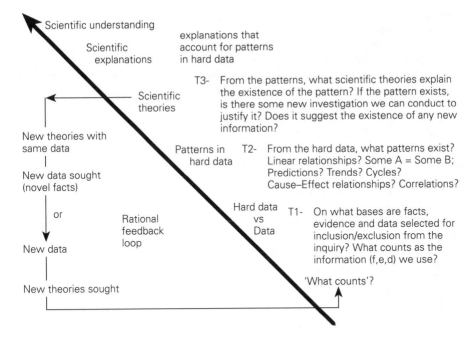

Figure 12.5:   Duschl's goal of science hierarchy
Source: Duschl, 1990

applying HPS to science education is one that rests squarely in the domain of constructing models to guide this practice.

In one sense, we are asking the same questions Schwab (1958) asked decades before. What is it that scientists do? But a focus on the development of theories and explanations moves us beyond the behavioural dimension into the cognitive and epistemic dimensions of that question. Contemporary HPS and forward looking applications of HPS to science education are exploring the details and consequences of both the behavioural and cognitive dimensions of science as they are practised. Through the careful consideration of cognitive theory we can effectively incorporate and coordinate information about the nature of science in science learning environments. We have argued that GKFs can be useful models to guide and assist teachers and students with this task.

References

Bransford, J.D. and Stein, B.S. (1984) *The IDEAL Problem Solver*, New York, Freeman.
Brown, A. and Campione, J. (1990) 'Communities of learning and thinking, or a context by any other name', in Kuhn, D. (Ed) *Contributions to Human Development Vol. 21: Development Perspectives on Teaching and Learning Thinking Skills*, Basel, Karger, pp. 108–26.

Bruer, J.T. (1993) *Schools for Thought: A Science of Learning in the Classroom*, Cambridge, MIT Press.

Clark, C. and Peterson, P. (1986) 'Teachers' thought processes', in Wittrock, M. (Ed) *Handbook of Research on Teaching*, (3rd edition) New York, Macmillan, pp. 255–96.

Cognition and Technology Group at Vanderbilt University (1992) 'Anchored instruction in science and mathematics: Theoretical basis, developmental projects, and initial research findings', in Duschl, R. and Hamilton, R. (Eds) *Philosophy of Science, Cognitive Psychology, and Educational Theory and Practice*, Albany, NY, SUNY Press, pp. 244–73.

Cohen, D.K., McLaughlin, M.W. and Talbert, J.E. (Eds) (1993) *Teaching for Understanding: Challenges for Policy and Practice*, San Francisco, Jossey-Bass Publishers.

Connelley, F.M., Finegold, M., Clipsham, J. and Wahlstrom, M. (1977) *Scientific Enquiry and the Teaching of Science*, Toronto, Ontario Institute for Studies in Education.

Doyle, W. (1983) 'Academic work', *Review of Educational Research*, **53**, pp. 159–99.

Duschl, R. (1990) *Restructuring Science Education: The Role of Theories and their Importance*, New York, Teachers' College Press.

Duschl, R. (1994) 'Research on the history and philosophy of science', in Gable, D. (Ed) *Handbook of Research on Science Teaching and Learning*, New York, Macmillan, pp. 443–65.

Duschl, R. (1995) 'Mas alla del conocimiento: Los desafios epistemologicos y sociales del la ensenanza mediante el cambrio conceptual' (Beyond cognition: The epistemic and social challenges of conceptual change teaching), *Ensenanza de las Ciencias*, **13**, 1, pp. 3–14.

Duschl, R. and Petasis, L. (1995, April) 'Discourse analysis as a window into the classroom', Paper presented at the annual meeting of the National Association for Research in Science Teaching, San Francisco.

Duschl, R. and Wright, E. (1989) 'A case study of high school teachers' decision making models for planning and teaching science', *Journal of Research in Science Teaching*, **26**, 6.

Erduran, S. and Duschl, R. (1995, April) 'Using portfolios to assess students' conceptual understanding of flotation and buoyancy', Paper presented at the annual meeting of the American Educational Research Association, San Francisco.

Gardner, H. (1991) *The Unschooled Mind: How Children Think and How Schools Should Teach*, New York, Basic Books.

Giere, R. (1988) *Explaining Science: A Cognitive Approach*, University of Chicago Press.

Gitomer, D. and Duschl, R. (1995) 'Moving toward a portfolio culture in science education', in Glynn, M.S. and Duit, R. (Eds) *Learning Science in the Schools: Research Reforming Practice*, Mahwah, NJ, Lawrence Erlbaum and Associates, pp. 299–326.

Glaser, R. (1994) 'Application and theory: Learning theory and the design of learning environments', Keynote address presented at the 23rd International Congress of Applied Psychology, July 17–22, Madrid, Spain.

Hodson, D. (1988) 'Toward a philosophically more valid science curriculum', *Science Education*, **72**, pp. 19–40.

Kitcher, P. (1993) *The Advancement of Science*, London, Oxford University Press.

Krajcik, J., Blumenfeld, P., Marx, R. and Soloway, E. (1994) 'A collaborative model for helping science teachers learn project-based instruction', *Elementary School Journal*, **94**, 5, pp. 483–97.

KUHN, T. (1962) *The Structure of Scientific Revolutions*, Chicago, University of Chicago Press.

MATTHEWS, M. (1994) *History Philosophy and Science Teaching*, London, Routledge.

MUNBY, H., ORPWOOD, G. and RUSSELL, T. (Eds) (1984) *Seeing Curriculum in a New Light: Essays from Science Education*, University Press of America, Lanham, Maryland.

NEWELL, A. and SIMON, H.A. (1972) *Human Problem Solving*, Englewood Cliffs, NJ, Prentice-Hall.

NOVAK, J.D. and GOWIN, D.B. (1984) *Learning How to Learn*, New York, Cambridge University Press.

RESNICK, L. (Ed) (1989) *Knowing, Learning and Instruction: Essays in Honor of Robert Glaser*, Lawrence Erlbaum Associates, Hillsdale, NJ.

RESNICK, L. and KLOPFER, L. (Eds) (1989) *Toward the Thinking Curriculum: Current Cognitive Research, 1989 Yearbook of the Association for Supervision and Curriculum Development*, Reston, VA, ASCD.

ROOT-BERNSTEIN, R. (1984) 'On defining scientific theory: Creationism considered', in Montagu, A. *Science and Creationism*, New York, Oxford University Press, pp. 64–94.

SCHWAB, J. (1958) 'The teaching of science as inquiry', *Bulletin of Atomic Scientists*, **14**, pp. 374–9.

SHAVELSON, R. and STERN, P. (1981) 'Research on teachers' pedagogical thoughts, judgments, decisions and behavior', *Review of Educational Research*, **51**, pp. 455–98.

SHULMAN, L. (1986) 'Those who understand: Knowledge growth in teaching', *Educational Researcher*, **15**, pp. 4–14.

THAGARD, P. (1992) *Conceptual Revolutions*, Princeton University Press, Princeton, NJ.

TOULMIN, S. (1958) *The Uses of Argument*, Cambridge, Cambridge University Press.

ZIMAN, J. (1984) *An Introduction to Science Studies*, Cambridge, Cambridge University Press.

13 Mental Modelling

Reinders Duit and Shawn Glynn

Abstract

Models may be viewed as representations of an object or of an event which are formed by the process of modelling. Modelling occurs both in science and in science education. Learning science has to do with modelling in at least two significant ways. Firstly, an important part of science education concerns the reconstruction of the product of modelling in science by the learner. Secondly, learning in general may be viewed as mental modelling. Analogies (and their relatives such as metaphors, similes, or allegories) are at the heart of modelling. What makes a model worthwhile, both in modelling in science and in mental modelling, is determined by the analogical relations between the 'original' and the model. The paper will analyse the issue of modelling in science education from the perspectives provided by research on the use of analogies in learning science. Learning of key science concepts and principles is viewed as construction of conceptual models.

Model: A Term with Manifold Meanings

It is quite common in science education that on the one hand terms used are somewhat vague, and on the other hand, basically the same concepts are communicated by different terms. Model is certainly among the terms that are used in such a broad spectrum of meanings that great care is necessary when that term is employed (Bunge, 1973, p. 91; Leatherdale, 1974, p. 41; Lind, 1980). Quite frequently not even single authors are consistent in the way they use this term. Nevertheless, there appears to be a common core across the different meanings in that models stand for something else, which they 'represent' in some way. Models may be viewed in terms of the structure presented in Figure 13.1. There are two domains, one may be called the 'source', the other the 'target'. The two domains share certain attributes or parts of structures but thus are different otherwise. Models then have to do with mapping the attributes and structures between different domains. The representation R stands for the attributes or structures that comprise the mapping. The term 'model' is used for different components in Figure 13.1. Quite frequently the source domain itself is called a model for the target domain. A well-known example in science education is the water model of the electric circuit. But also what is called representation R in Figure 13.1 is often viewed as a model in that it represents the mapping between the two domains, in other words, it models the mapping.

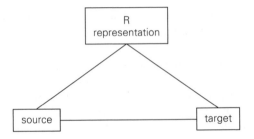

Figure 13.1: Structure of a model

Models and Analogies

Also the term analogy is used with manifold meanings, and care is necessary when this term is used. In this paper we use the term analogy in the following way. Analogies have to do with the situation portrayed in Figure 13.1. They stand for the mapping between two domains. Figure 13.2 outlines our view in a formal way (Duit, 1991). All boxes in Figure 13.2 stand for representations. There is a relation between the boxes R_1 and R_2: Certain attributes and parts of structures are the same in both representations, other attributes and parts of structures are only valid in R_1 or R_2 respectively. We call this kind of relation an 'analogy', or more precisely, an 'analogical relation'. We call R_1 the 'analog' (Glynn, 1991; Duit, 1991). R_2 is the 'target'. The representation R_m represents the attributes and parts of structures that are the same in R_1 and R_2. We call this kind of representation a 'model'. The term analogy then stands for an analogical relation between R_1 and R_2. R_1 and R_2 are analogous with regard to the structure presented in R_m. There may be analogical relations on different levels. If R_1 and R_2 are representations of two domains of reality (e.g., water and electric circuit), the analogy relation as portrayed in Figure 13.2 may be called an analogy of the first level. But analogical relations are also conceivable between two models.

As mentioned analogical relations may comprise identity of attributes (such as appearance, form, shape, or colour) between R_1 and R_2 and identity of parts of structures. The latter kind of relations are the more important ones, because only analogies that rest on deep structural identities between R_1 and R_2 have proven fruitful in constructing powerful models. Identities of attributes (i.e., 'surface' identities) appear to have mainly the function of leading attention to the deep structural identities (Gentner and Landers, 1985; Tenney and Gentner, 1985).

Our use of the term model clearly indicates that analogical relations are at the heart of models. It is the analogical relations that make a model a model. To learn a model in science instruction then means to learn the analogical relations that comprise that model. To construct a model, means to construct, and to create certain analogical relations. Glynn, Britton, Semrud-Clikeman, and Muth (1989, p. 385) call the representation R_m the 'superordinate concept' of analog and target. They see a creative function in the search for the superordinate: 'The identification and naming of the superordinate concept can suggest analogies, it also can stimulate students to generalize what they have learned and apply their learning to other concepts.'

There is a very important feature of the analogical relation that has far-reaching consequences for using analogies and hence for mental modelling: namely, the relation is symmetrical, because it is based on identities of attributes or structures. The terms analog and target do not therefore indicate some sort of logical hierarchy. They indicate the purpose of a particular use of an analogical relation. In other words, analog and target may change roles. As a consequence, whenever an analogy is employed in instruction students learn about analog *and* target as well. If, for example, the analog is viewed from the perspective of the target then new light is shed, so to speak, at the target and allows us to understand it in a more advanced way. In a sense the symmetrical relation allows mental bootstrapping. Very often, the analog is not as familiar to the learner as assumed by the provider of an analogy. From aspects that are already fairly familiar to the students, a first attempt can be made to construct the analogical relations. The first preliminary understanding of similarities between the two domains may be used to develop students' understanding of the analog which then leads to an enhanced understanding of the analogical relations.

Analogies and Their Relatives

In daily life comparisons are frequently drawn on something familiar when a new idea is explained to others. Also in the arts analogies and their relatives are employed to facilitate understanding by explicitly referring or implicitly alluding to similarities with something familiar (Sutton, 1978). Metaphors are used, for instance, if adequate vocabulary is missing, particularly when a totally new idea is being expressed. Metaphors, for this reason, very often have a certain surprise effect. The statement that 'education is sheep herding', which may serve as an example here, is in fact somewhat surprising when heard for the very first time. Literally taken, metaphors are often absurd. The analogical relation that is implicitly addressed by the metaphor has to be constructed by the addressee. Metaphors often aim at pictorial conceptions. Allegories explicitly try to explain something that is not accessible to our senses just by providing an image. Death embodied as the 'Grim Reaper' is a well-known example. Fables are also relatives of analogies. Here the behaviour of other creatures, for instance, animals have to be transferred to human behaviour. Parables like the similes that Jesus employs, in our view, can be seen in a similar way. They try to explain some abstract ideas that are inaccessible to our senses and our human understanding by comparing them to familiar events in daily life.

The very essence of employing analogies, and their relatives, is the analogical relation between an analog domain and a target domain. Sometimes these analogical relations are explicitly stated; very often, however, these relations have to be constructed by the person that tries to make sense of a comparison provided. With regard to mental modelling, which is the focus of this paper, the issue of analogies and their relatives indicates that there are at least two ways to use analogical relations in modelling processes: either provide students with the analogies or help students to construct their own analogies. In the first case, the analogical relations are explicitly worked out by a teacher or textbook author. The manner in which the two domains are similar, and also dissimilar, is explicitly listed. It appears that this is the common way of employing analogies at school such as the water analogy of the electric circuit. It will be mentioned in passing only that students' understanding of analogical relations is not sufficiently taken care of, and that at best the key similar features are highlighted when dealing with analogies at school and in textbooks (see Duit and Glynn, 1992; Treagust, Duit, Joslin and Lindauer, 1992). The other way of implying analogies is to let students construct an analogical relation with only little guidance. This leads to a way of analogy use that may be indicated by the term 'self-generated analogies' (Wong, 1993; Cosgrove, 1992).

Analogical Reasoning and Analogies as Learning Tools Provided for Students

Two points of view have to be clearly distinguished about the use of analogies and their relatives in the learning of science. There is the process of searching for similarities between the new and something already familiar that the learner, often initiated spontaneously when confronted with something unknown. The processes of mental analogical reasoning in the heads of the learners have to be differentiated from the use of analogies as teaching and learning tools provided by a teacher or a textbook. Analogies then are provided in order to incite processes of mental analogical reasoning (Duit and Glynn, 1992). The mental processes may be incited or not. In many cases reported in the literature these processes were obviously not incited because the analogies provided did not lead to the desired outcome (see reviews by Duit, 1991, and Duit and Glynn, 1992).

Mental Models and Conceptual Models

Modelling, the process of forming and constructing models is always a mental activity, of an individual or a group. Even if the product of a modelling process is a concrete object like a concrete model of the solar system, what counts is the mental representation of that object. The mental representation shared by a certain community allows the use of the product of the modelling process by those who gain the mental representation. This is a key reason why we use the term model here for representations, more precisely for mental representations in the way outlined in Figure 13.2.

Mental models refer to students' personal knowledge whereas conceptual models refer to scientifically accepted knowledge. We follow Norman (1983, p. 12) here:

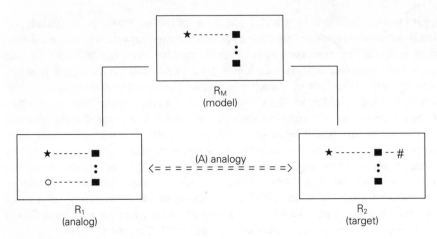

Figure 13.2:   Diagram of an 'analogical model'

Conceptual models are devised as tools for the understanding or teaching of physical systems. Mental models are what people have in their heads and what guides their use of things.

Mental models and conceptual models then are both representations of real-world processes or things. Conceptual models are shared by a community, mental models are personal (Glynn and Duit, 1995). Many learning difficulties in science instruction are caused by the fact that students' mental models and the conceptual models to be learned are grounded in significantly different general frameworks and are often contradictory. Learning science then means to develop students' pre-instructional mental models towards post-instructional models that share at least certain key facets with the conceptual model taught (see for more details Glynn and Duit, 1995).

Analogies and Learning

Our view of learning is in accordance with a constructivist perspective. In order to avoid misunderstandings that may come from a more philosophically oriented view on constructivism we want to point out that we have a 'psychological constructivism' in mind. We, therefore, like to avoid the term constructivist and speak about a 'constructive' view of learning instead (Glynn and Duit, 1995). This view is, as mentioned, psychological; it is practical (or even pragmatic) and not radical, and it is empirically based and focuses on the behaviour of students.

Briefly outlined, learning is seen as an active construction process on the basis of the conceptions (or mental models) the students already emphasize. We further hold that the function of cognition is adaptive and allows the learners to construct viable explanations of experiences. This is basically a Piagetian view of learning and the term 'viable' is used in the way as introduced by von Glasersfeld (1992). Learning is a personal, active construction of the individual and the process of constructing meaning is embedded in a social context.

As to the role of analogies in the learning process it is of key importance that learning is active construction on the basis of what is already known (Duit, 1991; Glynn, 1991; Glynn, Duit, and Britton, 1995). If a learner is confronted with something new, he or she scans, so to speak, the store of the already known to determine if there is something that is similar to the new. Piaget has called this mechanism assimilation. Wagenschein (1990, 1911), a German science educator, has pointed out that understanding always means to relate the so far unknown with something already familiar, and he claims that the new very often turns out to be a good old friend who is just a bit disguised. If in assimilation, the search for similarities, is not successful, in Piagetian terms processes of accommodation are necessary. Search for similarities, therefore, is a key process from the point of view of constructive learning that is taken here. It becomes obvious that what we have called analogical reasoning (seen as mental process) is at the heart of constructive learning.

Learning very often is conceptualized as conceptual change (Vosniadou, 1994; Duit, 1994). As there is some debate on what 'change' in conceptual change may mean, we want to point out that in our view change does not mean total extinction of the already existing conceptions, but that learning is a process of adaptive development that includes facets of change and enlargement in existing conceptions, and also a change in the situations and contexts in which the old and the new conceptions may be fruitful. The latter aspect has been recently emphasized by the 'situated cognition' perspective (Brown, Collins and Duguid, 1989; Resnick, 1991; Hennessey, 1993) but had been previously pointed out by science educators (e.g., by Jung, 1986). In our view this aspect of conceptual change highlights that learning is a complex process of adaptation that leads from students' conceptions (or in the terminology of modelling: from students' mental models) towards conceptual models. Analogies (seen as instructional tools) may be helpful aids in conceptual change oriented teaching and learning approaches (Glynn, Duit, and Thiele, 1995; Thiele and Treagust, 1991). It would appear that the mentioned symmetrical character of the analogical relations is of key importance with regard to conceptual change. Very often considerable reconstruction of the already known is necessary in order to develop understanding of the scientific point of view. Metaphors are also of particular interest in conceptual change learning. As mentioned, when they are literally taken, their absurdity provokes some sort of surprise. It appears that this aspect makes them valuable tools in learning because they may open up new perspectives: 'Something happens to us when we read a fresh metaphor. We are reorganising our patterns of previously organised meaning' (Gowin, 1983, p. 38).

Results of Research on Analogical Reasoning

Research on analogical reasoning in cognitive psychology and science education has shown that analogies can be fruitful tools in learning, in particular for learning science. But it has also become evident that attempts to support learning by analogies failed in several cases. We will now provide a brief overview of findings in the literature on analogical reasoning which predominantly follows the review by Duit (1991) because main lines of results reported there have also been supported by more recent research findings.

1 Spontaneous use of analogies

There are many research studies that clearly show that students frequently try to make sense of phenomena by employing analogies (see e.g., the studies in the bibliography by Pfundt and Duit, 1994). The studies confirm that analogies are common tools for explaining and trying to make sense of the unknown.

2 Analogies presented in science textbooks and instruction need considerable interpretation

When analogies are presented in textbooks or in instruction in order to ease or facilitate access to a certain science concept or principle, usually considerable interpretation is necessary if the analogy is to be used successfully (Harrison and Treagust, 1993). For analogies to be fruitful, learners require guidance.

3 Multiple analogies

Analogies often appear to facilitate or support learning only in specific areas of a target domain. Multiple analogies are, therefore, necessary in order to aid learning of broader domains.

4 Access to analogies via surface similarities — inferential power of analogies via deep structure similarities

It has already been mentioned above that similarities at the surface level of the analog and the target domains, such as shape or even colour, appear to ease and facilitate access to analogies. Such surface analogies usually have little if any inferential power.

5 The target domain must be viewed as sufficiently demanding

There is a tendency that analogies used as teaching and learning aids are only accepted by the learners if the task is sufficiently demanding for the learners. There is also the tendency that learners with lower 'ability levels' (e.g., in terms of Piagetian stages) are inclined to accept analogies as aids more readily than advanced students.

Very briefly summarized research has shown that analogies as teaching and learning tools have the following advantages and limitations (Duit, 1991, p. 666; see also Duit and Glynn, 1992).

6 Advantages

- They are valuable tools in conceptual change learning, because they may open new perspectives.

- They may facilitate an understanding of the abstract by pointing to similarities in the real world.
- They may provide visualization of the abstract.
- They may provoke students' interest and may therefore motivate them.

7 Disadvantages and potential dangers

Analogies are 'double-edged swords' (Glynn *et al.*, 1989, p. 387), which may totally mislead.

- An analogy is never based on an exact fit between analog and target. There are always features of analog structures that are different from those of the target. These features may mislead.
- Analogical reasoning is only possible if the intended analogies really are drawn by the students. If students hold misconceptions in the analog domain, analogical reasoning will transfer them into the target domain. It is therefore important to ensure that students in fact are sufficiently familiar with the analog domain and do not hold major misconceptions there; and, that the intended analogies really are drawn by the students.

Models and Modelling in Science Education Viewed from the Perspective of Analogical Reasoning

A key idea of this paper is to view the role of models and modelling in science education explicitly from an 'analogy perspective'. It has been argued that what makes a model a model is analogical relations, and that the processes of constructing (creating) analogies are at the heart of modelling. Modelling has a number of purposes in science education.

Firstly, it is an important task of both teachers and science educators to create new models that facilitate students' understanding of key features of science theories. The latest state of science knowledge is usually far too complex for beginners. Key ideas have to be structured so that they are accessible to the learners. Quite often models are created for such instructional purposes. The analogical perspective taken here endeavours to create the models in such a way that both similarities and dissimilarities are carefully worked out.

Secondly, models may be the content of science instruction in that the theory itself is not accessible to students of a particular age. Learning a model, in our view, means to learn the analogical relations that 'define' the model. Research on analogies as teaching aids may help to design instruction appropriately (see approaches in Glynn, Duit and Thiele, 1995). Also learning about the role and nature of models and modelling is an important issue for 'models in science education'. Students have to be introduced to the nature of the analogies that comprise models. They have to understand that analogies have a heuristic function only. Mach (1920) clearly stated analogies' function when he noted: 'Conclusions based on similarity and analogy are, strictly speaking, not a matter of logic, at least not of formal logic, but a matter of psychology' (the authors' translation).

Third, models may be used explicitly as teaching and learning aids. The paradigmatic example for science instruction is the use of the water model of the electric circuit in order to facilitate students' understanding of current flow. There is a certain irony with this classic example that students usually have similar misconceptions in the domain of the water circuit as compared to misconceptions in the domain of the electric circuit (Schwedes and Schilling, 1983; Schwedes and Dudeck, 1993). This example therefore clearly shows that it is important to inspect the analogical relations that are at the heart of the model carefully. However, when analogies are used as teaching aids it has to be remembered that a number of analogies provide a bridge towards understanding the conceptual models. Once they are learned the analogies are not needed any more, the bridge may be pulled down.

Modelling, finally, is a common behaviour of everybody in daily-life situations in that there is a search for similarities if something unknown has to be explained. But usually these spontaneous modelling processes lead to mental models that are not very close to the conceptual models of science. Guided modelling is in need of finding of a 'pathway' (Scott, 1992) from students' mental models towards the science conceptual model. This, so to speak, includes modelling the analogical relations that facilitate the pathway. Usually, there is no one-step path possible. Learning science should then be viewed as a chain of analogy construction processes from the initial mental model to a first intermediate model and so on towards the scientific model.

References

BROWN, J.S., COLLINS, A and DUGUID, P. (1989) 'Situated cognition and the culture of learning', *Educational Researcher*, **18**, pp. 32–42.

BUNGE, M. (1973) *Method, Model and Matter*, Dordrecht, The Netherlands, Reidel Publ.

COSGROVE, M. (1992) 'Teaching electricity — A place for learner-generated analogies', Occasional paper of the University of Sydney.

DUIT, R. (1991) 'On the role of analogies and metaphors in learning science', *Science Education*, **6**, 75, pp. 649–72.

DUIT, R. (1994, September) 'Conceptual change approaches in science education', Paper presented at the 'Symposium on Conceptual Change', Friedrich-Schiller-University, Jena, Germany, Sept. 1–3.

DUIT, R. and GLYNN, S. (1992) 'Analogien und Metaphern, Brücken zum Verständnis im schülergerechten Physikunterricht in Häußler', in HDUSSLER, P. (Ed) *Physikunterricht und Menschenbildung*, Kiel, Germany, Institute for Science Education, pp. 223–50.

GENTNER, D. and LANDERS, R. (1985) 'Analogical reminding: A good match is hard to find', Paper presented at the International Conference on Systems, Man and Cybernetics, Tucson, Arizona.

GLASERSFELD, E. VON (1992) 'A constructivist's view of learning and teaching', in DUIT, R., GOLDBERG, F. and NIEDDERER, H. (Eds) *Research in Physics Learning: Theoretical Issues and Empirical Studies*, Kiel, Germany, Institute for Science Education, pp. 29–39.

GLYNN, R. and DUIT, R. (1995) 'Learning science meaningfully: Constructing conceptual models', in GLYNN, S. and DUIT, R. (Eds) *Learning Science in the Schools: Research Reforming Practice*, Hillsdale, NJ, Lawrence Erlbaum, pp. 1–33.

GLYNN, S., DUIT, R. and BRITTON, B. (1995) 'Analogies: Conceptual tools for problem solving and science instruction', in LAVOI, D.R. (Ed) *Towards a Cognitive Science Perspective for Scientific Problem Solving, NARST Monograph*, Manhattan, KS, National Association for Research in Science Teaching, pp. 215–44.

GLYNN, S., DUIT, R. and THIELE, R. (1995) 'Teaching science with analogies: A strategy for constructing knowledge', in GLYNN, S. and DUIT, R. (Eds) *Learning Science in the Schools: Research Reforming Practice*, Hillsdale, NJ, Lawrence Erlbaum.

GLYNN, S.M. (1991) 'Explaining science concepts: A teaching-with-analogies model', in GLYNN, S.M., YEANY, R.H. and BRITTON, B.K. (Eds) *The Psychology of Learning Science*, Hillsdale, NJ, Lawrence Erlbaum, pp. 219–40.

GLYNN, S.M., BRITTON, B.K., SEMRUD-CLIKEMAN, M. and MUTH, K.D. (1989) 'Analogical reasoning and problem solving in science textbooks', in GLOVER, J.A., RONNING, R.R. and REYNOLDS, C.R. (Eds) *Handbook of Creativity*, New York, Plenum, pp. 383–98.

GOWIN, D.B. (1983) 'Misconceptions, metaphors and conceptual change: Once more with feeling', in HELM, H. and NOVAK, J. (Eds) *Proceedings of the International Seminar on Misconceptions in Science and Mathematics*, Ithaca, NY, Cornell University, pp. 39–46.

HARRISON, A.G. and TREAGUST, D.F. (1993) 'Teaching with analogies: A case study in grade 10 optics', *Journal of Research in Science Teaching*, **30**, pp. 1291–307.

HENNESSEY, S. (1993) 'Situated cognition and cognitive apprenticeship: Implications for classroom learning', *Studies in Science Education*, **22**, pp. 1–41.

JUNG, W. (1986) 'Alltagsvorstellungen und das Lernen von Physik und Chemie', *Naturwissenschaften im Unterricht — Physik/Chemie*, **13**, pp. 34, 2–6.

LEATHERDALE, W.H. (1974) *The Role of Analogy, Model and Metaphor in Science*, Amsterdam, The Netherlands, North-Holland Publication Company.

LIND, G. (1980) 'Models in physics: Some pedagogical reflections based on the history of science', *European Journal of Science Education*, **2**, pp. 15–23.

MACH, E. (1920) 'Ähnlichkeit und Analogie als Leitmotive der Forschung', in MACH, E., *Erkenntnis und Irrtum*, Leipzig, Germany, pp. 220–31.

NORMAN, D.A. (1983) 'Some observations on mental models', in GENTNER, D. and STEVENS, A.L. (Eds) *Mental Models*, Hillsdale, NJ, Lawrence Erlbaum, pp. 7–14.

PFUNDT, H. and DUIT, R. (1994) *Bibliography: Students' Alternative Frameworks and Science Education* (4th ed) Kiel, Germany, Institute for Science Education.

RESNICK, L.B. (1991) 'Shared cognition: Thinking as social practice', in RESNICK, L., LEVINE, J. and TEASLEY, S. (Eds) *Perspectives on Socially Shared Cognition*, Washington, DC, American Psychological Association, pp. 1–19.

SCHWEDES, H. and DUDECK, W.-G. (1993) 'Lernen mit der Wasseranalogie', *Naturwissenschaften im Unterricht Physik*, **16**, 4, pp. 16–23.

SCHWEDES, H. and SCHILLING, P. (1983) 'Schülervorstellungen zu Wasserstromkreisen', *Physica Didactica*, **10**, pp. 159–70.

SCOTT, P.H. (1992) 'Pathways in learning science: A case study of the development of one student's ideas relating to the structure of matter', in DUIT, R., GOLDBERG, F. and NIEDDERER, H. (Eds) *Research in Physics Learning: Theoretical Issues and Empirical Studies*, Kiel, Germany, Institute for Science Education, pp. 203–24.

SUTTON, C. (1978) 'Metaphorically speaking . . . The role of metaphor in teaching and learning science', Occasional Paper, Science Education Series, University of Leicester, UK, School of Education.

TENNEY, Y.J. and GENTNER, D. (1985) 'What makes analogies accessible: Experiments on that water-flow analogy for electricity', in DUIT, R., JUNG, W. and RHÖNECK, CH. VON (Eds) *Aspects of Understanding Electricity*, Kiel, Germany, IPN/Schmidt and Klaunig, pp. 311–18.

THIELE, R.B. and TREAGUST, D.F. (1991) 'Using analogies in secondary chemistry teaching', *The Australian Science Teachers Journal*, **37**, pp. 10–14.

TREAGUST, D.F., DUIT, R., JOSLIN, P. and LINDAUER, I. (1992) 'Science teachers' use of analogies: Observations from classroom practice', *International Journal of Science Education*, **4**, 14, pp. 413–22.

VOSNIADOU, S. (Ed) (1994) 'Conceptual change in the physical sciences', *Learning and Instruction*, **4**, pp. 121–3 [Special Issue].

WAGENSCHEIN, M. (1990) *Kinder auf dem Wege zur Physik*, Weinheim, Germany, Basel, Switzerland, Beltz.

WONG, E.D. (1993) 'Self-generated analogies as a tool for constructing and evaluating explanations of scientific phenomena', *Journal of Research in Science Teaching*, **4**, 30, pp. 367–80.

14   Texts and Contexts: Framing Modelling in the Primary Science Classroom

Carolyn Boulter and John Gilbert

Abstract

This paper presents models and modelling as critical aspects of the texts of primary science classrooms. It suggests that such texts are embedded in narratives which are made meaningful by the culture in which they are positioned. A framework is proposed to analyse the participation in modelling through argumentation and the modelling itself, taking the example of a class discussion about a forthcoming lunar eclipse.

The Nature of Science and Language in the Primary Classroom

Primary school science has been rather like the race between the hare and the tortoise. Only twenty years ago science had a low priority in the list of essentials for the primary school in the UK and was virtually invisible in the curriculum in practice (Clift, *et al.*, 1981). English and mathematics now contend with science for first place in the many revisions of the National Curriculum for England and Wales (DfE, 1995). Similar changes in the status of science can be distinguished worldwide (Meyer, *et al.*, 1992). In the intervening years the question has been largely whether content or process should drive the tortoise forward (Harlen, 1985). As the status of science has risen the process–content debate has become less of an issue and investigation has found an authorized place alongside content. It is now the nature of investigation that is the focus of the debate.

At the present time there is a wide variety of work being carried out into the nature of science, both within science as practised and in the science classroom. These studies look at how scientific knowledge comes into being in laboratories (Judson, 1995), how children perceive the processes that scientists go through (Driver, *et al.*, 1993), and how they operate as classroom scientists themselves (Lubben and Millar, 1994). In parallel with this work there has been an increasing interest in the language of science, its characteristics (Sutton, 1992) and its relationship to the language used in other parts of life. Much of this work has its foundations in the work of Piaget and Vygotsky and their concern to understand how language is used to build understanding. Such studies often concern themselves with sociolinguistic patterns deduced from speech in real situations (Edwards and Mercer, 1987) and with the analysis of printed scientific texts (Lowe, 1993).

Alongside these themes runs the continuing strand of research into the 'alternative conceptions' that children display and their relationship to official science. This is providing both an ever-increasing base of understandings across a wide range of content domains (Leach, *et al.*, 1992) and ways of approaching teaching and learning at all levels (Bell and Gilbert, [in press]).

Defining the Frame for Language and Science

As the status of science in primary school is enhanced, it is the nature of science and the language of science that form two major foci for research. It is the aim of this paper to sketch out the connections linking these two fields and to suggest ways of investigating them. In starting to paint such a picture it is necessary to define the frame into which it will fit and the materials with which it will be painted — the assumptions upon which our work into language and scientific understandings are based. Firstly, the work takes the view that it is of critical importance to be aware of the cultural influences on the construction of theory, both within official science and within science teaching and learning. Such cultural awareness includes an awareness of context on a large scale. Taking this stance, science cannot be seen as a plain and accurate representation of reality, for that is simply a 'culturally locked' position within one culture (Duncan and Ley, 1993). While accepting that culture is important, science can be seen as attempting culturally free exploration from a multiplicity of viewpoints producing a collage of theories, as the post-modernists would suggest (Gough, 1994). Alternatively it can be perceived as a culturally influenced interpretation of the world as we experience it in our culture. The texts that are produced by science must, from this perspective, be interpreted through the right glasses to be acceptable. We hold the latter, interpretative stance and are thus concerned to include analysis at the level of the cultural context in our framework.

The Importance of Context

The contextualist view of knowledge, that it is determined by the people and the time in which it is made, relates closely to the main tenets of the constructivist pedagogical approach. That is predicated on the belief that all knowledge is personally and socially constructed, such that all learners have prior conceptions and actively construct meaning, rather than on the belief that knowledge is the product of some innate individualistic mechanism (Gilbert and Watts, 1983). Meanings are therefore provisional and agreed through a process of individual reflection and social interaction. This tension between individual and social construction is thus seen as a structural feature of the development of understandings. These may be built in individual heads, in classrooms, in laboratories and many other social settings.

The building of meanings in such social settings is the domain of social semiotics, which deals with the systems of signs and symbols by which we make sense to, and of one another. Lemke (1990) makes it clear that each context can have different conventions for this meaning-making process. It is only when we belong to the same, or overlapping communities, that we share the conventions of meaning-making and the meanings themselves are thus shared. Each context is also contained within others producing a nested structure.

The Framework of Levels

The analysis framework we propose will therefore be capable of representing the contextualist view of knowledge and the nesting of the levels at which knowledge is built. It will allow the nature of such knowledge and its origins to be addressed within a variety of social settings.

The central importance of language in the building of understandings itself rests on a number of assumptions about the nature of language and the theories of how it operates. Following Lakoff and Johnson (1980), that all language is metaphorical and that one can only speak of one thing in terms of another, we seek to unravel the focal metaphors in contextual speech that serve as models. Research into models and modelling is a major theme in science education. It concerns itself with the way models are produced through analogy (Duit, 1991) and the ways in which they relate to aspects of the phenomenal world (Buckley, 1992). A fine-grained taxonomy of models, first attempted by Rouse and Morris (1986), is also of concern as it can give rise to an appreciation of critical aspects of the phenomenon modelled and of its mode of representation (Boulter, Buckley and Gilbert, forthcoming). In setting up the 'Models in Science and Technology — Research in Education' (MISTRE) Group in 1995 the rationale for the group was expressed as:

> A model is a representation of an object, event or idea. Such representations are used to summarize data, to make predictions, to guide enquiry, to justify outcomes, and to facilitate communication. Mental modelling is an activity undertaken by individuals whether alone or within a group. The results of the activity can themselves be expressed in the public domain, e.g., through speech or writing. Those expressed models which gain general acceptance to become consensus models play a central role in the conduct of both research and development. They also play a role in science and technology education.

Models and modelling are a tangible means of linking language and knowledge building. To encompass both this and the nested aspects of context, we propose a three-level framework of the level of the culture, the level of the narratives of social situations, and the level of the texts produced by, and within, such narratives. Each rests within the other and we should expect meaning to emerge from the effect of this nesting through metaphorical language which builds and uses models.

Science is talked about in many situations; laboratory, classroom, museum and home. In each situation events occur where groups of people talk together in different places and about different tasks. Any single event where the same people talk to each other in the same place and about the same things produces a text, a bounded piece of language of that particular event which may contain spoken, written and acted elements.

The social situation from which we have taken our data to illustrate how we set about analysing these texts is a primary classroom as the children investigate questions which they regarded as scientific. It involves a teacher and her class of 9–11-year-olds and the text that is produced as they interact sitting together in a corner of the classroom on the theme of a lunar eclipse that was about to happen at the time of the inquiry.

The Text Level in the Social Setting of the Classroom

Such texts provide us with data about which we can talk and about which we can build theories of the ways in which the types of model interact with each other and how modelling operates in this social setting. As mental modelling in the head is unavailable, the action or the speech in which individuals engage is the main data that we have as researchers. When individuals act, speak, draw, or write, they have voiced their representations of their mental models of the world as perceived. As they enter a place where they may be listened to, either there is a 'conversation with self' or with others, as texts are created or recreated (Figure 14.1).

There are several different models which may be embedded in these texts. The individual child's expressed models, the teacher's teaching models, and the consensus models, may be heard in the text of a class discussion event such as 'What do you think will happen in the eclipse tonight?'. There are other texts which may enter the discourse text, in particular the official curriculum, a text of official science such as a textbook or visiting expert, and the voice of the media. In each of these texts, the models of the domain may again be different and differently represented.

These spoken, written or action texts are the central data of classroom interaction. It is to these texts of classroom social events that we look in building theories of modelling and pose the question: How can the models and the process of modelling in the texts of the primary classroom engaged in science be analysed?

The Analysis of the Discourse Texts

It is important to theorize about how scientists model because it is this process which informs the curriculum and the teacher's approach to science teaching. However, it does not yet form a main focus of our research. Likewise, the processes by which teachers, both in training and in practice, develop models will be a critical part of a mosaic view of the classroom, but lies outside our present concerns. These are to focus on the models used by children and modelling processes that are taking

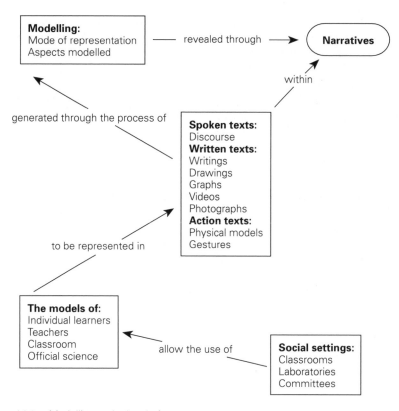

Figure 14.1: Modelling at the level of text

place in classroom interactions as the text of the discourse 'scrolls over', to adopt a word processing metaphor.

The Criteria for Rich Classroom Texts

Not all classroom texts will provide the richness of the data needed to hear the models which children themselves use. There are only certain conditions of participation in which children will voice their understandings and speak in a collaborative fashion (Boulter, 1992). The nature of the control of the language and the control of knowledge by the teacher are crucial. There needs to be a problem or task which can be debated and clear leadership from the teacher on what is expected in terms of interaction and the framing of the content. For children to deliberate on their own models in response to a task, they must both participate in the talk and in deciding the domain which is to be talked about. The prerequisite therefore for data in which children's models can be heard and deliberated in the text is a classroom situation in which the teacher is committed to this collaborative pedagogy.

In dialogue concerned with a prescribed but open-ended task being addressed by two individuals, Grannott (1993) showed that asymmetry in the perceived expertise in the knowledge domain, together with degree of collaboration (that is the extent to which the individuals work cooperatively together), were critical in determining the ways the meanings developed and the nature of the speech used. With equal expertise and low cooperation there was play and individuals developing independent meanings, speaking to themselves and occasionally to each other. With equal expertise and moderate cooperation there was group discussion, with individual voices listened to and responded to in turn, and separate meanings evolving which were partially shared. With an asymmetry of expertise and high cooperation, there was a scaffolding with some fragmented speech and the less expert person being engaged in sharing the meaning of the more expert person.

The text we chose to analyse initially was a whole class discussion of the teacher's question 'What do you think is going to happen during the eclipse tonight?'. The text was selected because the individual voices were distinct and the teacher did not show her domain specific knowledge, although she had close control of the management and interaction. There was an interplay of voiced and shared meanings on the eclipse over an extended discussion. This produces a coherent text with a rich depth of meanings which are formulated to persuade others in the class. A small section of this discussion gives a glimpse of Alex's explanation being challenged by another pupil who will later give his own explanation.

Teacher	Alex, would you like to draw for me here what's going to happen?
Child	He didn't —
Teacher	If there's something else you want to add just put your hand up.
Alex	This is the Sun.
Teacher	That's the Sun.
Alex	That's the Earth and that's the Moon. Light from the sun is going to the Earth but it won't reach the Moon because it's in the Earth's shadow.
Child	Sometimes it's the other way round.
Teacher	We haven't finished with Alex yet. So why is this happening tonight and not tomorrow night or last night?
Alex	It's because . . . um . . . the Earth moves around and the Moon moves round, so it only happens when they are in a line.
Teacher	So what shall we see tonight?

It was the persuasive nature of the text that led us to consider the nature and significance of the argumentation patterns between the utterances in the text. Kuhn (1992) considers that all thinking can be seen as argument and suggests that there are two forms that it takes: the didactic, where the audience is silent; and the dialogic, where each person gives their views. Kuhn's context was a series of interviews with adults. But for the context of the classroom, we proposed a third

Argument type / Aspect	Didactic Argument	Socratic Argument	Dialogic Argument
Intention	Meaning flows Transmission	Meaning mapped Discovery	Meaning shared Construction
Structure	Traditional rhetorical tools.	Initiation-Response-Evaluation. Reformulations. Implicit rules.	Framed deliberations. Rules explicit.
Audience	Pupils passive and receptive	A few questioned; others listen.	Wider audiences than the class.
	Cognitive conflict is unvoiced.	Cognitive conflict is unvoiced.	Cognitive conflict is voiced in deliberations.
Role of teacher	Representing the official understanding of science persuasively.	Providing a scientific concept map and access to it.	Mediating transitions within and between texts and narratives of science.

Figure 14.2: The characteristics of the types of argument

argument type, socratic, where the pupil audience was questioned (Boulter and Gilbert, 1995) (see Figure 14.2).

In Boulter and Gilbert (1995) we suggested that, in most classroom situations, the socratic form of argumentation by the teacher is the 'default' pattern (Janda, 1990) i.e., the one most commonly used. However, we expected that proficient teachers would be able to use and to assist their pupils to use the full range of argument patterns for different purposes and to switch easily between them. We expected that the deliberations between the children, recorded in a classroom committed to contextual, constructivist, science, would be of the dialogic kind.

Framing Modelling through the Analysis of Argumentation

By subjecting our classroom text of thirty-six minutes, now transcribed as utterances which we took as 'stretches of speech preceded and followed by silence or a change of speaker' (Crystal, 1985) to our framework for analysing argumentation, the way language structures knowledge through argumentation within a nested social context began to take shape. The utterances fall into sections which we called 'transactions' which are separated from each other by a change in focus and by 'boundary' utterances. So the transaction within the class discussion of 'What's going to happen in the eclipse tonight?', when the teacher focuses attention on the visible aspects of the eclipse with 'So what shall we see tonight?', starts off a series of boundary moves that lead into a debate about the colour of the moon during the eclipse. A child declares 'The moon's going to be a reddy colour', the teacher interrogates him, and the child explains, in a socratic exchange. Later, another pupil challenges this declaration with 'I just wouldn't have thought the moon was red' and he expands his challenge by 'I wouldn't have thought it because that would shine on there, it would cause a shadow over here so it would make the moon completely invisible'. This exchange is not completed until later in the discourse but it is a dialogic exchange, involving a challenge to a declaration. The third type of exchange is a didactic exchange where a declaration is followed by expansion and explanation, for example when the challenger expands his declaration that 'Its the same theory but with a different explanation', and goes on to say 'The light would go on this half of the Earth and the shadow that half and shadow the moon as well', and explains 'So it would completely shadow the moon', and expands the explanation 'So it wouldn't glow at all — you wouldn't see anything', and expands it again 'You see as it moves round its say a full moon, then it starts going half moon, quarter moon until one point tonight its completely invisible'.

These three types of argumentation all begin with a declaration. The boundary moves appear critical in the 'turning on' of one of the three pathways. We classify the boundary moves as 'procedural' or 'conditional', following Alexander and Judy's (1988) definitions:

> Declarative knowledge refers to factual information (knowing what), whereas procedural knowledge is the compilation of declarative knowledge into functional units that incorporate domain-specific strategies (knowing how). Conditional knowledge entails the understanding of when and where to access certain facts or employ certain procedures. (Alexander and Judy, 1988, p. 376).

Procedural strategies are concerned with how declarations enter the discourse and allow interaction, process and management strategies to emerge. They seem to allow the emergence of particular structures of investigative modelling, collaboration, and lesson form. Conditional strategies allow real life to enter the discourse. They seem to allow access to these out-of-class experiences of the pupils.

The argumentation in the discourse seems a useful way of looking at the patterns that may be present in the text of an event such as the class discussion described. It allows us to see how particular models of the eclipse or parts of these models are voiced in the children's talk and how they are responded to. The analysis of the argumentation thus frames the models and their use, showing how they persuade in this social context.

Towards an Analysis of Models and Modelling

In this collaborative classroom the models of the eclipse are voiced in response to the problem of explaining a phenomenon, the lunar eclipse that is not yet visible. Science deals with phenomena in the world. Any model is a representation of a phenomenon by another entity (source) and operates through analogy. Both the phenomenon and the entity which represents it may be made up of parts. During representation some aspects of the source of the model are transferred to the target phenomenon (Norman, 1983). In analysing models and modelling in the text it is therefore important to be able to distinguish between phenomena and the parts of which they are made, and the entities and their separate parts. In addition, the other aspects of the phenomenon; the relationship (or function) of the parts, the way they behave through time, and the causal mechanisms (Buckley, 1992), which are focused upon in the dialogue allow us to analyse how that particular piece of analogy is operating. Moreover, the mode of representation of the model is significant.

In the text described the first model voiced by Alex gives a description of the eclipse which relates the parts of the phenomenon, the sun, the Earth and the moon in a straight line. He draws these as circles suggesting that the analogical source may be spherical balls. The aspect he chooses first in his model is the relationship between the structural aspects of the parts of the phenomenon.

He then draws the light from the sun (another part of the phenomenon) by tracing the line from the sun circle to the Earth circle, suggesting that the analogical source may be the straight line travelling of an object such as a bullet. The aspect here that he focuses on is behavioural, what happens through time to this part of the phenomenon.

He finishes his response, to the first teacher prompt to explain what will happen in the eclipse, by giving a structural (what it looks like) description of the Earth's shadow.

In response to the teacher's prompt to explain why it will happen tonight, he indicates through circling hand movements that the Earth will go round the sun and the moon round the Earth suggesting that here he has an analogy of objects travelling in a circle like a ball on a spun string. In doing so he focuses on the cause of the eclipse and its mechanism.

The teacher presses him to expand and he focuses on a new structural aspect of the phenomenon, that the moon will be red in colour. Here it is hard to see the analogy and it becomes clear later that this image has been seen on television news coverage. Finally, he concludes his contribution in response to her request for a reason by drawing a line which bends round the Earth and hits the moon, suggesting again an analogy of a travelling object, and how it behaves.

In this example we have various parts of the eclipse phenomenon represented; the sun, Earth and moon, the light both white and coloured and the Earth's shadow. As each part is mentioned it allows the focus to fix on different aspects of the parts of the phenomenon, the structure, relationship, behaviour and mechanism. The usefulness of this analysis is that it provides opportunities for looking at the patterns in the appearance of these aspects and the mismatch that might arise between the teachers' expectations and the pupil's use.

The Narrative Level in Learning Science

We have attempted to show that the texts of classroom events are critical data for an analysis of the relationship between the building of understanding and language use through argumentation and modelling. These texts do not exist in a vacuum, but are part of a narrative which (in this case) is being told through the entire classroom discourse. The transaction which has been illustrated is part of a sequence of transactions during the class discussion which may be seen as a mini-narrative of model building in this social setting. The story has a sequence of phases with critical points, tensions, and resolutions, and the individuals seek to take from this common story meaning for their individual stories. It is a narrative of model building about the nature and operation of the eclipse.

Scientific narratives about the world are seldom seen as they are being built, as in this classroom, but are implicitly present in various sequences of texts from which the developing narrative has been rendered invisible. The curriculum document and the scientific paper are important examples in science education. Others such as the museum exhibit and the video may also be relevant. They may contain an implicit narrative structure themselves which the producers expect to be read in a certain order to produce a particular learning outcome. However, they may not be read in this way nor seen as being a product of a contextual process in which people have developed an understanding through time. So, in approaching such undisclosed narratives for the first time, many are read as simple statements. The texts are not related to each other in ways which enable the reader to progress and to relate the developing meaning to her own understanding. So for instance, in the museum, the narrative linking the texts of individual exhibits is seldom explicit and the visitor will either not seek a connection or will construct a personal narrative. The narrative level thus gives meaning to the texts which comprise it and itself has meaning within the culture in which it was constructed.

The Cultural Level

The framework discussed above contains an attempt to link the narrative level to the textual level. The narratives exist with cultures. Within 'British culture' there exist many cultures, so unravelling the ways in which they give meaning to narratives is a formidable task. In any culture, models and modelling are both produced by, and represent, that culture. They are taken up and influence decision making and the imaging of possibilities and prescriptions for action and reform. The ideology of the present political culture in British education is described by the image of the market-place. Within this way of perceiving education, the science education of pupils in primary schools is a subject of major concern and debate because the purpose of schooling in this framework is seen as determined by its success in providing industry with scientifically informed workers. On finishing school, the potential of pupils is seen in terms of their suitability for the job market. This market-place image is the guiding metaphor imposed on education in Britain in the 1980s and 1990s. It is within this general culture, influenced by such metaphors, that the various educational situations, both formal and informal, in which children learn are situated. These situations range from the classroom to the museum, from the living room to the field trip. Insofar as an ideology is accepted by the readers of the story it will give meaning to contemporary narratives of classroom, museum, and curriculum. Insofar that ideological metaphors are not accepted as adequate, there will be a mismatch between writers and readers of narratives.

This paper suggests that modelling and persuasive argumentation are critical aspects of the nature of science education and shows ways in which texts of classroom situations can be analysed for argumentation and modelling patterns. It describes a framework in which the texts forming the data for the analysis are contextualized by the narratives within which they are positioned and the culture which interacts with these narratives.

References

ALEXANDER, P. and JUDY, J. (1988) 'The interaction of domain-specific and strategic knowledge in academic performance', *Review of Educational Research*, **58**, 4, pp. 375–404.

BELL, B. and GILBERT, J. (in press) *Teacher Development: A Model from Science Education*, London, Falmer Press.

BOULTER, C. (1992) 'Collaborating to investigate questions: A model for primary science', PhD thesis, Reading University.

BOULTER, C., BUCKLEY, B. and GILBERT, J. (forthcoming) 'Approaching a typology of models'.

BOULTER, C. and GILBERT, J. (1995) 'Argument and science education', in COSTELLO, A. and MITCHELL, S. (Eds) *Competing and Consensual Voices*, London, Multilingual Matters.

BUCKLEY, B. (1992) 'Multimedia, misconceptions and working models of biological phenomena: Learning about the circulatory system', PhD thesis, Stanford University, USA.

CLIFT, P., WEINER, G. and WILSON, E. (1981) *Record Keeping in Primary Schools*, London, Macmillan.

CRYSTAL, D. (1985) *A Dictionary of Linguistics and Phonetics*, Oxford, Blackwell.

DfE (1995) *The National Curriculum (England)*, London, HMSO.

DRIVER, R., LEACH, J. and SCOTT, P. (1993) *Students' Understanding of the Nature of Science*, CLIS Research Group, Leeds University, Science Education Group, York University.

DUIT, R. (1991) 'On the role of analogies and metaphors in learning science', *Science Education*, **75**, 6, pp. 649–72.

DUNCAN, J. and LEY, D. (Eds) (1993) *Place/Culture/Representation*, London, Routledge.

EDWARDS, D. and MERCER, N. (1987) *Common Knowledge: The Development of Understanding in the Classroom*, London, Routledge Kegan Paul.

GILBERT, J. and WATTS, D.M. (1983) 'Concepts, misconceptions and alternative conceptions: Changing perspectives in science education' *Studies in Science Education*, **10**, pp. 61–98.

GOUGH, N. (1994) 'Regarding nature in new times: Reconceptualising studies of science and environment', Annual Conference of the Australian Association for Research in Education, Newcastle NSW, 27 November–1 December.

GRANNOTT, N. (1993) 'Separate minds, joint efforts, and weird creatures: Patterns of interaction in the co-construction of knowledge', in WOZNIAK, R. and FISCHER, K. (Eds)

Development in Context: Acting and Thinking in Specific Environments, New Jersey, Erlbaum.

HARLEN, W. (Ed) (1985) *Primary Science: Taking the Plunge*, London, Heinemann.

JANDA, M. (1990) 'Collaboration in a traditional classroom environment', *Written Communication*, **7**, 3, pp. 291–315.

JUDSON, H. (1995) *The Eighth Day of Creation*, London, Penguin.

KUHN, D. (1992) 'Thinking as argument', *Harvard Educational Review*, **62**, 2, pp. 155–78.

LAKOFF, G. and JOHNSON, M. (1980) *Metaphors We Live By*, London, Chicago University Press.

LEACH, J., DRIVER, R., SCOTT, P. and WOOD-ROBINSON, C. (1992) *Progression in Understandings of Ecological Concepts by Pupils aged 5 to 16*, Leeds, CLIS Research Group.

LEMKE, J. (1990) *Talking Science: Language, Learning and Values*, Norwood, NJ, Ablex.

LOWE, R. (1993) *Successful Instructional Diagrams*, London, Kogan Page.

LUBBEN, F. and MILLAR, R. (1994) *A Survey of the Understanding of Children Aged 11–16 of Key Ideas about Evidence in Science (PACKS Research paper 3)*, University of York, University of Durham.

MEYER, J., KAMENS, D. and BENAVOT, A. (Eds) (1992) *School Knowledge for the Masses, Studies in Curriculum History*, London, Falmer Press.

NORMAN, D. (1983) 'Some observations on mental models', in GENTNER, D. and STEVENS, A. (Eds) *Mental Models*, Hillsdale, NJ, Erlbaum, pp. 7–14.

ROUSE, W. and MORRIS, N. (1986) 'On looking into the black box: Prospects and limits in the search for mental models', *Psychological Bulletin*, **100**, 3, pp. 349–63.

SUTTON, C. (1992) *Words, Science and Learning*, Buckingham, Open University Press.

YOUNG, M. (1971) *Knowledge and Control: New Directions for a Sociology of Education*, London, Macmillan.

Part III
Approaches to Science Instruction

Introduction
Geoff Welford

This collection of papers ranges across a broad spectrum of topics. Miller and Lubben tackle the often overlooked question of how to teach and learn about the methods and procedures of scientific enquiry. It is perhaps only recently, in the UK anyway, that there has been a realization that this area of students' understanding is not 'caught', but has to be taught and learned. Benloch and Pozo examine the persistence of students' conceptual understandings and ideas with age in the face of instruction. Watson and Dillon also look at the issue of the progression in pupils' understanding describing changes over time in their explanations of phenomena and the concomitant pedagogical implications. Lichtfeldt looks at the development of pupils' cognitive structures and understandings as they progress along the 'pathways to an atom-idea'. Mashhadi takes us into the area of students' comprehension of the concepts of quantum physics asking mind-numbing questions like 'Are electrons particles or waves?' Lock picks up the gauntlet of the teaching of controversial issues, looking at student attitudes to, and knowledge about, the highly current areas of Biotechnology and Genetic Engineering. These papers are perhaps bound together by their obvious enthusiasm for the search for ways to make things clearer for students — and for us, the readers of their work.

15 Knowledge and Action: Students' Understanding of the Procedures of Scientific Enquiry

Robin Millar and Fred Lubben

Abstract

It is widely agreed that learning about the methods and procedures of scientific enquiry is an important element of science education. There is less agreement, however, about what these methods are and what might be taught to young people. The Procedural and Conceptual Knowledge in Science (PACKS) project documented school students' (aged 9–14) perform-ance of practical investigation tasks and probed their understandings and reasons for specific actions.[1] To interpret this data, the project developed and used a model linking specific aspects of understanding to performance. Four areas of understanding are seen to influence actions: understanding of relevant science content, of the use of relevant equipment, of the purpose of the task (frame) and of the collection and evaluation of empirical data. The latter appears to play a particularly important role in determining the quality of students' responses to a task. In a second phase of the project, a survey of students' ideas about measurement revealed a wide range of understandings, some of which differ markedly from the accepted scientific view. This suggests that more explicit teaching interventions may be required to develop more fruitful understandings of measurement and of empirical data in general.

Introduction

It is a truism to say that the central aim of science education is to help the learner acquire 'an understanding of science'. This is widely taken to include at least three elements:

- an understanding of some parts of the corpus of accepted science know-ledge (knowing some facts, laws, theories about the natural world);
- an understanding of the methods and procedures of enquiry used in science (knowing something about how the body of scientific knowledge has been established and how it is still being extended);
- some understanding of science as a social enterprise (the inter-relationships between science and the wider society; issues concerning the application of science; the internal social structures of science; issues and disputes within science and their resolution).

This paper focuses on the second of these elements. Both the practice and the rhetoric of science education suggest that teachers, and science educators more generally, give some importance to this aspect of science learning. We use practical work extensively to support our teaching of science ideas: seeing for yourself is a warrant for accepting the ideas the teacher wants to get across; or the teacher may use practical work to challenge students' expectations and, hopefully, to encourage them to develop their ideas and conceptions so that they move closer to the accepted scientific view. Also, from their experience of practical work, students are expected to learn something about the methods and procedures of scientific enquiry. It is this aspect of science learning that we wish to explore in this paper.

In some countries, curricula have given explicit emphasis to teaching and learning about the methods of scientific enquiry. In the English National Curriculum, one of the four Attainment Targets deals with 'Experimental and Investigative Science' (DFE/WO, 1995). This part of the curriculum sets out a programme of work, and identifies specific learning targets, for helping students to develop their understanding of the scientific approach to enquiry — the procedures involved in carrying out an enquiry 'scientifically'. Pupils are required to carry out scientific investigations; in this context, an 'investigation' is a practical task in which the pupil decides the strategy and methods to be adopted, chooses the apparatus to use, decides what measurements to take, and interprets the results. At various stages during their school career, their performance of an investigation task of this sort is assessed against a set of specified criteria.

Teaching and assessing scientific investigations is the aspect of the Science National Curriculum which has caused the greatest difficulty — both for teachers and for the curriculum planners. It has been changed considerably in the two revisions of the National Curriculum since 1989, and many problems have arisen as teachers try to understand and to implement the curriculum requirements. We would suggest that these difficulties are not simply an artefact of poor curriculum design and implementation — but that the challenge of deciding what we might want to teach, and what we could possibly teach, pupils about the methods of science is real and difficult.

Our starting position, then, is that it is important that we teach pupils something about the nature of scientific enquiry. Indeed it is impossible to teach any science content without implicitly conveying some messages about the methods of enquiry which have led us to this knowledge. The argument which is sometimes put forward, that an understanding of science method is valuable because it provides students with a powerful and generally applicable method of 'finding things out', is largely rhetorical; there is little evidence that any such method exists, or that it would be appropriate to use it in non-scientific contexts even if it did. Rather, it is important that we teach about scientific enquiry in order that students might understand the warrants for accepting the scientific account and appreciate both the power and the limitations of scientific knowledge.

The key problem in teaching students about science method is that there is little agreement about what this 'method' is. No one can describe it satisfactorily in detail. Curriculum prescriptions have tended to present distorted and misleading images of scientific enquiry. The English National Curriculum is a case in point: it implies that one-off, and relatively quick, individual investigations lead to knowledge of the natural world, if undertaken appropriately and with sufficient care. This is the empiricist fallacy writ large: that knowledge comes directly and easily from data collected by unbiased observation of the world. In addition, driven by the pressure to devise a progressive hierarchy of investigation performance for assessment purposes, the National Curriculum has tended to narrow the notion of a 'scientific investigation' to an exploration of the effect of one or more independent variables on a dependent variable. In this, it has been influenced by the working definition of a science investigation adopted by the Assessment of Performance Unit in the 1980s (APU, 1984a,b), which in turn drew on the role of the ability to control variables in Piagetian stage theory. The net outcome has been to narrow the range of tasks which can count as 'investigations' — so that the sort of work undertaken bears little relation to the real work of scientists. This unnatural constriction of the notion of 'scientific investigation' is most marked in biology and chemistry contexts, where little work can be devised which fits the teaching and assessment model.

Again, though, we would emphasize that these problems are not exclusively related to the English National Curriculum. Teachers in most countries subscribe to the idea that teaching pupils about scientific method is important and so the issues about what this 'method' is, and how it might be taught, inevitably arise.

Teaching and Learning about Scientific Enquiry

One response to the difficulty of specifying what we mean by 'science method' is to appeal to the holistic nature of scientific enquiry and to treat understanding as largely tacit (Woolnough, 1989). Students simply come to understand how to design and carry out a scientific investigation from experience of seeing, and doing, a large number of science practical tasks. If we take it as given that most science graduates *have* come to an understanding of how to investigate a practical problem scientifically, and acknowledge that most have followed courses which involved no explicit teaching of science methods, then it seems clear that some 'learning from experience' of just this sort must indeed be occurring.

But, if learning can result from a relatively unplanned sequence of experiences (i.e., unplanned in relation to learning about science method; the sequence is based on other criteria), then surely it ought to be possible, in principle at least, to plan the sequence to achieve this learning more efficiently and effectively. To do this, however, we would first need to clarify what we mean by an understanding of scientific method.

A Knowledge-based Model of Investigation Performance: The Packs Project

In the Procedural and Conceptual Knowledge in Science (PACKS) project (Millar *et al.*, 1994), we took the view that carrying out an investigation task is largely a

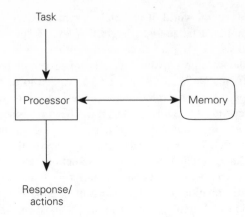

Figure 15.1: Responding to an investigation task: A simple model

knowledge-based (as opposed to a skill-based) performance. That is, better performance results from having access to more knowledge (or better structured knowledge) which is relevant to the task in hand. Faced with a new investigation task, which is different from anything encountered before, a student must draw on existing knowledge about previously encountered situations which he or she considers to be similar in some way. The task is a trigger to actions; the decisions about exactly which actions are taken draw on ideas held in long-term memory. We might envisage a model like that of Figure 15.1.

Key questions to ask then are:

- What kinds of knowledge are recalled and why are these selected?
- How are these memorized elements stored (for example, are they stored as rules, or concepts, or as whole episodes)?
- Are the recalled items *generic* (widely applicable to many, or even most, investigation tasks), or are they largely *investigation-specific* (or specific to investigations in particular domains)?

In the PACKS project we asked small groups of pupils to carry out an investigation task. We used seven different tasks, though each pupil group did only one. We observed the groups closely as they did the task. In total, we observed sixteen groups of 9-year-olds, thirty-two groups of 10–11-year-olds and thirty-two groups of 14-year-olds on each task. The pupils were asked to predict the expected outcome before beginning, and a record was kept of their actions as they carried out their investigation. We asked them to keep a written record of their results and to write down their conclusion at the end. After they had finished, we asked them to answer some diagnostic questions which probed their understanding of the science ideas involved in the investigation and their ability to interpret data of the sort they might have produced in the investigation, presented in various formats. We also talked to them as they worked on the task, trying to explore their reasons for some

TASK

Cool drink

If you want a cold drink on a hot day, you can use a 'coolbag'. First you cool your bottle of drink in the fridge, then you put it into the coolbag. The coolbag is designed to keep it cold for as long as possible.

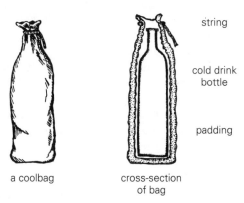

string

cold drink
bottle

padding

a coolbag   cross-section
of bag

How well does a coolbag work? This depends on:

- the material you use for the padding
- the thickness of the padding

You are given some pieces of material (bubble wrap, foam and fleece). Any of these could be used to make the padding for a coolbag. Whichever one you choose, you can use one layer, two layers, or more to make the coolbag.

Your job is to find out how the thickness of the padding affects how well the coolbag works. Use the response sheet to record what you do and what you find.

Figure 15.2:   The 'cool drink' task

of the decisions they took. (For a fuller discussion of the research strategy and methods, see Millar *et al.*, 1994)

One PACKS task concerned heating and insulation. Pupils were asked to think about the effectiveness of a coolbag, designed to keep a cold drink cool on a hot day. They were asked to investigate how the thickness of the insulating material used affected the performance of the coolbag. The task is shown in Figure 15.2. Sheets of three different wrapping materials were provided: fleece, plastic foam and bubble wrap. Measuring cylinders and 100 ml beakers were made available, as were thermometers, stopclocks, a supply of ice-cold water, scissors, string, sticky tape and rubber bands. We will use the *Cool drink* task to illustrate the kinds of understandings which students appeared to draw upon in their responses. A detailed account of students' responses to this task can be found in Lubben and Millar (1994); here there is space only to state the main findings.

As we might expect, students' understanding of thermal processes influenced their actions. Those who had a secure mental model of heat flowing from high to low temperature, through a 'barrier' (of insulating material), were able to carry out more coherent and purposeful investigations and to interpret their results than students with a view of materials as 'intrinsically' hot or cold, or with a rather hazy and insecure mental model. Differences were apparent in the inclusion, or not, of an initial temperature measurement, the period over which temperature change was measured and the perceived importance of the type of insulation material. Specific manipulative skills such as knowing how to use and take readings from a thermometer also, as one might expect, made a significant difference to actions. All of these are domain-specific understandings, useful in carrying out investigations of thermal processes but not applicable in other science domains.

For some students, the conceptual understanding they drew upon in relation to this particular thermal process seemed to be embedded in their recall of a similar investigation — of rates of cooling and the drawing of cooling curves. Several groups based their experimental designs closely on this previously encountered investigation, taking readings at fixed, and unnecessarily short, intervals and drawing graphs. This type of memory element is very clearly investigation-specific and would only be accessed in response to a task with clearly perceived similarities to a previously encountered task.

Two other types of understandings, however, seemed less clearly task- or domain-specific. First, we found that many students, in responding to this and other tasks, re-interpreted the given task and carried out what was, in effect, a quite different task. The forms which this re-interpretation took fell into a number of categories. Some students seemed to want to construct a physical model, in this case, of a coolbag, and to show that it worked. Others did not focus on making their container look like a coolbag but still wanted simply to produce the desired effect — keeping water cool for as long as possible. (Some even thought that a good coolbag would make the temperature of the water fall further.) We described these students as adopting a *modelling frame* for their investigation. Some students re-interpreted the task as varying parameters (the insulating material or its thickness) in order to optimize the performance of their model coolbag. This approach does involve changing one or more variables and observing the outcome, but the emphasis is on improvement and optimization, and not on exploring a relationship between two factors, as the given task asked. Students who re-interpreted the task in this way we described as adopting an *engineering frame*. Those who interpreted the task as we intended, and compared different thicknesses of insulation, looking for a trend in performance with thickness, or a relationship between thickness and rate of temperature change, we saw as adopting a *scientific frame*. In using the terms engineering and scientific to label these frames, we are drawing on parallels between our findings and those reported by Schauble *et al.* (1991) who saw a similar distinction in students' responses to investigations in computer-based microworlds, and on the laboratory bench. The understandings which underpin students' frame choices may be generic, though the engineering/scientific distinction only applies to investigations of relationships between variables.

Finally, and perhaps most importantly, we found that the quality of investigation performance of many students was strongly influenced by their understanding of the nature of empirical evidence. A poor understanding of how to collect reliable data, and of how to evaluate the reliability of the data collected, seemed to stand in the way of satisfactory performance for many groups of students. Many groups collected sketchy data, of low accuracy and uncertain reliability, yet were prepared to draw firm conclusions from this. Across the range of investigation tasks we used, we observed groups of students:

- taking only one measurement of a quantity when it would have been easy to repeat it several times to check (and to estimate) its reliability;
- repeating a measurement only if some very obvious mistake had occurred in an earlier reading;
- making a small change in a factor, rather than a larger change which might produce a more clearly observable difference in outcome;
- attributing significance to small differences between single measurements made under different conditions;
- considering that the quality of data (or of observed patterns in data) can only be assessed by checking against an external authority (the teacher, or a book), but not 'from inside', using the collected data alone.

All practical investigation in science involves making observations or measurements. The reliability, or trustworthiness, of this empirical data is always an issue. An understanding of empirical evidence — of the procedures which should be followed in order to collect reliable data, and why these are necessary; and of ways in which the quality of collected data can be evaluated from the data set itself — is a generic aspect of procedural understanding.

The types of understanding, then, which we see as influencing students' performance of investigation tasks are summarized in Figure 15.3.

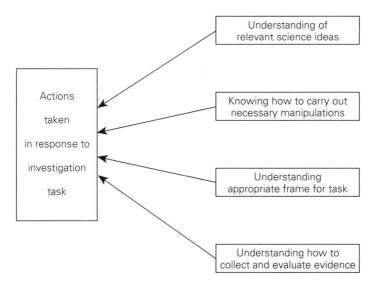

Figure 15.3: Types of understanding which underpin observed performance on investigation tasks

PACKS Phase 2: Probing Understandings of Empirical Evidence

Because of its crucial role in investigation performance, we decided, in a second phase of the PACKS project, to explore students' understandings of empirical evidence further, using a written survey instrument. This consisted of a set of diagnostic questions, probing students' ideas about measurement and evidence. Questions looked at students' ideas about whether to repeat measurements, and about what to do with the data generated, about how to deal with apparently anomalous data, and about judging the reliability of data from the spread of values in a set of repeated measurements. The results of a survey involving over 1000 pupils are reported and discussed by Lubben and Millar (forthcoming). Among the more interesting findings are that:

- significant numbers of 11-year-olds (about 35 per cent) think that the purpose of repeating measurements is to obtain two identical results, which can then be taken as the 'true value'. Only 18 per cent of 11-year-olds see repeating as a means of dealing with the inevitable uncertainty in measurements, though this rises to 50 per cent by age 14 and reaches 70 per cent at age 16.
- Almost no 11-year-olds identified and rejected a clearly anomalous value in a set of five readings. By age 16 only 13 per cent rejected the anomalous value and averaged the remaining four.
- When asked to consider two sets of five measurements with the same average, one widely spread and the other much more narrowly, only 23 per cent of 11-year-olds thought the more narrowly spread set was more 'trustworthy', whilst 42 per cent thought both were equally 'trustworthy' because they had the same average. By age 16, 39 per cent thought the narrowly spread set was more 'trustworthy', but 40 per cent still considered both to be equally 'trustworthy'.

On the basis of the survey findings, we have proposed (Lubben and Millar, forthcoming) a hierarchy of levels of students' understanding of how to collect, deal with, and evaluate the reliability of empirical evidence.

In general terms, the survey responses corroborate our view from the main phase of the PACKS work that an understanding of empirical evidence is a critical aspect of understanding the scientific approach to enquiry. Little work has been carried out on this aspect of science understanding, and that which has been reported deals with much more advanced students (Séré *et al.*, 1993). We would see this as an important area for further research, to improve our understanding of students' responses to practical tasks in science and hence our ability to plan improved forms of instruction.

Note

1   The Procedural and Conceptual Knowledge in Science (PACKS) project was supported, between October 1991 and March 1994, by Economic and Social Research Council

(ESRC) grant no. L208 25 2008 as part of the research programme Innovation and Change in Education: the Quality of Teaching and Learning. It was based at the Universities of York and Durham. We acknowledge the contributions of Richard Gott and Sandra Duggan, our co-researchers in the PACKS project, to discussions about the issues developed in this paper.

References

ASSESSMENT OF PERFORMANCE UNIT (APU) (1984a) *Science Assessment Framework Age 13 and 15: Science Report for Teachers (2)*, London, Department of Education and Science/Welsh Office/Department of Education for Northern Ireland.

ASSESSMENT OF PERFORMANCE UNIT (APU) (1984b) *Science Assessment Framework Age 11: Science Report for Teachers (4)*, London, Department of Education and Science/Welsh Office/Department of Education for Northern Ireland.

DEPARTMENT FOR EDUCATION/WELSH OFFICE (DFE/WO) (1995) *Science in the National Curriculum (1995)*, London, HMSO.

LUBBEN, F. and MILLAR, R. (1994) *Children's Responses to the 'Cool drink' task and Probes*, PACKS Research Paper 1, Department of Educational Studies, University of York/School of Education, University of Durham.

LUBBEN, F. and MILLAR, R. (forthcoming) 'Children's ideas about the reliability of experimental data', *International Journal of Science Education* (in press).

MILLAR, R., LUBBEN, F., GOTT, R. and DUGGAN, S. (1994) 'Investigating in the school science laboratory: Conceptual and procedural knowledge and their influence on performance', *Research Papers in Education*, **9**, 2, pp. 207–48.

SCHAUBLE, L., KLOPFER, L.E. and RAGHAVAN, K. (1991) 'Students' transition from an engineering model to a scientific model of experimentation', *Journal of Research in Science Teaching*, **28**, 9, pp. 859–82.

SÉRÉ, M.-G., JOURNEAUX, R. and LARCHER, C. (1993) 'Learning the statistical analysis of measurement errors', *International Journal of Science Education*, **15**, 4, pp. 427–38.

WOOLNOUGH, B.E. (1989) 'Towards a holistic view of processes in science education', in WELLINGTON, J. (Ed) *Skills and Processes in Science Education: A Critical Analysis*, London, Routledge, pp. 115–34.

16   What Changes in Conceptual Change?: From Ideas to Theories[1]

Montserrat Benlloch and Juan Ignacio Pozo

Abstract

There are numerous studies that have demonstrated the persistence of alternative conceptions. There is a tendency, in some studies, for these conceptions to be implicitly considered
as isolated units of knowledge. An idea or conception is considered to be persistent when it
appears in subjects of different ages and levels of instruction but, although the idea persists,
its meaning and relevance may change with the acquisition of new ideas. In fact some cognitive theorists conceive of students' representations as implicit theories, in which change
should be observed, not so much in each of the ideas or units making up the theory, as in
the organization adopted by these ideas. The same idea may appear persistently within
different implicit theories. In order to confirm this hypothesis, fifty-one children, aged from
9 to 14, and divided into four groups, were confronted with a simple task in which a balloon
attached to a bottle was inflated when the bottle was heated. We analysed separately each one
of the ideas expressed by the subjects and through a Multiple Correspondence Analysis, the
relationships between these ideas making up the different implicit theories. We thus identified fifty-four different ideas organized around five different implicit theories. Results show
that there were no differences between age groups in their ideas, when these were analysed
separately. However, the theories change with age and instruction. Thus, whilst students'
conceptions, taken as isolated ideas, persist, the relationships between these ideas change
substantially, giving rise to new forms of organization. The persistence of a conception does
not necessarily imply that no conceptual change has occurred.

Theoretical Framework

According to the research on children's alternative ideas (or misconceptions) about
scientific phenomena, these ideas are not only ubiquitous and pervasive, but also
persistent. The same ideas have been identified in subjects of different ages and
with different science instruction. For instance, several studies (e.g., Seré, 1985;
Stavy, 1988) have shown that students' ideas about gases, and more specifically
about air and its properties, are relatively persistent, and contradictory to those
maintained by scientific theories.

Research has mainly concentrated on those conceptions that are most persistent and resistant to change, and show that if change is achieved, it tends to be
fleeting and rarely generalized. Although the representational nature of alternative
conceptions tends not to be explicit in most studies, it can be observed implicitly in the methodology employed. The diverse conceptions held by students are

frequently analysed as isolated ideas, but without explicitly studying the relationships between a conception or idea and other ideas maintained by the same subject in this or other contexts. Thus, in many studies these conceptions are understood, or studied, as isolated units of knowledge, as diSessa (1993) maintains. Alternative conceptions studied as isolated ideas scarcely change. Indeed, even studies that attempt to promote conceptual change barely manage to move really deep-rooted alternative conceptions: at best, what is achieved is that students' initial ideas become mixed with the new ideas, but are not abandoned (Duit, 1994).

However, several cognitive theorists suggest that children have alternative theories, with some level of internal organization, and that these theories undergo some radical changes in their development. According to this approach, students have theories about different domains, and not only some dispersed ideas (e.g., Carey, 1985; Chi, Slotta and de Leeuw, 1994; Pozo and Carretero, 1992; Pozo *et al.*, 1992; Vosniadu, 1994). For these authors, children's theories are submitted to conceptual changes as a consequence of cognitive development and/or instruction. They maintain that students' alternative conceptions do indeed constitute theories, implicit in nature and based on certain epistemological suppositions and developed through basic cognitive processes. Thus, intuitive models or implicit theories would differ from scientific models not only in their content, but also in the processes upon which they are based and in the way they are organized. Scientific theories would be the product of theoretical reflection, of research, and of the links made between theories and between theories and data, and would be guided by criteria of explanational coherence. Implicit theories, on the other hand, would be based on the establishment of regularities in the environment, they would result from the use of simplified heuristic rules, they would be more unstable and dependent upon the context, and would be less concerned with explanational coherence than with predictive efficacy (e.g., Pozo *et al.*, 1992).

From this perspective it is assumed that these theories change with age and instruction. However, they would not be attributable to a general cognitive development, but rather to an increase in knowledge in each specific scientific domain, a consequence of school instruction: strong conceptual change would not involve the mere substitution of some concepts for others, but a profound theoretical restructuring (Carey, 1985; Vosniadu, 1994).

For instance, according to Chi's theory of conceptual change, the learning of scientific theories may involve a radical conceptual change, as it requires the student to employ new and different ontological categories (Chi, Slotta and de Leeuw, 1994). From Chi's perspective, a strong conceptual change occurs when a concept is assigned to a new ontological category. Learning that heat is a process and not a substance — removing it from its category and placing it in a new one — involves a strong ontological reassignation. At a more elemental level, the difficulties encountered by both children and adults in understanding the characteristics and nature of air would result from the fact that air, in intuitive theories based on the everyday investigation of the environment, belongs to the 'matter' category but has many features characteristic of non-material entities: air cannot be seen or held, and has no defined shape, etc.

Thus, in students' theories about the nature of air and other gases, air is not conceived as a material entity with the same properties as those ontologically attributed to matter (weight, shape, size . . .). Conceptual change thus involves a theoretical restructuring, more than a change in specific ideas about gases, so that students must accept that air is an entity with all the properties of a material, but must later understand that heat is a process, and not a substance or material entity.

Objectives and Hypotheses

The principal objective of this study is to demonstrate the relevance of analysing students' conceptions about a specific phenomenon (in this case the expansion of air with a rise of temperature) not only as isolated *ideas* or units of knowledge, but also as part of certain implicit conceptual structures, which we call 'implicit theories', that can be deduced from the relations existing between these ideas. Whilst many of these ideas, taken separately, may persist with age and instruction, their relationship to other ideas may change in the framework of these implicit theories. Thus, different theories could share ideas or units of knowledge, but within different conceptual organizations.

In this study, we assume a theoretical and methodological approach that allows us to differentiate between the ideas held by students (or each one of their verbal expressions with a semantic content regarding the expansion of air) and the theories in which they are embedded, analysed on the basis of the relationships implicitly existing between these ideas. We study children's ideas about air and its properties, in the context of their implicit theories about the expansion of air with temperature increase.

Methodology

Subjects

Subjects were fifty-one pupils from the same state school, with ages ranging from 9.7 to 14.4 years. They were divided into four groups, corresponding to different educational levels: fifth grade (9.7–11.7 yrs), sixth grade (11.7–12.4 yrs), seventh grade (12.6–13.6 yrs) and eighth grade (13.7–14.4 yrs).

Task and Procedure

Each subject was asked to observe individually a situation of the expansion of air and to undergo a semi-structured interview carried out using a clinical methodology. A transparent and thermo-resistant vessel, containing only air and sealed with a balloon, was placed on a gas ring. The air expanded and the balloon inflated as shown in Figure 16.1 below.

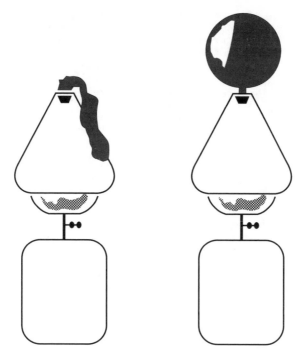

Figure 16.1: Apparatus used in the study

When the balloon had inflated to a volume not greater than that of the vessel, the subject was asked about the reasons for the increase in size — Why does the balloon inflate? — and about the presence of the air inside the vessel and the balloon: What is there in the bottle now? What is there in the balloon? Is there the same quantity of air in the whole bottle? Is there the same quantity of air in the upper part of the bottle as in the rest of it? And in the lower part of the bottle?

Finally, the subject was asked about the consistency or variation of the quantity, weight and volume of the air before and after the lighting of the gas ring. Pointing to the vessel and the balloon, he or she was asked: Inside the whole of the apparatus, is there now more, less or the same amount of air than there was before the heating process? Does the air inside the vessel now weigh more, less or the same as it did before being heated? Does the air now occupy the same amount of space as it did before being heated, or does it occupy more or less? Subjects were asked to give reasons for their answers.

Data Analysis and Results

To facilitate understanding of the analyses carried out, we shall present the analysis criteria and the results obtained for each aspect analysed. These aspects are: the ideas about air and heat; the ideas about the position of the air; the ideas about the

Table 16.1: Examples of scientific and non-scientific ideas about air and heat

Categories 1	**The heat produces a change in the properties of the receptacle or in its inner content.**
Examples of ideas	The heat causes the balloon to inflate/go down, stretch/retract, increase/decrease. The hot air makes the balloon inflate, stretch or increase. The heat warms the air. *The heat warms the molecule or atom particles. *The heat changes the weight of the air. The air fills more/less space. It becomes larger or smaller. The air expands/contracts. The air spreads in. *The air loses/increases strength. *Some component of the air has/loses/increases strength. A pressure is present or made present on the air.
Categories 2	**The changes occurring in the inner receptacle are due to an increase of matter which comes from the outside.**
Examples of ideas	*The bottle or the balloon is pierced with holes. *The heat transmits matter as heat. *Matter is transmitted, enters or escapes as air. *Matter is transmitted, enters or escapes as: steam, oxygen, gases, substances. *Matter is transmitted, enters or escapes as molecules.
Categories 3	**The changes occurring in the receptacle are due to an increase of the matter which comes from the inside.**
Examples of ideas	*More air is formed, converted, destroyed, composed or appears. *Steam, smoke, gas or a substance is formed or destroyed, composed or appears. *Some particles form, break down or multiply.
Categories 4	**The changes occurring in the inner receptacle are due to a change in the position of the air and/or the heat.**
Examples of ideas	*The heat changes its position. *The heat changes the position of the air. *The heat changes the position of some of the components of the air. The heat changes the position of the particles or molecules.
Categories 5	**The changes of temperature causes a change in the nature of movement of the air.**
Examples of ideas	The air moves around with more or less force. The air moves around more/less rapidly. The particles move around more/less rapidly. The air collides with more/less force. Some component collides with more/less force. The particles collide with more/less force. The air expands/contracts, moves around more. The particles expand/separate more in the space.
Categories 6	**Statements that refer to logical reasons: Identity, compensation, inversion.**
Examples of ideas	The receptacle is a closed system. Nothing goes in or out. Principle of conservation of matter. Nothing happens. The present effects balance the initial situation.

Asterisks (*) mark the non-scientific ideas. The rest is correspondent with scientific ideas.

conservation of quantity, weight and volume; and finally the implicit theories held about the inflation of the balloon.

Ideas about Air and Heat

Interviews were taped and codified, taking as the unit of analysis the sentences produced by the subjects. Each sentence was codified according to the semantic content of the verbs used for talking about the air and the change of temperature (Hickman, 1990). Thus, we obtained in total fifty-four different ideas about air and heat contained in the students' sentences. These ideas, which could be either scientifically accepted or alternative, were then categorized, according to their content, in six main categories (see Table 16.1).

We computed the frequency of scientifically accepted and alternative ideas for each age group. According to a Kruskal-Wallis statistical test there were no differences between groups in their ideas about air and heat ($H = 1.38$; $p = 0.09$). These ideas are persistent and do not change significantly with age and instruction.

Ideas about the Position of the Air

These ideas were categorized in the following three main categories:

1 Only balloon: All the air is in the balloon; the bottle is empty.
2 Bottle empty in part: The balloon is full of air and a part of the bottle is also occupied by air.
3 Both bottle and balloon with air: The entire apparatus, both the balloon and the bottle, is occupied by air.

After computing the frequency of each answer category by group (see Figure 16.2), results did not show differences between groups in their conceptions about the position of the air when the balloon is inflated ($H = 1.38$; $p = 0.71$). In this case ideas are also persistent.

Ideas about the Conservation of Quantity of Matter, Weight and Volume

We made a similar analysis for the ideas about quantity of matter, weight and volume before and after the change. We identified five answer categories:

1 All Change (change without differentiation of properties): The same change in the three properties: quantity, weight and volume increase or decrease together.
2 No change (conservation without differentiation): Quantity, weight and volume do not change.

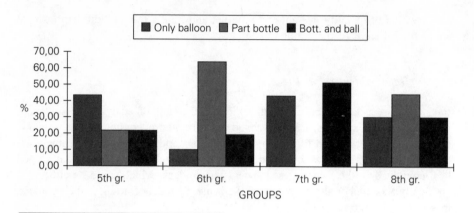

1. Only the balloon contains air
2. One part of the bottle is empty
3. The balloon and the bottle contain air

Figure 16.2:   Ideas about the position of the air

3   Change with covariation: Two properties change but the other is con-
    served. Changes are always in the same direction.
4   Change with compensation: Volume increases, but quantity and weight
    decrease.
5   Change with differentiation: Volume increases, but quantity and weight do
    not change.

Having classified all the answers obtained according to these categories (see
Figure 16.3), we did not find significant differences between groups in their ideas
about conservation (H = 6.46; p = 0.09). Once again, ideas are persistent across age
groups.

Implicit Theories

Once categorized, all the ideas (about air and heat, position of the air and conservation of properties) were submitted to a Multiple Correspondence Analysis (MCA),
in order to find patterns of implicit organization for them, or the implicit theories
about air and its properties maintained by children. MCA is a factorial analysis
applied to discrete or categorial variables. It is a descriptive analysis that operates
with active variables, which define the space, and supplementary variables, which
are projected upon it. MCA reduces a multidimensional space to factorial planes
(see Figure 16.4). The classifications obtained demonstrate the relations of dependence and independence between variables (Greenacre, 1984).

The MCA described five different implicit theories. Figure 16.4 represents
graphically the distribution of different ideas between theories. As it can be seen,

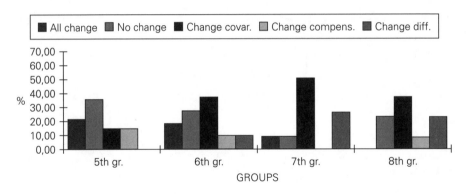

1. The three properties change.
2. No properties change.
3. Changes with covariation.
4. Changes with compensation.
5. Changes with differentiation.

Figure 16.3: Ideas about the quantification of properties

certain ideas are common to more than one theory, the pattern of relations between ideas being that which determines the theory held by a subject. The five theories identified can be summarized as follows:

Theory 1 The inflation of the balloon is caused by an increase in the quantity of air from outside of the apparatus. The system is open, and there is an increase of material from outside.

Theory 2 The inflation of the balloon is caused by a change in position of the air which, on being heated, moves up from the bottle into the balloon. The system is closed, and there is neither an increase nor a decrease in the amount of material, nor of any of the three properties.

Theory 3 The inflation of the balloon is caused by a process of transmutation that affects the air inside the apparatus. The system is closed, and there is an interior increase of material.

Theory 4 The inflation is caused by a change in some property, such as weight and/or volume of the air contained in the interior of the apparatus. The system is closed and some property increases, excluding that of quantity.

Theory 5 The inflation of the balloon is the result of an increase in the volume of air inside the apparatus. It is a closed system with increase in volume of air exclusively.

We computed the number of subjects using each theory (see Figure 16.5) and performed a Kruskal-Wallis test with the data obtained. Results showed a significant effect for the group variable ($H = 11.07$, $p = 0.01$). When differences between pairs of groups were analysed with a Chi-square test, we found differences between

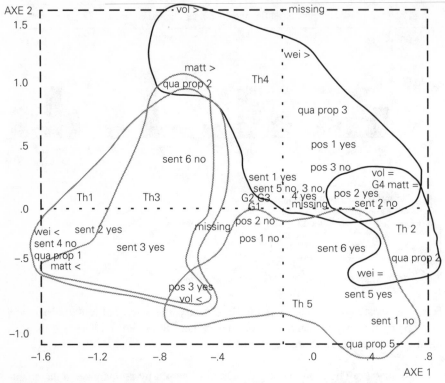

TH = theory; VOL = volume; MATT = quantity of air; WEI = weight; QUA PROP = quantity of properties; SENT = sentences about scientific and non-scientific ideas

The cluster groups the different ideas giving rise to the implicit theories. Thus, for example, theory 2 brings together ideas about the constancy of the properties of air, regardless of temperature increase (QUA PRO 2; WEI =; MATT =; VOL =) incorporates no sentence related to the entry of external air (SENT 2 — NO), and combines logical reasons (SENT 6 — YES) justifying the absence of changes.

Figure 16.4: Graphical representation of theories according MCA

fifth graders and seventh graders (z = 2.90; p < 0.001), and fifth graders and eighth graders (z = 2.83; p < 0.001), demonstrating a gradual change of the implicit theories held, which gives rise to their restructuring with age and instruction.

Conclusion

The results obtained in this study support our hypothesis that changes in scientific knowledge take place within the framework of the implicit theories held by students. In fact, there were no differences in the individual ideas held by the different groups about air and heat. Neither were there differences in the ideas about the position of the air and the conservation of quantity, weight and volume. Thus, we

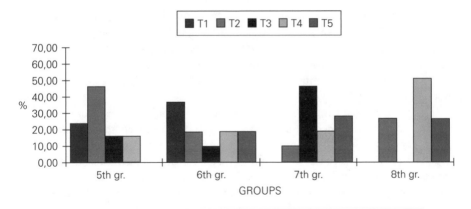

1. The system is open. There is an increase of matter coming from outside.
2. The system is closed. None of the properties of the air changes.
3. The system is closed but there is an increase of matter coming from inside.
4. The system is closed and some properties increase but never the quantity of air.
5. The system is closed and only the volume of the air increases.

Figure 16.5: Theories about expansion of the air

can conclude, in line with many studies about alternative conceptions, that students' ideas in this domain are really persistent.

However, through a qualitative analysis, based on a Multiple Correspondence Analysis, we have also been able to identify different answer patterns, or implicit theories, representing different ways of organizing these individual ideas. The MCA detected five types of theory about the expansion of the air due to a change in temperature. These theories improve significantly with age. Each theory presents a different version of the concept of air. Ideas about air and heat do not change significantly with age. Some are expressed in a similar way in different theories, but change their meaning. Beyond their apparent similarity, the statements 'the air rises because the heat pushes it', or 'the air rises because it has got hot and lost weight' involved different reasoning. The first causal relation contains a concept of heat as a force; the second presents the heat as being responsible for a change in temperature that indirectly modifies the position of the air. In each case the affirmation that 'the air rises' refers to a unidirectional change in the position of the air, implicitly conferring upon it a solidity that is not a quality of fluids.

Thus, different theories share some common ideas. Although ideas persist, their meaning changes as the conceptual structure (or theory) in which they are embedded changes. Conceptual development only can be traced when we consider concepts within theoretical structures. Indeed, while we did not find significant changes in ideas, if we consider them separately, there were significant differences with regard to the theories held by different age groups. Theories 4 (change of properties of the air) and 5 (expansion of the air), closer to the scientifically accepted explanations, were more frequent in older subjects. As predicted by Chi's theory of conceptual change, children's theories about air and heat are subject to diverse ontological changes. The principal difficulty for students, in the age groups we studied, is to accept that air is a material entity with all the properties of a substance; this will later make possible a further and more difficult ontological change, whereby heat will be conceived, not as a substance, but as a process of interaction, which will bring the students closer to the scientific theories in this domain (Benlloch, 1993).

In sum, we can conclude that students have implicit theories about the nature of air, and that these theories do change, even though their component ideas seem to be persistent. Just like particles, ideas are not independent components added together to form a broader and coherent system, but elements that interact and combine according to complex patterns of dynamic relationships. Ideas, then, work in some ways like particles, whose interaction occasionally changes the appearance of reality, forming new substances or, in this case, more advanced theories.

The study of conceptual change should not be directed towards the analysis of ideas or elements, one by one, but towards dynamic relationships, sometimes implicit, between those elements. As is the case with recent cognitive theories of mental models or parallel processing, it is the activation patterns and the connections between the elements that form the structure of implicit theories held by students at any given moment. If we want to understand the nature of conceptual change, we must concentrate our efforts in the study, not of the elemental particles of knowledge, but of the way they relate to one another. Just like the students trying to understand the nature of matter, we must realize that knowledge is not a static substance, but a process of interaction in continuous movement.

Note

1   This paper is based on the doctoral dissertation of Montserrat Benlloch supervised by Juan Ignacio Pozo. We are grateful for the methodological assistance provided by Francesc Martinez and Roser Riu. We would also like to thank David Weston, who helped us to prepare the English version.

References

BENLLOCH, M. (1993) 'La génesis de las ideas sobre la composición de la materia' (The development of ideas about the composition of matter), Unpublished Doctoral Thesis, University of Barcelona, Faculty of Psychology.

CAREY, S. (1985) *Conceptual Change in Childhood*, Cambridge, MA, MIT Press.

CHI, M.T.H., SLOTTA, J.D. and de LEEUW, N. (1994) 'From things to processes: A theory of conceptual change for learning science concepts', *Learning and Instruction*, **4**, 1, pp. 27–43.

DISESSA, A. (1993) 'Towards an epistemology of physics', *Cognition and Instruction*, **10**, 2–3, pp. 105–225.

DUIT, R. (1994, September) 'Conceptual change approaches in science education', Paper presented at the Symposium on Conceptual Change, Jena, Germany.

GREENACRE, M.J. (1984) *Theory and Applications of Correspondence Analysis*, New York, Academic Press.

HICKMAN, M. (1990) 'Coding manual for narrative utterances', Unpublished research report, Nijmegen, Max Planck Institute for Psycholinguistic.

POZO, J.I. and CARRETERO, M. (1992) 'Causal theories and reasoning strategies by experts and novices in Mechanics', in DEMETRIOU, A., SHAYER, M. and EFKLIDES, A. (Eds) *Neopiagetian Theories of Cognitive Development: Implications and Applications*, London, Routledge and Kegan Paul.

POZO, J.I., PÉREZ ECHEVERRÍA, M.P., SANZ, A. and LIMÓN, M. (1992) 'Las ideas de los alumnos sobre la ciencia como teorías implícitas' (Students' ideas about science as Implicit Theories), *Infancia y Aprendizaje*, **57**, pp. 3–22.

SERÉ, M. (1985) 'The gaseous state', in DRIVER, R., GUESNE, E. and TIBERGHIEN, A. (Eds) *Children's Ideas in Science*, Milton Keynes, Open University Press.

STAVY, R. (1988) 'Children's conception of gas', *International Journal Science Education*, **10**, 5, pp. 533–60.

THAGARD, P.R. (1992) *Conceptual Revolutions*, Cambridge, MA, Cambridge University Press.

VOSNIADU, S. (1994) 'Capturing and modelling the process of conceptual change', *Learning and Instruction*, **4**, 1, pp. 45–69.

17 Development of Pupils' Ideas of the Particulate Nature of Matter: Long-term Research Project

Michael Lichtfeldt

Abstract

In the long-term research project 'pathways to an atom-idea', pupils in a Berlin Gymnasium are observed from grade 7 through grade 11. The development of pupils' cognitive structures and understanding in the topics in 'the microworld' is described. Using different research methods (e.g., analysis of outcomes from all pupils out of four parallel courses, observing of single pupils in detail, video-recordings in lessons and interviews outside school) the following problems are addressed: What influences do physics and chemistry lessons have on science thinking and learning? What is the influence of pupils' alternative frameworks within the cognitive field of particles and atoms?

Design of the Research Project 'Pathways to the Atom-idea'

A continuing long-term research project designed to address this problem has been under way since 1993. Beginning at grade 7 in one Berlin Gymnasium (grammar school), pupils are being observed through to grade 11. In grade 7 the pupils do not study physics or chemistry, both subjects starting in grade 8. From grades 8 to 11, the pupils cover themes such as the particulate structure of matter (solids, liquids and gases), electrons in metals as carriers of electric current, radioactivity and the use of atomic models for understanding compounds in chemistry.

The design for this long term study had two main lines:

1 A sample of twenty-four pupils from four different courses at the same level were observed in detail. During this time, they answered questionnaires, they were video-recorded during the lessons in which these topics were taught, and they came to the university to be interviewed and to perform experiments. From the resulting data, frameworks of conceptions (mappings of each pupil's ideas and their meanings) were constructed. Step by step series of different frameworks can show the development of ideas as a model for an individual learning process.

2 Every half year all the pupils from all four different courses (about 100 pupils in all) completed the same questionnaire and were video-recorded during their lessons. These less detailed data provided a baseline of pupils' main ideas and have been compared with the results of international

research studies which are all focused on single aspects of the particulate structure of matter and atoms. A comparison with the historical development and growth of what is known about the atom's structure is also in progress. The complete results are intended to form the basis for the construction of a new curriculum for the teaching and learning of the particulate structure of matter and atoms in levels 8 to 10. The whole research programme is summarized in Figure 17.1.

A pilot study was done in January 1993 at a Gymnasium in Berlin. At the same time a questionnaire was given to all pupils (N = 334) in the lower level (grades 7 to 10). The questions were different in type, some examples of which are shown here:

Page 1 thirteen sentences with the choice 'yes', 'no', 'I don't know':
- e.g.,: There is nothing between the particles of gas

Page 2 open questions:
- e.g.,: What does an atom look like?
- e.g.,: Write down other words which you associate with the given word 'atom'
- e.g.,: Tell me the situation and date when you first heard the word 'atom'

Page 3 drawings to underline the written words
- What does an atom look like?
- Make a description!
- Draw a picture!

Page 4 and 5 twenty-four sentences with the choice 'yes', 'no', 'I don't know, it could be':
- e.g.,: If an atom has a positive charge it has only protons and neutrons
- (including four sentences about pupils' opinions about physics lessons and physics teaching)

Page 6 logical connections of thirteen given words about the 'atom-world'
- connect these words to sentences!

The results of the pilot-study showed that pupils used different forms of language: either 'everyday-language' or 'subject-language'. The main everyday-language did not depend on the time spent studying physics or chemistry. Pupils tried to integrate the new ideas from school lessons into their private networks of ideas with the school learning sometimes later obscured by the old networks of ideas. Figure 17.2 shows the thinking world of a young boy in grade 7. His drawing was a mixture of different models given by science textbooks.

This is an example to show the difficulties teachers face in their lessons about the particulate nature of matter. The boy already seemed to know a lot about atoms, molecules, shells and so on. The question to be asked is how did he react to teachers' presentations in school? Was he able to integrate the new information about atoms or did he select only some details for integration into his picture?

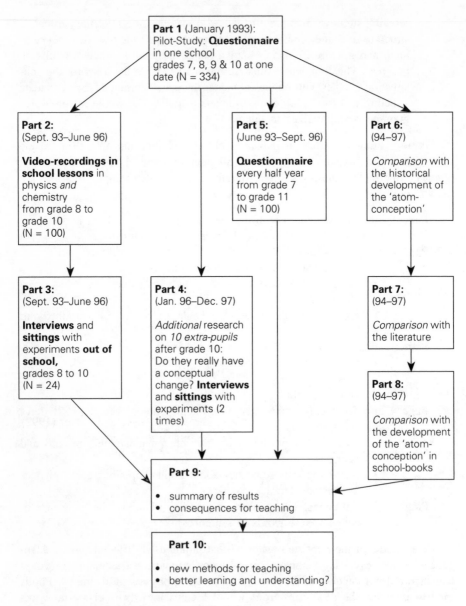

Figure 17.1: Research programme 'pathways to the atom-idea'

Theoretical Frame of the Long-term Research

On the basis of pupils' alternative frameworks the observation of learning pathways covers a wide field involving research topics in psychology of learning, linguistics and neuronal research. To handle this huge area, it was necessary to concentrate research on single goals. Therefore the research reported here does not address

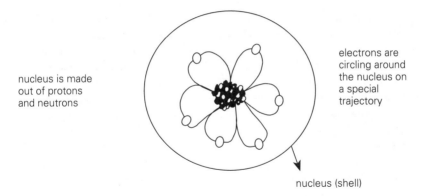

nucleus is made
out of protons
and neutrons

electrons are
circling around
the nucleus on
a special
trajectory

nucleus (shell)

Figure 17.2:   Boy, 13 years old, grade 7 (without having physics or chemistry in school)

learning processes on the basis of neuronal structures (Edelman, 1987; Calvin, 1990).

Growth of knowledge depends on pupils' prior knowledge, their alternative every day frameworks. It affects learning in physics and chemistry either by pushing knowledge forward or acting like a barrier (Reinmann-Rothmeier and Mandl 1995). Carey (1986) describes such processes as 'conceptual-change-models' whereas Strike and Posner (1985) and Posner *et al.* (1982) call them 'accommodation'. This could be seen as a reference to Piaget (1985). Strike and Posner mark a conceptual change as a break between pre-lesson-knowledge and science-knowledge, whereas Stavy (1991) declares that change of knowledge can only result from conceptual growth. This is the constructivist basis of the work of the 'Children's Learning in Science Project' Driver (1989), Scott (1992) and Scott, Asoko and Driver (1992).

There are other relevant influences on learning: motivation is necessary and science ideas have to be fruitful and plausible (Posner *et al.*, 1982). Additionally, pupils' cognitive structure of their science world has to have a dimension of truth (Schaefer, 1990). If pupils are to accept the science view they must recognize that their former knowledge is not wholly correct. Thus it is possible that even where pupils are well motivated they do not change their concepts and often only pick out bits of knowledge to integrate into their previous understanding (Stavy, 1991).

There is a chance of conceptual change if pupils see the problem to solve in science lessons as like their problems in everyday life (Resnick, 1987, pp. 13–20). On the other hand school knowledge can be likened to 'dead knowledge' in pupils' minds and they do not tend to use it in everyday life (Mandl, Gruber and Renkl, 1993a). It is suggested that science learning is always a conclusion of cognitive, social and cultural aspects from outside and within school (Shuell, 1993). Otherwise knowledge exists on different 'islands' (Mestre, 1991) which are also called 'compartments of knowledge' (Mandl, Gruber and Renkl, 1993b) or 'scripts' (Gardner, 1991). Teaching in school should aim to develop pupils to become 'skilled problem solvers' (Mestre, 1991) able to link different 'scripts'.

The growth of knowledge is based on the language which is used by pupils. Like Vygotsky (1962), linguistic researchers distinguish between a language of science and a language of everyday life (Ingendahl, 1975). In their early childhood children are able to learn the different terms in use for the same concept, whereas in school they have to work with a single, precisely defined term for an idea or concept. Seiler (1985) describes in his theory about 'growth of ideas and their meanings' the difference between everyday- and school-knowledge. It is useful to define levels in learning: intuitive learning out of school which will be represented by the everyday-language and school-learning which is represented by the science-language. In Seiler's theory the 'growth of ideas and their meanings' depends on an existence of a meta-learning strategy. For a learner, this makes the process of learning and understanding visible, plausible and open to new interpretations. Out of this theory, it is possible to define consequences for learning of the use of language (Wagenschein, 1976). It is not enough to learn science simply by doing. Learners must know what they are doing. School learning must become a conscious process.

Starting from this viewpoint, we used a model of learning involving a scheme of three learning levels. This idea corresponds with Gardner (1991) who postulates that besides the intuitive-learner and the school-learner, there is the learning-expert who is able to link the intuitive learning level with the school learning level. In the phases of expert learning pupils are able to discover new ideas in science.

Results on Alternative Frameworks in Particle and Atom Theory, Given by Other Research Projects

In the lower level of Gymnasium (grade 8 through 10), pupils get a 'picture' of the microscopic world of the particle. Particles are presented as continuous bodies explained in a discontinuous way. Particles do not have the attributes (e.g., temperature, colour, etc.) of the whole body, but often pupils do not accept this. Additionally, teachers often do not differentiate between particles and atoms or molecules, so that pupils are unable to differentiate between different models. Atoms become particles with the same attributes as the whole body. Research projects (see, for example, Pfundt, 1982a; Novick and Nussbaum, 1978 and 1981; Brook, Briggs and Driver, 1984; Wightman, Green and Scott 1986; Johnston and Driver, 1987; Lijnse *et al.*, 1990) have shown that pupils' learning pathways starting from the point of a continuous body to the microscopic world of particles are very complicated. Pupils do not accept the particle model in the way intended by scientists, not seeing the function of the model to act as a bridge between explanation of phenomena and experiments (Duit, 1992). The real character of science models does not exist in their minds.

Mostly, research projects are short-term, or they are concentrated on special questions and topics from the atom- and particle-theory (Pfundt, 1982a; Duit, 1994). Results from these studies influence teachers' teaching methods since it is suggested the teaching and learning process will become more straightforward. However, there is no evidence to suggest that long-term retention by children of their learning has happened.

Recent projects or re-interpretations of classical or empirical studies show a strong linkage between models and everyday experience in pupils' knowledge and comprehension (Todtenhaupt, 1995; Löffler, 1992 and 1994). The learned science explanation is usually absent from pupils' responses. This has led us to question the use of models of the atom in teaching at the lower grades of Gymnasium (Buck and Mackensen, 1994; Buck, 1987).

This 'real-world-view' acts as a barrier to the effective learning of science in the upper grades of Gymnasium (grades 11 through 13). In physics and chemistry pupils' old fixed ideas dominate their understanding of modern science. Pupils cannot develop a feeling for such subjects as quantum physics (Fischler and Lichtfeldt, 1992 a and b). They try to merge the new ideas into their old meanings and interpret experimental results from their existing points of view. After the teaching of the accepted science explanations of recent scientific discoveries, pupils fall back into their old macroscopic views of knowledge. What they have learned for written tests and other school examinations could be likened to learning in a foreign language. In most cases, pupils did not see the differences between the scientific point of view and their own, causing their school learning to fade and be forgotten after a comparatively short time (Bethge, 1988; Fischler and Lichtfeldt, 1992 a and b; Lichtfeldt, 1992 a, b and c; Petri and Niedderer 1994 and 1995).

Results of the Long-term Study

The long-term project has been under way for two years enabling a good overview of individual learning and alternative frameworks. This will be demonstrated in what follows with:

- some examples from school lessons;
- answers and drawings from the questionnaire, especially from responses of two girls during the two years; and
- networks constructed in a way to show progression.

Pupils have significant difficulties especially with changes of state, the transformation of bodies from solids to liquids and from liquids to gases. They develop their own models of the particulate structure of matter on the basis of their ideas and they try to hold onto this model even as the teacher presents a physical elicitation. For example, water is a familiar everyday substance to all pupils, but its atomic structure can be difficult for them to understand. Simply watching water boil elicits several questions: Where do the gas bubbles come from when water boils? What kind of material are these bubbles made of? Is it, possibly, that water is a gas made of particles?

The following describes a discussion in a grade 8 physics lesson. The teacher wants to develop a model of the particulate nature of water. He presents a beaker filled with little glass marbles and another beaker filled with water.

T	We should have a look at the water. How could it be constructed?
11	There are a lot of water-molecules inside.
21	Two-thirds are made of hydrogen and one-third is made of *oxygen*.
23	*Oxygen* has to be inside.
30	Why must *oxygen* be inside? In real water in a lake perhaps but not in this one (pointing to the beaker with the marbles).
24	There is *space* inside!
T	Why could it be?
19,24	Oh yes, there is *air* between!
T	The little particles are moving all the time. Why should there be anything between them? (while speaking he is shaking the beaker.)
21	They must have *space* for moving. If there isn't any *oxygen* between they couldn't move.
04	That's like mineral water! There are also little *gas-marbles*. And in real water it must be *oxygen* because fishes have to live there.
22	But in reality you can't see the particles!
32	But they must be inside! The fishes are swimming and breathing!

Pupils' answers are a mixture of everyday- and school-knowledge. The teacher is not able fully to understand pupils' meaning. The model becomes reality for the pupils, the space between the particles is necessary for their understanding that water is composed of oxygen and hydrogen. The particles themselves are identified as hydrogen, the space between them as oxygen. Both kinds of material are seen to be inside the water, but for them the water is synonymous with the particles of hydrogen with all the attributes water has. Pupils are still thinking of water as a continuous material. They were only able to add the new idea of gas-particles in solution or as composing water having seen them portrayed as marbles. These results are quite similar to the results of the CLIS research group. (For a brief description see Scott, 1992)

The teacher presented the model in order to generate better understanding in the pupils. He thought that it could have a bridging-function, but his strategy went wrong: pupils picked up little pieces of knowledge to join them into their ideas and the 'model of the model' became the reality for pupils' discussion. Figure 17.3 shows the interdependence between the 'world-views' which are very different. It thus becomes necessary to construct situations in school-lessons which takes account of this misinterpretation of the models used by teachers by encouraging pupils to discuss their ideas not only of the abstract model of particles, but of the concrete model used to portray it.

In the long-term study, the same questionnaire was given to all participating pupils every six months.

The following examples are the responses of two girl-pupils during the time from June 1993 to January 1995. Comparison of 'written answers' and 'drawings' (see Figures 17.4 and 17.6) shows that different parts of pupils' knowledge is tapped by the different methods used and each can give additional information about pupils' thinking. The use of 'mappings' especially can shed light on how

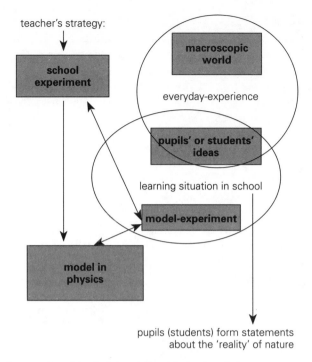

teacher's strategy:

school experiment

macroscopic world

everyday-experience

pupils' or students' ideas

learning situation in school

model-experiment

model in physics

pupils (students) form statements about the 'reality' of nature

Figure 17.3: Cooperation of 'reality', 'model' and 'school-reality'

pupils really think. In this method, pupils have to construct a network of their understanding of meaning using given words (see Figures 17.5 and 17.7). Between these words there is one line for each connection of meaning given after every half-year. The connection itself is divided into:

R represents a relation between general ideas and subideas. Features which form a general idea can be transferred to the subidea.

CM typifies characteristic features of an idea through the use of other ideas.

AM typifies active features which characterize an idea.

ZO illustrates a relation between two ideas without itself being a characteristic feature.

Descriptions of an Atom by Pupil 7319

This girl has given the following answers at various points in the progression study:

grade 7 An atom is composed of little bricks (made out of matter) which are indivisible. Many atoms compound to molecules or

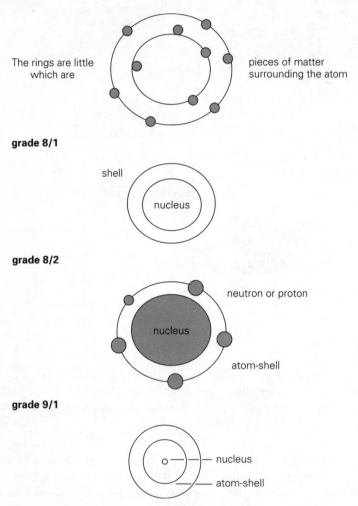

grade 7

The rings are little which are

pieces of matter surrounding the atom

grade 8/1

shell

nucleus

grade 8/2

neutron or proton

nucleus

atom-shell

grade 9/1

nucleus

atom-shell

Figure 17.4: Drawings of an atom given by girl-pupil 7319

crystals. The molecules are made of two pieces of matter which exist of two or more atoms. These atoms could be equal or unequal. The molecules of a compound are the smallest units of particles which have the important attributes of the matter they are made of.

grade 8/1 The atom is like a bullet. It has a nucleus and a shell.

grade 8/2 The atom possesses a nucleus and a shell. The shell is packed with protons and neutrons.

grade 9/1 An atom has a body like a bullet. It exists out of a nucleus and an atom-shell.

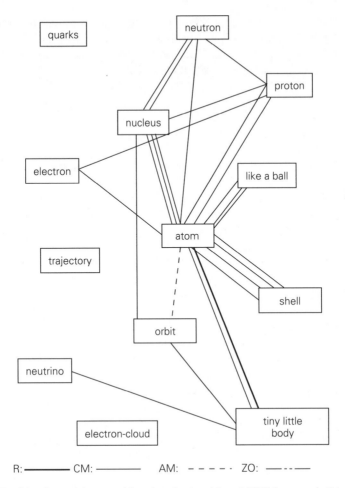

Figure 17.5: Mappings of the atom idea given by the girl-pupil 7319 from grade 7 to grade 9/1

Before lessons in physics and chemistry had begun pupil 7319 had already developed her own meaning of the structure of an atom. She was drawing little bits of material from which the body is made and attached them to shells. After six months at school she drew a different picture, this time the simplified model given by teachers in physics and chemistry. But had the girl really changed her structure of meaning? By the end of grade 8 she had reverted to drawing the old picture again, but with new words given as an explanation. However, these words were integrated into the former idea of an atom with now these 'little bits' called neutrons and protons.

This is a good example of the research assumption that there are two learning levels. By grade 9, the school-picture replaces the everyday-picture again, although to complicate matters she was taught about atoms and molecules just before answering the questionnaire! Her written answers confirm the explanation depicted by her pictures. The mapping of relations between given words shows the strong meaning structure in the area 'like a ball-atom-shell-nucleus'. Later on, she added the idea of 'an atom is like a tiny little body' into her knowledge structure, most likely because of her teacher's presentation of the microscopic world (see below again).

Descriptions of an Atom by Pupil 7414

This girl has given the following answers at various points in the progression study:

grade 7   no answer — (see Figure 17.6)

grade 8/1   Atoms — you can't see them with your eyes and therefore you can't describe them exactly. Atoms are particles and the matter is made of them.

grade 8/2   What do atoms look like? . . . You can't describe them exactly. We say they are like little marbles. The atoms are so small that you really can't see them. Until now, a special apparatus is not constructed which makes atoms visible.

grade 9/1   An atom is very small. It is made of one nucleus and shells. In contrast to the outside-shell the nucleus is so tiny. We could imagine: if a nucleus has a size of 2 millimeters the shell has a diameter of 200 meters. The atom itself is made of protons in the nucleus and electrons in the shells.

This second girl (pupil 7414) answered quite differently. At first she had no idea what an atom looked like and her written answers underline this position (see Figure 17.6). Her main reason given is that nobody is able to see an atom yet, but

grade 7   I have no idea how does an atom look like.
grade 8/1   – no drawing –
grade 8/2   – no drawing –
grade 9/1

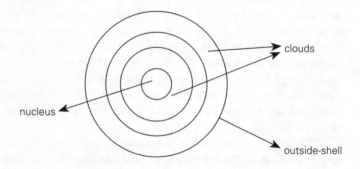

It is a model

Figure 17.6: Drawings of an atom given by girl-pupil 7414 from grade 7 to grade 9/1

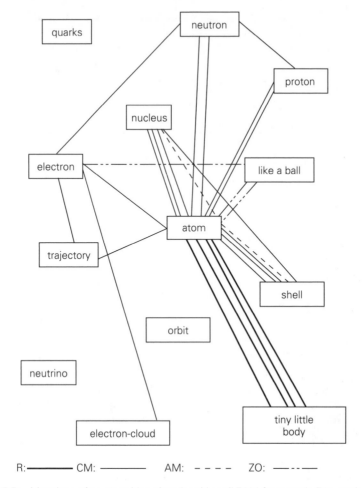

Figure 17.7: Mappings of an atom idea given by girl-pupil 7414 from grade 7 to grade 9/1

she already had some knowledge about the atom. The mapping showed that she thought that of an atom as a tiny little body with a fixed shell and a nucleus (see Figure 17.7).

An additional question was given to the pupils to get more information about their thinking about atoms and particles: Do particles or do atoms have the same attributes as the whole body which is made out of these pieces?

Figure 17.8 shows the influence of school-lessons on pupils' answers. At the beginning of physics- and chemistry-lessons both girls (7319 and 7414) did not imagine that particles had the same attributes. Later on particles were seen as little pieces of a continuous body. The same result is given by girls' answers about atoms: the main line rises up but not straight ahead. Deriving from those parts of lessons in which the teacher explained the difference between atoms and the whole body, the girls have given answers from a 'physical' point of view.

Michael Lichtfeldt

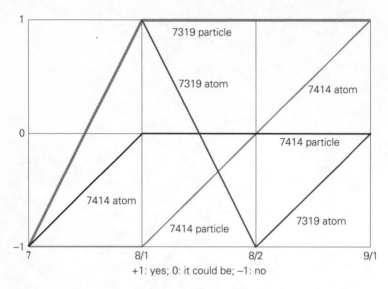

+1: yes; 0: it could be; −1: no

Figure 17.8: Particles and atoms have the same attributes as the whole body

Summarized, the information about atoms, molecules and particles in school-lessons caused the girls to modify their understanding. The question then arises of what exactly causes this to happen? Probably, the art of teachers' presentations has a large influence on the development of knowledge and cognitive meaning. Otherwise in the case of our two girls, they already had a formulated structure of meaning about particle-atom-*reality* which made the change possible. Thus it is necessary to develop situations in lessons in which pupils can consciously develop their meanings and structures of knowledge — see Figure 17.9.

Interviews and Observations of Experiments

Pupils' behaviour during experiments at the university was quite different from their behaviour during the lessons. Even very quiet pupils were able to talk continuously about the experiments. It is thus possible to collect a lot more elaborate information about their learning and understanding. In these more contrived situations pupils were able to connect intuitive-knowledge and school-knowledge to new ideas and to develop a better understanding (see Figure 17.9). The pupils — working in groups of two or three together — were given enough time for discussion and for the realization of their ideas in the experiments. This is what I am calling a 'special area of learning', a situation where pupils were more able, indeed encouraged, to develop new ideas as an 'expert-learner' (see Gardner, 1991). This could be seen during the interview: At the beginning both the girls showed their old fixed main ideas again e.g.:

a heavy body has heavy atoms or molecules;
a heavier body is a bigger body;

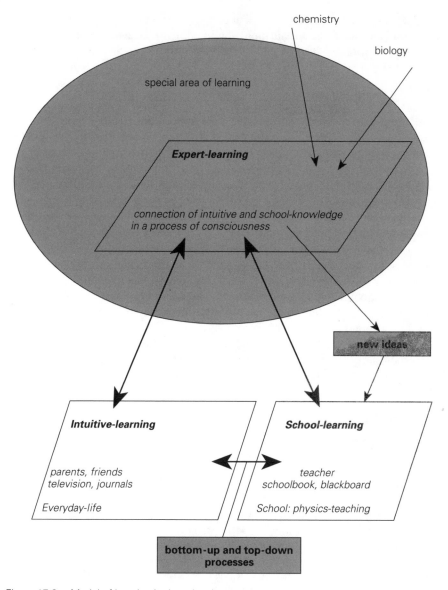

Figure 17.9: Model of learning in three levels

the particles inside the water must be blue, otherwise the water couldn't be blue.

However, after nearly half an hour of discussion they began to show an appreciation of the taught-science (physics) point of view. They drew on the school-knowledge, were able to compare it with their own previously held ideas about the phenomenon, to see differences in the viewpoints and to articulate their understanding well enough to show that they had constructed a new meaning structure. Just how stable in all situations this modified conception will prove to be, should be seen in 1997 at the end of the long-term research.

Conclusion

It has only been possible here to give a flavour of the results coming out of a longitudinal study. The main point has been to show and explain the different knowledge-levels using examples. One problem of normal school lessons is that pupils and their teacher do not have enough time to develop a discussion between them and so arrive at a closer understanding of both the learners' difficulties and the teacher's grasp of where the pupils have reached in their thinking. Often pupils learn the school-knowledge scheme only for presentation in home- or class-work. Outside school, they fall back onto their everyday-knowledge which has picked up scraps of school-knowledge compatible with their preconceptions. Therefore it is necessary to know pupils everyday-knowledge structures before teaching and to develop only those models in school which are quite similar to pupils' ideas. Otherwise pupils are not able to see the different and special explanatory potential characters of a model, tending to see the model as the reality (Figure 17.3) rather than as a device to aid their grasp of sometimes abstract phenomena. Finally, teaching has to point out the different viewpoints of everyday-life and science-world step by step establishing comprehension at each point, which takes time and skill so that most of the pupils will be able to become expert-learners.

References

BETHGE, TH. (1988) 'Aspekte des Schülervorverständnisses zu grundlegenden Begriffen der Atomphysik', University dissertation, Bremen.

BROOK, A., BRIGGS, H., and DRIVER, R. (1984) *Aspects of Students' Understanding of the Particulate Nature of Matter*, Leeds Centre for Studies in Science and Mathematics Education, University of Leeds.

BUCK, P. (1987) 'Der Sprung zu den Atomen', *Physica Didacta*, **14**, 1/2, pp. 41–5.

BUCK, P., and MACKENSEN, M.v. (1994[5]) 'Atomistische Modellvorstellungen und ihre Wirkungen auf das Naturverständnis (junger Menschen)', in *Naturphänomene Erlebend Verstehen*, Aulis, Köln.

CALVIN, W.H. (1990) *The Cerebral Symphony: Seashore Reflections on the Structure of Consciousness*, Bantam Books, New York.

CAREY, S. (1986) *Conceptual Change in Childhood*, Cambridge, MA.

DRIVER, R. (1989) 'Changing conceptions', in ADEY, P. (Ed) *Adolescent Development and School Science*, London, Falmer Press, pp. 61–84.

DUIT, R. (1992) 'Atomistische Vorstellungen bei Schülern', in FISCHLER, H. (Ed) *Quantenphysik in der Schule*, Kiel, pp. 201–14.

EDELMAN, G.M. (1987) *Neural Darwinism — The Theory of Neuronal Group Selection*, Basic Books, New York.

FISCHLER, H., and LICHTFELDT, M. (1992a) 'Modern physics and students' conceptions', *International Journal of Science Education*, **14**, 2, pp. 181–90.

FISCHLER, H., and LICHTFELDT, M. (1992b) 'Learning quantum mechanics', in DUIT, R., GOLDBERG, F. and NIEDDERER, H. (Eds) *Research in Physics Learning: Theoretical Issues and Empirical Studies*, IPN, Kiel, pp. 240–51.

GARDNER, H. (1991) *The Unschooled Mind: How Children Think and How Schools Should Teach*, Basic Books, New York.

HEWSON, P.W., and HEWSON, M.G. (1992) 'The status of students' conceptions', in DUIT, R., GOLDBERG, F. and NIEDDERER, H. (Eds) *Research in Physics Learning: Theoretical Issues and Empirical Studies*, IPN, Kiel, pp. 59–73.

INGENDAHL, W. (1975) *Sprechen und Schreiben*, Heidelberg, Quelle und Meyer.

JOHNSTON, K., and DRIVER, R. (1987) *A Case Study of Teaching and Learning about Particle Theory*, CLIS Project, Leeds.

KNOTE, H. (1975) 'Zur Atomvorstellung bei 13 bis 15-Jährigen', *Der Physikunterricht*, **9**, 4, pp. 86ff.

LICHTFELDT, M. (1992a) *Schülervorstellungen in der Quantenphysik und ihre möglichen Änderungen durch Unterricht*, Westarp Wissenschaften, Essen.

LICHTFELDT, M. (1992b) 'Erprobungen der "Einführung in die Quantenphysik": Lernprozesse und Veränderungen von Vorstellungen', in FISCHLER, H. (Ed) *Quantenphysik in der Schule*, IPN, Kiel, pp. 253–69.

LICHTFELDT, M. (1992c) 'Schülervorstellungen als Voraussetzung für das Lernen von Quantenphysik', in FISCHLER, H. (Ed) *Quantenphysik in der Schule*, IPN, Kiel, pp. 234–44.

LIJNSE, P.L., LICHT, P., DE VOS, W. and WAARLO, A.J. (Eds) (1990) *Relating Macroscopic Phenomena to Microscopic Particles: A Central Problem in Secondary Science Education*, Utrecht Centre for Science and Mathematics Education, University of Utrecht.

LÖFFLER, G. (1992) 'Piagets und Inhelders Interviews zum kindlichen Atomismus unter einem phänomenologischen Gesichtspunkt reinterpretiert', *Chimica Didactica*, **18**, 2, pp. 85–99.

LÖFFLER, G. (1994) 'Analyse und Reinterpretation von Untersuchungen von Novick und Nussbaum zum Teilchenbild der Materie', *Chimica Didactica*, **20**, 1, pp. 5–34.

MANDL, H., GRUBER, H., and RENKL, A. (1993a) 'Das träge Wissen', *Psychologie Heute*, Septemberjournal, pp. 64–9.

MANDL, H., GRUBER, H., and RENKL, A. (1993b) 'Misconceptions and Knowledge Compartmentalization', in STRUBE, G., and WENDER, K.F. (Eds) *The Cognitive Psychologie of Knowledge*, Amsterdam a.o., pp. 161–76.

MESTRE, J.P. (1991) 'Learning and instruction in pre-college physical science', *Physics Today*, **9**, pp. 56–62.

NOVICK, S., and NUSSBAUM, J. (1978) 'Junior high schools pupils' understanding of the particulate nature of matter', *Science Education*, **62**, pp. 273ff.

NOVICK, S., and NUSSBAUM, J. (1981) 'Pupils' understanding of the particulate nature of matter', *Science Education*, **65**, pp. 187ff.

PETRI, J., and NIEDDERER, H. (1994) 'Eine Fallstudie zur Veränderung von Schülervorstellungen in der Atomphysik (S II)', in BEHRENDT, H. (Ed) *Zur Didaktik der Physik und Chemie*, Leuchtturm, Alsbach, pp. 298–300.

PETRI, J., and NIEDDERER, H. (1995) 'Learning pathways in atomic physics (grade 13)', Paper presented at the European Conference in Science Education Research, Leeds April 1995.

PFUNDT, H. (1981) 'Das Atom — Letztes Teilungsstück oder erster Aufbaustein?: Zu den Vorstellungen, die sich Schüler vom Aufbau der Stoffe machen', *Chimica Didactica*, **7**, 2, pp. 75–94.

PFUNDT, H. (1982a) 'Untersuchungen zu den Vorstellungen, die Schüler vom Aufbau der Atome haben', *Der Physikunterricht*, **16**, 1, pp. 51ff.

PFUNDT, H. (1982b) 'Ein Weg zur Atomhypothese', *Chimica Didactica*, **8**, 3, pp. 143–56.

PFUNDT, H., and DUIT, R. (1994[4]) *Students' Alternative Frameworks and Science Education*, IPN, Kiel.

PIAGET, J. (1985) *The Equilibration of Cognitive Structure*, Chigago.

POSNER, G.J., POSNER, G.J., STRIKE, K.A., HEWSON, P.W. and GERTZOG, W.A. (1982) 'Accommodation of a scientific conception: Toward a theory of conceptual change', *Science Education*, **66**, pp. 211–27.

REINMANN-ROTHMEIER, G., and MANDL, H. (1995) 'Wissensvermittlung: Ansätze zur Förderung des Wissenserwerbs', in KLIX, F., and SPADA, H. (Eds) *Enzyklopädie der Psychologie; Themenbereich C: Theorie und Forschung; Serie II: Kognition; Band G: Wissenspsychologie; 14 Kapitel* (in press).

RESNICK, L.B. (1987) 'Learning in school and out', *Educational Researcher*, **16**, pp. 13–20.

SCHAEFER, G. (1990) 'Zickzack-Lernen als Methode oder: Kann aus Wirrwarr Ordnung entsehen?', in OEHMIG, B. (Ed) *Erleben, Beobachten, Untersuchen: Zur Didaktik von Untersuchungen*, FU, ZI Fachdidaktiken, Berlin, pp. 110–31.

SCOTT, P.H. (1992) 'Pathways in learning science: A case study of the development of one student's ideas relating to the structure of matter', in DUIT, R., GOLDBERG, F. and NIEDDERER, H. (Eds) *Research in Physics Learning: Theoretical Issues and Empirical Studies*, Proceedings of an International Workshop, Bremen 1991, IPN, Kiel, pp. 203–24.

SCOTT, P.H., ASOKO, H.M. and DRIVER, R. (1992) 'Teaching for conceptual change. A review of strategies', in DUIT, R., Goldberg, F. and NIEDDERER, H. (Eds) *Research in Physics Learning: Theoretical Issues and Empirical Studies*, Proceedings of an International Workshop, Bremen 1991, IPN Kiel, pp. 310–29.

SEILER, TH.B. (1985) 'Sind Begriffe Aggregate von Komponenten oder idiosynkratische Minitheorien?', in SEILER, TH.B. and WANNENMACHER, W. (Eds) *Begriffs und Wortbedeutungsentwicklung*, Berlin, Heidelberg, pp. 105–31.

SHUELL, T.J. (1993) 'Towards an integrated theory of teaching and learning', *Educational Psychologist*, **28**, 4, pp. 291–311.

STAVY, R. (1991) 'Using analogy to overcome misconceptions about conservation of matter', *Journal of Research in Science Education*, **28**, 4, pp. 305–13.

STRIKE, K.A. and POSNER, G.J. (1985) 'A conceptual change view of learning and understanding', in WEST, L.T.H. and PINES, A.L. (Eds) *Cognitive Structure and Conceptual Change*, New York.

TODTENHAUPT, S. (1995) *Zur Entwicklung des Chemieverständnisses bei Schülern*, Lang, Frankfurt/M. a.o.

VYGOTSKY, L.S. (1962) *Thought and Language*, MIT Press, Cambridge, MA.

WAGENSCHEIN, M. (1976) *Die pädagogische Dimension der Physik*, Vieweg, Braunschweig.

WIGHTMAN, T., GREEN, P. and SCOTT, P.H. (1986) *The Construction of Meaning and Conceptual Change in Classroom Settings: Case Studies on the Particulate Nature of Matter*, Leeds.

18 Biotechnology and Genetic Engineering: Student Knowledge and Attitudes: Implications for Teaching Controversial Issues and the Public Understanding of Science

Roger Lock

Abstract

Knowledge and attitudes towards biotechnology and genetic engineering have been invest-igated in a questionnaire study of 188 14–15-year-old students (112 males, seventy-six females) drawn from six schools in England.

One-third of the sample, and more males than females, did not know what biotechno-logy or genetic engineering was, and nearly half the sample could not give examples of biotechnology or genetic engineering. Internal consistency of response to attitude questions was high. Attitudes of students were context-dependent: there was broad approval of genetic engineering applied to microbes and plants but not of genetic engineering applied to animals; females were particularly unsupportive of genetic engineering applied to farm animals.

Teaching about genetic engineering increased student knowledge levels and reduced uncertainty of attitudes leading to increased approval of genetic engineering in all contexts surveyed.

This paper concludes with a consideration of the implications of these findings for the teaching and learning of controversial science–society issues and for a curriculum which ad-dresses the public understanding of science.

Introduction[1]

Durant (1990) has summarized the rationale for concern about the public under-standing of science into three categories; cultural, practical and political. The first concerns the need for people to be informed about what are probably the most important achievements of our time, that is science as part of our heritage. The second concerns the need to understand how some of the everyday science-based technologies work at the functional level. The political rationale is about demo-cracy in action. Effective citizen participation depends upon access to particular knowledges such that policy issues may be subjected to critical scrutiny. It is in this respect that the importance of disseminating scientific and technological informa-tion is accentuated. (Dickens, 1992; Science Museum, 1994; Ziman, 1980). The significance of the public understanding of science and the related public attitudes is demonstrated by the activities of science/technology-based commercial concerns, The Royal Society and various social action groups (e.g., animal rights, conservationists etc.) (Irwin, 1995; Roberts, 1988; Turney, 1994). In general, these activities aim to close the gap between scientists and general public. They include the staging of special events, innovative ways of presenting science to the public through day conferences, exhibitions, etc. or concentrate upon developing scientists' communication skills.

Formal education offers unique opportunities for the improvement of the public understanding of science. Not only can it provide access to the majority of the future population, it has an explicit function to develop pupils' scientific understanding in preparation for adult life (Husen, 1991; Jenkins, 1990). It is also the main arena in which non-scientists learn about science (Cross and Price, 1991; Dixon, 1989). This potential within schools has been capitalized by various anti-science pressure groups. Schools and students are the principal targets for publicity and promotion of their views (Parliamentary Office of Science and Technology, 1992). Despite this, the significance of particular school experiences for the development of conceptual understanding of, and attitudes towards, science-society issues has only rarely been empirically investigated. (Lock and Miles, 1993; Lock, Miles and Hughes, 1995.)

The place in formal education of biotechnology, and in particular genetic engineering, has been reinforced in syllabuses offered to 14–15-year-old students by the prominent position afforded to such work by the Science National Curriculum in England (Department for Education and Science (DES), 1991). However, inclusion in the National Curriculum is not the only rationale for studying such a topic in schools. For most people the period of formal education is the major lifetime opportunity for understanding the science that will impact on their lives and lifestyle. Adults are expected to play a full and responsible role in society which includes applying the knowledge, understanding and attitudes gained from their study of science to their everyday life. Biotechnology and genetic engineering are aspects of science content rich with opportunities for work of this kind.

Method

This study was carried out using 188 students (112 males, seventy-six females) drawn from six schools in central England. The schools involved included independent and maintained, selective and comprehensive, single sex and coeducational. They had urban, suburban and more rural catchment areas and contained students from a wide range of religious and cultural backgrounds. Students were involved over a period of at least two double lessons. Although this was an opportunity sample and not one selected at random to be representative of 14–15-year-old students in general, it was a carefully constructed group of schools and classes, and there was no reason to suspect any specific bias in the sample.

The study involved three elements:

1   a questionnaire that measured knowledge and attitude (15/20 minutes);
2   teaching materials/activities (90/120 minutes); and
3   questionnaire (15 minutes).

The first questionnaire was developed to measure the level of knowledge and type of attitudes that students held about biotechnology and genetic engineering before any teaching on the topic. It used open-ended questions to assess knowledge and a Likert-type scale to measure attitude. It was administered by the teachers involved in the study during science lessons at the schools in the survey.

Activities in the lessons were student-centred and involved comprehension exercises and small group discussion of ethical and moral issues involved in genetically engineering bacteria, tomatoes and sheep.

The second questionnaire was administered at the end of the second/third lesson devoted to the study. It replicated the knowledge questions involved in the first questionnaire and a selection of the attitude scale items as well as including open questions that explored what students thought about the materials and the work they had studied.

The questionnaires had been subjected to small-scale trials prior to the study and validation by administering in one-to-one situations with students as semi-structured interviews.

Earlier papers have reported on student knowledge and attitude (Lock and Miles, 1993) and on the influences of teaching upon them (Lock *et al.*, 1995). Here a brief synopsis of the earlier findings is presented to set the context for the bulk of this paper which focuses on the implications of the findings for teaching controversial science and society issues.

Results

Student Knowledge

Responses were categorized and then collated for gender and the whole sample from questionnaires administered *before* (B) the teaching materials and activities were introduced and *after* (A) the exercise was completed. Results are presented in Tables 18.1 and 18.2. Both tables show a low 'no response' rate, possibly indicating a high level of student interest and involvement in the study. The rate of 'no response' to examples of genetic engineering (Table 18.2) was higher than for the other question, possibly because this included candidates who knew no examples but did not respond rather than stating 'don't know'. This view is substantiated by the drop in 'no response' to this question after the exercise, whereas in Table 18.1 there is an increase in such responses. This latter pattern is expected from some adolescents in a study of this kind where a near-identical questionnaire is administered twice with only a short time interval.

About 50 per cent of the sample did not understand what genetic engineering meant prior to the teaching. The exercise has increased understanding of the term.

Table 18.1: What does genetic engineering mean?: Pupil knowledge before and after related teaching (Multiple responses recorded)

Response	Before Males (n = 112)	After Males (n = 108)	Before Females (n = 76)	After Females (n = 71)	Before All (n = 188)	After All (n = 179)
No Response	5	9	9	4	14	13
Don't know	33	5	17	7	50	12
Manipulate/change /alter genes	32	47	20	20	52	67
Manipulate/change /alter organisms	24	37	18	11	42	48
Reference to DNA or gene	19	21	17	13	36	34
Transfer genes	7	10	2	9	9	19
Genes from one organism to another	1	2	5	0	6	2
Select characteristics	1	4	2	10	3	14
Make genes	2	3	3	0	5	3
Cure diseases	1	1	0	2	1	3
Precise form of selective breeding	0	13	0	19	0	32
Cells from one organism to another	0	4	0	0	0	4
Reference to genetic fault	0	1	0	2	0	3
Reference to specific product	0	4	0	1	0	5
Other	0	7	0	6	0	13

Source: Lock et al., 1995

More students understood that the process involves manipulating, changing and/or altering genes and/or organisms. Others preferred a less specific definition that saw it as a precise form of selective breeding or as selecting characteristics. There were about 20 per cent of the sample who referred to DNA or genes but who did not show an understanding of how they were involved in genetic engineering. The percentage of such students was the same both before and after the lessons. After teaching, about 10 per cent saw the process as involving transfer of genes, but few

Table 18.2: Pupil knowledge of examples of genetic engineering before and after related teaching (Multiple responses recorded)

Examples	Before Males (n = 112)	After Males (n = 108)	Before Females (n = 76)	After Females (n = 71)	Before All (n = 188)	After All (n = 179)
No response	13	12	13	9	26	21
Don't know	42	2	30	4	72	6
Humulin	16	3	15	6	31	9
Cystic fibrosis	10	2	3	1	13	3
Genetic fingerprinting	0	0	7	4	7	4
Improve cereals	4	4	0	8	4	12
Cross breeding	5	1	2	2	7	3
Prevent diseases	3	2	0	2	3	4
Improve food taste	1	6	1	3	2	9
Improve food quality	1	2	1	2	2	4
Genetic implants	1	0	2	1	3	1
Develop flower colour	3	0	1	0	4	0
Pharmaceuticals from sheep milk	0	24	0	11	0	35
Reduce tomato spoilage	0	50	0	16	0	66
Authentic human milk from cows	0	6	0	6	0	12
Tomato shape	0	0	0	2	0	2
Improve yield	0	3	0	1	0	4
Increase plant variety	0	3	0	6	0	9
Other (not genetic engineering)	12	13	4	10	16	23

Source: Lock *et al.*, 1995

made the specific point that the transfer of genes is from one organism to another. Some misconceptions occur as the result of teaching with, for example, the belief that whole cells are transferred from one organism to another.

Table 18.2 shows that the teaching reduced the size of the 'don't know' response considerably. It appears that knowledge of examples of genetic engineering was enhanced more successfully than knowledge of the definition. As with

Table 18.3: Reliability of student response (Percentages to nearest whole number)

Statements	M	F	T	M	F	T	M	F	T	M	F	T	M	F	T
	I strongly agree			I agree			I'm not sure			I disagree			I strongly disagree		
*8. Altering the genes in fruit to improve their taste is not acceptable to me.	3	7	4	15	24	19	11	13	12	52	55	53	19	1	12
*28. Altering the genes in fruit to improve their taste is not acceptable to me.	2	8	4	15	17	16	8	22	14	52	49	51	23	3	15

* Numbers show positions of statements in the questionnaire
Key: M = males, F = famales, T = total sample
Source: Lock et al., 1995

Table 18.1, there are data that suggest students learn incorrect information. For example, in Table 18.2 an increase in examples which are not genetic engineering is shown.

Overall, the increase in student knowledge of examples of genetic engineering was pleasing.

Student Attitudes

Attitude measurement is fraught with a number of difficulties, not least their ephemeral and vacillating nature, sometimes even within the timespan taken to complete a questionnaire. With this in mind, the first questionnaire had a statement repeated in exactly the same format within its structure (statements 8 and 28). The results from these questions are given in Table 18.3 and show the high level of consistency of response achieved.

There are a range of factors that could influence pupil attitudes in an exercise such as this. It is tempting to ascribe changes in attitude, if any, solely to the work involved in this study. This will be a factor, and a major one at that, but it will not be the only factor involved; other factors may emanate from outside the science laboratory.

The attitude statements used before the teaching and learning activity showed broad approval for genetic engineering applied to microbes and plants but more disagreement with applying the process to animals (Table 18.4). Females were particularly unsupporting of genetic engineering in animals.

The major change that is noted between the two questionnaires is in the reduction of the 'I'm not sure' responses. Such a pattern was seen in response to ten of the thirteen statements (Lock et al., 1995). Two of the statements where such a decrease was not observed related to genetic engineering of animals.

A reduction in the uncertain response is heartening as it suggests that teaching and learning about controversial issues can help students to clarify their position.

Table 18.4: Attitudes to genetic engineering (Percentages to nearest whole number)

(a) Microbes

Statements	I strongly agree			I agree			I'm not sure			I disagree			I strongly disagree		
	M	F	T	M	F	T	M	F	T	M	F	T	M	F	T
*19. Microbes should be genetically engineered to make them more efficient at decomposing human sewage.	23	20	22	59	58	59	14	20	17	2	1	2	2	0	1

(b) Plants

	M	F	T	M	F	T	M	F	T	M	F	T	M	F	T
*9. Altering the genes of plants so that they will grow better in salty soils is acceptable to me.	22	5	15	61	61	61	9	21	14	7	11	9	1	1	1
*18. We should not alter the genes in plants to get them to make more oils useful in manufacturing.	5	1	3	19	25	21	17	16	17	47	55	51	13	1	8

(c) Animals

	M	F	T	M	F	T	M	F	T	M	F	T	M	F	T
*2. Changing the genetic make up of farm animals should be banned by law.	21	26	23	26	34	29	31	28	30	16	12	14	6	0	4
*13. Inserting genes from human cells into the fertilized eggs of sheep is acceptable to me.	3	0	2	7	0	4	21	18	20	31	46	37	38	34	36

* Numbers show position of statements in the questionnaire
Key: M = males, F = females, T = total sample
Source: Modified from Lock and Miles, 1993

However, reduction in the level of uncertain response is not the only indicator that attitudes had changed over the period of study.

There was also a reduction in the level of disagreement with genetically engineered changes in over half of the statements (Lock *et al.*, op. cit.). Such a finding suggests that most of those who changed their responses from 'I'm not sure' moved to a position where they supported genetic engineering.

Table 18.5 shows that more females than males disapproved of genetic engineering, particularly in contexts where animals are involved but levels of disapproval between the genders differ little with respect to manipulation of microbes. Table 18.5 further shows that a majority approved of genetic engineering in animals where drug production was involved for the treatment of human or animal conditions.

Table 18.6 highlights the influence that teaching activity or style may have on student attitudes. In one school (Y) the majority of students disagreed with the statement that changing the genetic make up of farm animals should be banned by

Table 18.5: Attitudes to genetic engineering for pharmaceutical and veterinary products before and after related teaching (Percentages to nearest whole number)

Attitude to Genetic Engineering for Human Medicines

(a) Microbes

Statements Before/After Teaching		I strongly agree			I agree			I'm not sure			I disagree			I strongly disagree	
	M	F	T	M	F	T	M	F	T	M	F	T	M	F	T
*14. I am against changing the genes of microbes so that they make medicines for humans. Before	2	0	1	7	7	7	10	26	17	54	49	52	28	17	23
*6. After	2	1	1	7	6	7	10	16	12	50	59	54	31	18	26

(b) of Animals

	M	F	T	M	F	T	M	F	T	M	F	T	M	F	T
*20. Genetically engineering cows to produce life saving drugs for humans is not acceptable to me. Before	6	5	5	19	24	21	18	22	20	39	37	38	19	9	15
*10. After	3	4	3	9	17	13	16	21	18	49	44	46	23	14	20
*23. Using genetically engineered sheep to produce medicines for humans is a good idea. Before	13	7	11	39	32	36	19	29	23	22	25	23	7	7	7
*12. After	15	10	13	45	49	47	16	21	18	19	16	18	5	4	4

Attitude to Genetically Engineering Animals for Veterinary Products

	M	F	T	M	F	T	M	F	T	M	F	T	M	F	T
*18. Changing the genes of animals to produce vaccines to treat animal diseases is not acceptable to me. Before	4	1	3	18	17	18	16	29	21	44	43	44	18	8	14
*9. After	4	3	3	16	18	17	18	27	22	44	45	45	8	7	13

* Numbers show positions of statements in the questionnaires
Key: M = males, F = females, T= total sample
Source: Modified from Lock et al., 1995

Table 18.6: Differences in attitude between students in two schools after teaching activity

	I strongly agree	I agree	I'm not sure	I disagree	I strongly disagree
School X (n = 46)	6	9	11	15	5
School Y (n = 50)	4	21	12	13	0

Source: Lock *et al.*, 1995

law, a position that reflected the combined views of the whole sample. In contrast, students in school X had a wider spread of views with more, in real and proportionate terms, in disagreement. This difference between schools cannot be explained in terms of gender differences between the groups and is more likely to be influenced by factors associated with the teaching and learning styles to which pupils were exposed.

Implications

The findings presented in preceding sections raise implications for the approaches used in teaching and learning about controversial science–society issues and for the curriculum. Each of these issues will be addressed in turn.

Teaching and Learning about Science: Society Issues

Firstly, it is important to include controversial science–society issues as an integral part of workschemes; not as an 'add on' at the end of a unit, nor as an extra for homework or as the final element of extension material for fast workers, but as a central theme covered by *all* students. Having included such work, it is vital that the specific contribution that science makes to work with these issues is a key element of the teaching and learning strategies used; for it is not only science teachers who include topics like genetic engineering in their workschemes. In essence this means adopting a 'scientific' approach, one where students are asked to distinguish between fact and opinion and to determine if data support the interpretation that is presented. Students should be encouraged to show scientific attitudes such as curiosity, open mindedness and respect for evidence. They should be willing to tolerate uncertainty. Above all, it is knowledge rather than hearsay upon which sound opinion is based. Student views should respect and not contradict or conflict with the evidence. This places the onus on teachers to provide the accurate data and information that might underpin student opinion.

Having included controversial issues in a work scheme, the next priority is to ensure that a balanced approach is provided in dealing with such issues. This means that students should be exposed to a range of views and value positions, not just indoctrinated with the view held by the teacher.

Teachers could present a range of possible views without indicating which they personally support, or alternatively, resource material representing a wide range of different view points could be presented to students (for further details on teaching strategies see, for example, Bridges, 1986). However balanced a teacher attempts to be, there will almost certainly be some influence on students' views, but if there has been an even-handed lesson, conducted in a scientific manner, then students will have a model of how to approach similar controversial issues in contexts outside the school environment. Having been exposed to a range of viewpoints, students should be encouraged to discuss their views with their peers. Through such discussion students should be encouraged to make up their own minds about an issue. The significant elements here are that students come to their views through a critical evaluation of the evidence which should be seen to support rather than contradict their opinion. The nature of the student opinion should not be important, but the fact that it has been gained through critical reflection and respect for evidence is important.

A further teaching approach is to ask students to justify and defend their position to their peers. In such activities students may reveal inconsistencies in their views and attitudes. Drawing apparent inconsistencies in attitudes to students' attention and seeking explanations can be a powerful classroom strategy to generate debate, discussion and further clarification and justification of attitudes. The materials, used by all students involved in this study, described the production of pharmaceuticals in sheep milk through insertion of human genes into fertilized sheep eggs. While the majority of students approved of using genetically engineered sheep to produce human medicines, they disapproved of the insertion of human genes into fertilized sheep eggs which makes this possible. Facing students with dilemmas such as this leaves them to question and clarify their position. At what point, if at all, does it become acceptable to insert genes into sheep? For what purpose would this process be acceptable/not acceptable?

It is important to include the teaching of controversial science–society issues in order to develop student understanding of science, scientists and the dilemmas faced by scientists through their professional activities. There may be some teachers who consider teaching about controversial issues to be the preserve of English or Religious Education teachers and those concerned with personal and social education. In my opinion this is a mistaken view. It is vital that such issues are covered in cross-curricular contexts, as in this way the distinctive contribution that science and scientists make becomes evident. The way that science interacts with daily life is made explicit and applied issues such as those relating to food production and food labelling can be drawn to the attention of future citizens in an objective and unsentimental manner. Equally, it is vital that we illustrate that the moral high ground is not the exclusive preserve of the non-scientists. In such ways teachers can make a major contribution to the public perception of science and scientists. By exposing students to the kind of moral and ethical issues that face scientists working in the field of genetic engineering they may come to challenge the media stereotyped view of a scientist as a hard, uncaring and unsympathetic individual.

Curriculum

The first implication for the curriculum is that of including the content relating to controversial science–society issues in a prominent enough position. While there have been many attempts to marginalize and exclude such content, the current discussion will focus on two of the more recent examples.

In 1983 discussions were well underway about the development of an examination that combined the, then largely separate, General Certificate of Education Ordinary Level (GCE) from the less highly rated Certificate of Secondary Education (CSE). Physics was one of the first new syllabuses to be submitted to, and considered by, the then Secretary of State for Education, Sir Keith Joseph. The proposals, which came from Her Majesty's Inspectorate and a host of professional associations, industrialists and curriculum development bodies, argued that the syllabus should emphasize the wider social and economic implications of the subject. The rejection of the proposals by Sir Keith led the editor of the Times Educational Supplement (TES) to comment (TES, 1983):

> Where Sir Keith has gone farther out on a limb is in his total rejection of anything which suggests that the study of physics should include any consideration of the social and economic issues which arise from the application of scientific knowledge. Thus, while he insists that pupils must learn about the technological applications, he believes they must be rigorously steered away from the interesting questions of value, morality and expediency, of which (it is to be hoped) scientists have become increasingly aware. Although social and applied issues were included in the syllabuses for the new GCSE courses, the examinations assessed only fact and not views and opinions.

A further bid for status for social and ethical issues in science was made in one of the early drafts of a science national curriculum (DES, 1988). In the proposed attainment target 21, Science in Action, it was suggested that

> Pupils should develop a critical awareness of the ways that science is applied in their own lives and in industry and society, of its personal, social and economic implications, benefits and drawbacks.

By proposing to devote a complete attainment target to such issues a clear signal could be given to students and teachers about the status of such work. However, by the time a statutory version of the curriculum had been produced, not only had the attainment target been deleted, but references to ethical issues were confined to the statements of attainment and the Programme of Study. The position with respect to genetic engineering is shown in the following extract from the Key Stage 4 (14–16-year-olds) Programme of Study (DES, 1991).

> Using sources which give a range of perspectives, they (pupils) should have the opportunity to consider the basic principles of genetic engineering, for

example, *in relation to drug and hormone production*, as well as being aware of any ethical considerations that such production involves.

Not only were the social, moral and ethical issues marginalized in terms of their status within the National Curriculum, but they were often included in a position which suggested that study of such issues was only appropriate for the most able students. The statement of attainment related to the part of the Programme of Study quoted above was located at level 10 and hence deemed only suitable for students who would attain the highest standards; only a tiny proportion of the grade A candidates. It is hardly surprising, therefore, that in many schools study related to this specific statement of attainment was not included.

In an interesting way the same statement of attainment illustrates the further progressive marginalization of ethical issues. In the equivalent component of the *revised* National Curriculum (DFE, 1995) it reads

Pupils should be taught the basic principles of cloning, selective breeding and genetic engineering.

The consideration of ethical issues has gone! It may be that some of the concepts involved in the understanding of genetic engineering are only accessible to the most able 16-year-old, but this does not mean that others should be excluded from considering the moral and ethical issues arising from such work. That social and ethical issues are not the preserve of able students was illustrated by the interest and understanding reported by students of all abilities involved in this study (Table 18.7).

Table 18.7: The level of student interest and understanding in work on genetic engineering

Was the work interesting?	
YES	58%
SOME OF IT	25%
NO	16%
Did you understand the work?	
YES	51%
SOME OF IT	42%
NO	6%

Notes: (n = 179)

The National Curriculum appears to have taken an entrenched view of science locked in the tradition of abstract concepts and rote-learning. There is something at odds with a compulsory curriculum that makes students do more work of the type that they opted to avoid when this was possible.

Figure 18.1 shows the contrast between the science that students meet in school and that in the world outside it. It compares the list of topics, representing a term's work, taken from the exercise book of a 15-year-old student working

School Science topics January to March 1995 Year 10, 14/15-year-olds	Terrestrial Television — science topics March 27th — April 1st 1995
In this ten week period there were eleven topics studied.	In this one week period there were twenty-five programmes on factual science, i.e., excluding drama involving hospitals.

These were:

- Gases

- Why are solids solid?
- Gas Laws
- Structure of atoms
- Differences between elements
- Periodic table

- Separating techniques

- Emulsions and foams
- Radioactivity
- Making materials stronger
- Oxidation

Including:

- The information war on international battlefields
- Urban foxes
- The third sex
- Greyhound euthanasia
- Sex, slugs and the speed of light
- Vets talking with distressed owners of dying pets
- Organic meat — are customers paying over the odds?
- Chemical warfare
- Extracting DNA from Pharoahs
- Eco-terrorism
- Control of Nuclear Weapons

Figure 18.1: Science subjects from school science and terrestrial television

towards a double award science examination with the evening output, in a single week, of the four terrestrial television channels.

There is a clear mismatch between the world of science portrayed in school and the science that students may meet in their leisure time. As science teachers we should be concerned at this gulf and should spend some time offered by the current five-year period of curriculum stability in redressing such an imbalance in the curriculum as preparation for students of the next millennium.

The case for including more work with a social, moral and ethical base is strong. Such a curriculum change would be popular, particularly with female students, and could lead to more positive views of science. Not only could this contribute to increased numbers studying science beyond the compulsory years of schooling, but also it could do much to develop and enhance the public understanding and perception of science and scientists.

It is not too dramatic to stress the importance of including work on controversial science–society topics in the curriculum for *all* 11–16-year-olds. An informed population with an objective view of science and scientists based on experience, knowledge and understanding is a prerequisite for progress in the twenty-first century.

Note

1 My thanks to Mairéad Dunne and Allan Soares who contributed to the introduction.

Roger Lock

References

BRIDGES, D. (1986) 'Dealing with controversy in the curriculum: A philosophical perspective', in WELLINGTON, J.J. (Ed) *Controversial Issues in the Curriculum*, Oxford, Basil Blackwell.

CROSS, R.T. and PRICE, R.F. (1991) *Teaching Science for Social Responsibility*, Sydney, St. Louis Press.

DES (1988) *Science for Ages 5–16: Proposals of the Secretary of State for Education and Science and the Secretary of State for Wales*, London, HMSO.

DES (1991) *Science in the National Curriculum* (1991), London, HMSO.

DFE (1995) *Science in the National Curriculum*, London, HMSO.

DICKENS, P. (1992) *Society and Nature: Towards a Green Social Theory*, London, Harvester Wheatsheaf.

DIXON, B. (1989) *Society and Science: Changing the Way We Live*, London, Cassell.

DURANT, J.R. (1990) 'Copernicus and Conan Doyle: Or, why should we care about the public understanding of science?', *Science and Public Affairs*, **5**, 1, pp. 7–22.

HUSEN, T. (1991) 'Opening statement', in HUSEN, T. and KEEVES, J.P. (Eds) *Issues in Science Education: Science Competence in a Social and Ecological Context*, Exeter, Pergamon.

IRWIN, A. (1995) 'Scientists call for genetic freeze', in *The Times Higher Education Supplement*, January 20, p. 2.

JENKINS, E. (1990) 'Scientific literacy and school science education', *School Science Review*, **71**, (256), pp. 43–51.

LOCK, R. and MILES, C. (1993) 'Biotechnology and genetic engineering: Students' knowledge and attitudes', *Journal of Biological Education*, **27**, 4, pp. 101–6.

LOCK, R., MILES, C. and HUGHES, S. (1995) 'The influence of teaching on knowledge and attitudes in biotechnology and genetic engineering contexts', *School Science Review*, **76**, 276, pp. 47–59.

PARLIAMENTARY OFFICE OF SCIENCE AND TECHNOLOGY (1992) *The Use of Animals in Research, Development and Testing*, London, POST.

ROBERTS, G.G. (1988) 'At home with science and technology', *Science and Public Affairs*, The Royal Society, London, 3, pp. 53–72.

SCIENCE MUSEUM (1994) *UK National Consensus Conference on Plant Biotechnology*, Final Report, 2–4 November 1994, Regents Park, London, London Science Museum/BBSRC.

TES (1983) 'To the barricades', in *The Times Educational Supplement*, 18 March, p. 16.

TURNEY, J. (1994) 'The Public understanding of genetics: Where next?', *Bio-technology Education*, **3**, 3, pp. 16–21.

WOOD, N. (1983) 'Sir Keith tries to keep new exams nuclear free', in *The Times Educational Supplement*, 18 March, p. 1.

ZIMAN, J. (1980) *Teaching and Learning about Science and Society*, Cambridge, Cambridge University Press.

19 Progression in Pupils' Understanding of Combustion

Rod Watson and Justin Dillon

Abstract

Progression in pupils' ideas about the process of burning was investigated using open-ended and structured response questions. A questionnaire was administered to a sample of 299 pupils aged 14 and 15, of whom thirty-three were followed up using the same questionnaire sixteen months later. Pupils' responses were analysed using operational definitions of different types of explanation developed in an earlier study (Prieto, Watson and Dillon, 1993). The way in which the pupils' explanations changed over the sixteen-month period and the pedagogical implications are discussed.

Introduction[1]

A number of studies have been carried out to try to understand pupils' conceptions of combustion (Andersson, 1986a, 1990; BouJaoude, 1991; Meheut, Saltiel and Tiberghein, 1985; Meheut, 1982; Ross, 1991; Schollum and Happs, 1982).

In some studies, categories were developed to code the responses to particular questions (Driver, Child, Gott, Head, Johnson, Worsley and Wylie, 1984) whereas other authors have developed more general categories which can be applied to the responses to a variety of questions (Andersson, 1986a, 1990; Meheut *et al.*, 1985; Meheut, 1982; Pfundt, 1981). In previous studies of pupils' understanding of combustion (Prieto, Watson and Dillon, 1993; Watson, Prieto and Dillon, 1995) the general categories proposed by Andersson (1986a, b and 1990) were adapted and developed to categorize patterns of pupils' thinking across a range of questions, and a model to describe pupils' progression in understanding of combustion was proposed. In this paper a follow up study of thirty-three of the original 299 pupils, after sixteen months is described, and conceptual difficulties are identified and discussed.

Models of Pupils' Understandings of Combustion

Andersson (1990) describes a categorization system which can be applied to pupils' responses to a variety of questions concerning both chemical and physical change.

The categories are disappearance, displacement, modification, transmutation and chemical interaction. These are illustrated as follows:

1 Disappearance

When asked about the weight of exhaust gases produced when petrol is burnt in a car (Andersson and Renstrom, 1983) some pupils answer that some of the petrol is used up in the car and disappears.

2 Displacement

An example is of pupils explaining the disappearance of water from a puddle on the floor, by saying that the water had penetrated the floor, i.e., it had been displaced to a different place (Bar and Travis, 1991).

3 Modification

Meheut *et al.* (1985) give examples of pupils explaining the burning of alcohol and the boiling of water in terms of modification of liquid alcohol to alcohol vapour and liquid water to steam.

4 Transmutation

Some changes are described in terms of transmutation of substance into energy, of energy into substance or of one substance into a new one, for example iron wool being transmuted into carbon during combustion (Osborne and Cosgrove, 1983).

5 Chemical change

Ideas of chemical change are applied correctly to examples such as petrol burning, but also incorrectly to physical changes (Osborne and Cosgrove, 1983). Some pupils think that the bubbles of steam are oxygen and hydrogen gases.

Meheut *et al.* (1985) developed an alternative way of categorizing the responses of pupils (aged 11–12) based on their ideas of conservation. There are some similarities with Andersson's categorization scheme, but Meheut *et al.* also incorporate the nature of the combustible material in the categorization system. Pupils' responses can be divided into two groups according to the nature of the combustible substance. The first group includes responses about metals, wax, water, and alcohol, which are said to melt or evaporate, rather than burn, or using Andersson's terminology, are modified. The second group includes responses about wood, cardboard, paper, alcohol and air, which are seen to burn and be changed into another substance or nothing. Using Andersson's categories, these substances disappear or are transmuted. An important feature in the categorization of Meheut *et al.* is that during transmutation each substance is transmuted separately. Pupils often fail to realize that matter is interacting.

The role of oxygen and air is not dealt with completely in either Andersson's or Meheut's model. One difficulty in interpreting responses to questions designed to elicit pupils' understandings of the role of oxygen and air in combustion is that not all pupils have a good understanding of the nature of air (Russell, Longden

Table 19.1: Operational definitions of pupils' explanations

Role of oxygen/interaction

C: Pupils recognize that the combustible substance and oxygen/air interact. The reaction is irreversible.

T: There is no interaction between the combustible substance and the oxygen/air. Oxygen/air may be or may not be recognized as necessary for combustion to take place. Burning is a destructive process. The destructive process may release or liberate substances from the combustible substance. It is irreversible.

M: Oxygen/air is not involved in the change. The change is reversible.

Flame/fire

C: Energy changes may be observed but are not explained. The flame/fire is evidence of a chemical reaction. The flame contains both the combustible substance and oxygen/air reacting.

T: The flame/fire is an active agent of change. Air/oxygen may be needed to 'feed the flame' or 'keep it alive'. Air/oxygen is transmuted by the flame/fire or is consumed by it. Matter may be transmuted into heat. Flames contain only the combustible substance or oxygen/air or possibly both but with no interaction.

M: The flame/fire is a source of heat to make the modification occur.

Products and reactants

C: The products contain the reactants in a different chemical combination. Mass is conserved provided that pupils think that gases weigh something or gas is not 'lost' to the atmosphere. Properties are not conserved.

T: Substance is changed from one substance to another or into nothing during combustion. Oxygen/air may be transmuted separately into a product. It may be needed but does not interact in a chemical sense. Mass may increase, decrease or stay the same (because the transmuted products are different from the reactants). Properties are not conserved. Properties may be substantialized and therefore be involved in the transmutation.

M: One substance changes to a different form of the same substance. Substance is conserved. Mass may be conserved, but this depends on whether different forms of the same substance are considered to weigh the same. Some properties are conserved.

and McGuigen, 1991; Brooks and Driver, 1989). Brooks and Driver reported that even at age 16 about three-quarters of pupils thought that air had zero or negative weight. Pupils may, therefore, conserve matter in their explanations of chemical reactions, but not necessarily conserve mass.

The ideas in these papers and the results from a questionnaire study were used as a basis for developing operational definitions (see Table 19.1) of categories of pupils' explanations (Prieto *et al.* 1993, Watson *et al.* 1995). These operational definitions treat the 'Disappearance' category as a limiting form of 'Transmutation', where the material is transmuted to nothing, as in Andersson's earlier paper (1986a) and Pfundt (1981). The 'Displacement' category did not seem to fit the data in the present study. The analysis of pupils' explanations in this paper has, therefore, been based on the categories 'Chemical reaction' (C), 'Transmutation' (T), and 'Modification' (M).

In addition to the three explanatory categories described in Table 19.1 there is a further category of description (D) in which pupils simply describe what they see. There may be a limited amount of inference based on the observations, but no underlying explanation of combustion, e.g., if a substance burns away to an ash, they may expect it to weigh less because it looks as though it should.

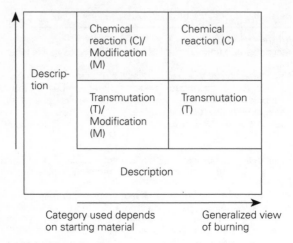

Figure 19.1:   A model of progression

A Model of Progression

A model of progression proposed by Prieto *et al.* (1993) is shown in Figure 19.1. Progression is represented by moving diagonally from the bottom left of the figure to the top right. The model proposes two dimensions to progression: the development of ideas about interaction (vertical axis), and the development of a generalized view of burning (horizontal axis). The category of description has been placed around the left and bottom of the figure to indicate its lowest level in the hierarchy of explanations.

In this study the responses of pupils to the same questionnaire, given at age 14–15 and then sixteen months later, are analysed in order to determine how their explanations had changed and whether their explanations shifted in the directions expected from the model proposed in Figure 19.1.

Methodology

A questionnaire containing nine mainly open questions, was developed through a series of four pilot studies which involved about 150 pupils. The main foci of the questionnaire were: the requirements; the process; the products (including aspects of conservation) and the examples of combustion that pupils give. The questionnaire was administered to 299 pupils aged 14 and 15. The questionnaire was administered again to a subsample of thirty-three of the same pupils sixteen months later.

The studies of the younger pupils had included practical work to identify a range of gases (including carbon dioxide, hydrogen and oxygen); observation of demonstrations of the change in volume of air during burning of copper and various other substances; weight changes on heating, due to evolution of, or reaction with, gases. They had also heated a range of metals in air or oxygen and seen demonstrations of more reactive metals being heated in air or oxygen. At the later age they were nearing the end of their science course leading to a national external examination to be taken at age 16, and had studied a wide range of chemical reactions, including use of chemical equations, structure and bonding.

The responses to five of the questions were categorized using the operational descriptions in Table 19.1. The other questions mainly provided qualitative data showing the contexts in which pupils thought about combustion. Sometimes it was only possible to distinguish whether pupils were using (M) or (C)/(T), but not to distinguish between (C) and (T). These responses were categorized as (X). Some pupils provided a lot of information in their responses which enabled their responses to be allocated to a particular category with confidence. For other pupils less information was available for categorization. Nevertheless the operational definitions in Table 19.1 allowed the responses to be categorized reliably. The responses to thirty of the 299 questionnaires administered to pupils aged 14–15 were categorized by two independent coders and an agreement of 94 per cent was achieved.

The next stage in the analysis was to look for patterns in the responses of individual pupils across questions at the two ages, and to use these patterns to categorize pupils on the two dimensions indicated in Figure 19.1. This was followed up by an analysis to identify conceptual problems which may have been inhibiting progress.

Results

Progression in Developing a Generalized View of Burning

At both ages many pupils seemed to confuse change of state with burning. At the older age, seventeen pupils gave the same number of modification responses whereas seven gave more and 9 gave fewer than at the younger age. The most common of these was in question 6 where many pupils considered the wax of the candle to be simply melting. Another example was in question 9 where some pupils thought that the magnesium had simply changed from a 'solid' form to a 'powdered' form and so had not changed weight.

Progression between Descriptive, Transmutation, Intermediate and Chemical Reaction Models of Burning

In order to determine which model of combustion pupils were using, responses to individual questions have been combined to form four groups of pupils: 'Chemical reaction', 'Intermediate', 'Transmutation' and 'Descriptive'. If a pupil used Chemical reaction explanations on three or more occasions, or used Chemical reaction explanations on two occasions and did not use Transmutation explanations, the pupil was placed in the Chemical reaction group. Pupils categorized in the Transmutation group use Transmutation explanations and no Chemical reaction

Table 19.2: Combined classification of pupils with respect to Transmutation/Chemical reaction

Chemical reaction
   (i) 3 or more C + others, (ii) 2 or more C + others which are only X

Intermediate
   (i) 2C + some T, (ii) 1C + some T, (iii) Some C + some X + some T, (iv) 3 or more X

Transmutation
   (i) All T (3T, or 4T, or 5T, or 6T) (ii) Mixed T and X (3T + 1X, OR 4T + 1X, 2T + 2X)

Descriptive
   2 or fewer explanations: responses mainly descriptive

Table 19.3: Progression after 16 months
(number of pupils at two ages)

Age 15/16	Age 14/15				
	Descriptive/ Other	Transmutation	Intermediate	Chemical reaction	Total
Descriptive/Other	4	0	0	0	4
Transmutation	1	2	2	0	5
Intermediate	1	5	14	1	21
Chemical reaction	0	0	2	1	3
Total	6	7	18	2	33

explanations, whereas the Intermediate group uses both kinds of explanations or Intermediate explanations only. The fourth group contains pupils who rarely explain, but simply describe. The combined categories are summarized in Table 19.2.

The combined categories were used to classify the same pupils at age 14–15 and again 16 months later as shown in Table 19.3. For the majority of pupils (twenty-one pupils) there is no overall change in classification over the period of 16 months and so these pupils appear on a diagonal line going from top left to bottom right of Table 19.3. Three pupils seem to have regressed and so appear above this diagonal, whereas nine pupils seem to have progressed and appear below the diagonal.

The patterns of change between ages 14–15 and ages 15–16 within the four categories of pupils will now be discussed with respect to the changes in pupils' conceptions of the reactants involved, how the transformation takes place, and the products of the reaction.

Reactants

The open format of the questionnaire allowed pupils to mention oxygen as a reactant in response to up to six of the questions. The need for oxygen for burning was recognized by the vast majority of pupils at both ages. Almost all the pupils in the other categories mentioned the need for oxygen at both ages, in at least one example, except those categorized as 'Descriptive' at age 14–15. There was an overall increase with age, in the numbers of times that oxygen or air was mentioned as taking part in combustion in response to the whole questionnaire. The mean number of times mentioned at age 14–15 is 3.0 and at age 15–16 is 4.0 (significant at 0.001 level). There was, however, some uncertainty about the role of the oxygen. Many pupils described oxygen as interacting with the flame. For example, in a question about a candle in a gas jar with its lid on:

Q6b   What has happened to the air inside the gas jar?
P15   It has been used up by the flame.
P44   The flame used it up as if it was breathing.

About half of the Intermediate group (i.e., eight out of seventeen at age 14–15 and nine sixteen months later) and a few in the Transmutation group (i.e., one at age 14–15 and three sixteen months later) and none in the Descriptive group gave this kind of explanation. It is interesting to note that this kind of alternative explanation did not decrease with age.

How the Transformation Takes Place

The data from this study indicate that most pupils have difficulty in envisaging how combustion occurs. Pupils spontaneously made a small number of references to particles or microscopic structure. In all eleven pupils at age 14–15 and twelve pupils at age 15–16 used ideas of particles and structures in their explanation, but only three of these used the ideas at both ages. A typical example is:

Q4   Why do you think some substances can burn and others cannot burn?
P42, age 15   Some substances can burn because of the number of electrons in their outer shell. If the shell needs to dispense (lose) or get one or two electrons it gets very reactive and reacts with heat.
P42, age 16   Some substances can burn because they are a fuel, but others like metal are not a fuel and therefore cannot burn.

A similar spontaneous low level of use of the particulate theory is also noted in a study by Abraham, Williamson and Westbrook (1994).

Many pupils used causal reasoning (see Andersson, 1986b) to explain the process of combustion. The idea of the flame consuming oxygen is mentioned above. Many pupils also considered the flame or heat to actively change the combustible substance:

Q6(a) What do you think has happened to the wax of the candle?

P42, age 16 It has been melted by the flame and when the flame goes out the wax will reset hard.

Q6(b) What has happened to the air in the gas jar?

P42, age 16 It has been burned up by the flame and changed into carbon dioxide.

Q6(c) Is anything formed that you cannot see? Please explain your answer.

P42, age 16 Yes, carbon dioxide. The oxygen cannot simply disappear because there is something there to start with (oxygen) or air, then it must have been changed or converted to something else. We know this even though we cannot see it.

This kind of explanation is a characteristic of Transmutation explanations and was used by about three-quarters of the Transmutation respondents at both ages. A small number of Intermediate respondents also used this kind of explanation and there are indications that this kind of explanation may be used more by the older group (i.e., by seven pupils at age 14–15 and twelve at age 15–16). Similar explanations were given in response to question 8 about putting out a fire. One-third of the pupils used very active language and talked about 'killing the fire', 'smothering' or 'suffocating' the flame, 'drowning' the oxygen, or 'drowning' the fire.

Products

The number of products of combustion that were named correctly increased with age and is one of the factors that accounts for the shift from the categorization of pupils in the Transmutation category towards the Chemical reaction category. For example thirteen pupils correctly identified one product in at least one reaction at age 14–15 and twenty-four did so at age 15–16.

In spite of the general trend with age to identify more products correctly, heat is also viewed by about a third of all respondents as a product. These respondents are spread fairly equally between the different categories of pupils. Some of these pupils also view heat as a substance. For example:

Q7 When a match burns, you see a flame. What is the flame made of?

P43, age 16 'Pure heat energy'.

The idea of heat as a substance increased with age in all categories of pupils with a total from four pupils to fourteen in the older age group. Three of these pupils used this idea consistently at both ages.

Conclusion

The model of progression presented in Figure 19.1 divides progression into two dimensions. There appears to be no discernible pattern relating the progress of the pupils in this study along the two axes. They appeared to make no progress along the horizontal axis, i.e., in distinguishing reversible and irreversible changes, but some made progress along the vertical axis from Transmutation to Chemical reaction explanations.

The model of progression indicates that as pupils progress along the horizontal axis they will become better at recognizing examples of heating which involve combustion (irreversible change) and those which do not (reversible change). The lack of progress on this dimension reported in this study, has also been noted in studies by Boo (1994) and Abraham *et al.* (1994) where the problem was found to persist with more advanced students. A possible explanation for this may lie in the curriculum structure of chemistry: chemistry tends to be organized around topics such as combustion, oxidation, or fuels, rather than focusing on transformations which to the pupils appear very similar. An assumption in such a curriculum is that distinguishing between physical and chemical change is largely unproblematic and that pupils do not need to be taught specifically how to do this. Perhaps more attention needs to be given to comparing explicitly such processes as burning alcohol and boiling water, or burning wax and melting ice. It appears that pupils need more help in identifying what aspects of such processes are similar and what aspects are different, so that they are better able to distinguish between chemical and physical change.

Progress along the vertical axis on Figure 19.1 was more substantial. Pupils identified more reactants and products correctly and appeared to be moving towards chemical change as a process of interaction. The pupils, however, seemed to have difficulties in envisaging how the reaction takes place and particularly the role of oxygen. The particle model of chemical reactions is notable for its absence. A minority of pupils used this model spontaneously and most of those that did, only used it at one of the two ages, suggesting that the model was not well integrated into their explanatory frameworks.

Pupils tended to use causal explanations (Andersson, 1986b) for chemical change. In their explanations some identified the flame as interacting with oxygen or the combustible substance, and the type of interaction was one where the flame acted upon these substances to change them. Many pupils seemed to view the flame almost as a living entity: it is often referred to as such in everyday language. A fundamental confusion here was between things and processes. The flame was being viewed as an entity in itself, rather than a manifestation of a process of interaction. Similarly heat was often viewed as a product rather than as a process of energy transfer, and the idea of heat as a substance increased with age. It may be that formal study of exothermic and endothermic reactions had reinforced the concept of heat as a substance, for example by writing enthalpy changes as part of chemical equations. Transmutation explanations are clearly able to incorporate the idea of matter being transmuted into heat during combustion.

The difficulties that pupils have in distinguishing substances and properties has been noted in a number of studies in the areas of physics and chemistry (Chi, 1991; Paris, 1992; Reiner, Chi, and Resnick, 1988; Sanmartí, 1989, Sanmartí, Izquierdo and Watson, 1995). Chi maintains that it is psychologically impossible to change pupils' conceptions which incorrectly identify processes as substances to the scientifically acceptable ones. Instead teachers should aim to build a competing scientifically acceptable framework of explanation which pupils will eventually see as more useful than their alternative explanations; or at least which they will be able to apply in appropriate situations.

This leads again to a consideration of the curriculum structure. A fundamental assumption in secondary chemistry curricula is that pupils know the differences between substances and properties. One problem, however, in changing from a Transmutation framework to a Chemical reaction framework is the difficulty of differentiating between processes of interaction and their manifestations, and substances. It appears that the curriculum should explicitly address the problem of how pupils can distinguish between substances and properties.

This study shows that pupils are insecure in their understanding of some fundamental concepts and indicates a need to re-assess chemistry curricula in the light of what is now known about pupils' difficulties in understanding.

Note

1 This work is part of larger study and we would like to thank Dra. T. Prieto of the University of Malaga, for help in developing the questionnaire and the analytical frameworks used, and the British Council which funded the collaboration with the University of Malaga through the Acciones Integradas programme.

References

ABRAHAM, M.R., WILLIAMSON, V.M. and WESTBROOK, S.L. (1994) 'A cross-age study of the understanding of five chemistry concepts', *Journal of Research in Science Teaching*, **31**, 2, pp. 147–65.

ANDERSSON, B.R. and RENSTROM, L. (1983) *How Swedish Pupils, Age 12–15 Explain the 'Exhaust' Problem*, EKNA group, Department of Education and Educational Research, University of Gotenborg, Box 1010, S-431, 26 Moindal, Sweden.

ANDERSSON, B.R. (1986a) 'Pupils' explanations of some aspects of chemical reactions', *Science Education*, **70**, 5, pp. 549–63.

ANDERSSON, B.R. (1986b) 'The experiential gestalt of causation: A common core to pupils' preconceptions in science', *European Journal of Science Education*, **8**, 2, pp. 155–71.

ANDERSSON, B.R. (1990) 'Pupils' conceptions of matter and its transformations (age 12–16)', *Studies in Science Education*, **18**, pp. 53–85.

BAR, V. and TRAVIS, A.S. (1991) 'The development of the concept of evaporation', *J. Res. Sc. Teach.*, **28**, 4, pp. 363–82.

BOO, H.K. (1994) 'A-level chemistry students' conceptions and understandings of the nature of chemical reactions and approaches to the learning of chemistry content', Doctoral thesis, King's College London, University of London.

BOUJAOUDE, S.B. (1991) 'A study of the nature of students' understandings about the concept of burning', *Journal of Research in Science Teaching*, **28**, 8, pp. 689–704.

BROOKS, A. and DRIVER, R. (1989) *Progression in Science: The Development of Pupils' Understanding of Physical Characteristics of Air across the Age Range 5–16 Years*', CLISP, Centre for Studies in Science and Mathematics Education, University of Leeds.

CHI, M. (1991) 'Conceptual change within and across ontological categories: Examples from learning and discovery in science', in GIERE, R. (Ed) *Cognitive Models of Science: Minnesota Studies in the Philosophy of Science*, Minneapolis, MN, University of Minnesota Press, pp. 133–90.

DRIVER, R., CHILD, D., GOTT, R., HEAD. J., JOHNSON, S., WORSLEY, C. and WILEY, F. (1984) *Science in Schools: Age 15, Research Report no. 2, APU*, Dept. of Education and Science, London.

MEHEUT, M. (1982) 'Combustion et réaction chimique dans l'enseignement destiné à des élèves de sixième', These de doctorat de troisième cycle, Université de Paris.

MEHEUT, M., SALTIEL, E. and TIBERGHIEN, A. (1985) 'Pupils' (11–12 year olds) conceptions of combustion', *European Journal of Science Education*, **7**, pp. 83–93.

OSBORNE, R.J. and COSGROVE, M.M. (1983) 'Children's conceptions of the changes of state of water', *Journal of Research in Science Teaching*, **20**, pp. 825–38.

PARIS, M. (1992) 'Pupils' understanding of the concepts of acid and base', Doctoral thesis, Universitat Autonoma de Barcelona.

PFUNDT, H. (1981) 'Pre-instructional conceptions about substances and transformations of substances', *Proceedings of the International Workshop on 'Problems Concerning Students' Representation of Physics and Chemistry Knowledge'*, Ludwigsburg, pp. 320–41.

PRIETO, T., WATSON, J.R. and DILLON, J.S. (1993) 'Pupils' understanding of combustion', *Research in Science Education*, **22**, pp. 331–40.

REINER, M., CHI, M. and RESNICK, L. (1988) 'Naive materialistic belief: An underlying epistemological commitment', *Tenth Annual Conference of the Cognitive Science Society*, Montreal, Canada, pp. 544–51.

ROSS, K. (1991) 'Burning — A constructive not destructive process', *School Science Review*, **72**, 251, pp. 39–50.

RUSSELL, T., LONGDEN, K. and MCGUIGEN, L. (1991) *Materials*, Primary SPACE Project Research Report, Liverpool University Press.

SANMARTI, N. (1989) 'Difficulties in understanding the difference between the concepts of mixture and compound', Doctoral thesis, Universitat Autonoma de Barcelona.

SANMARTÍ, N., IZQUIERDO, M. and WATSON, J.R. (1995) 'The substantialisation of properties', *Science and Education*, **4**, 4, pp. 1–21.

SCHOLLUM, B. and HAPPS, J.C. (1982) 'Learners' views about burning', *The Australian Science Teachers' Journal*, **28**, 3, pp. 84–8.

WATSON, R., PRIETO, T. and DILLON, J. (1995) 'The effect of practical work on students' understanding of combustion', *Journal of Research in Science Teaching*, **32**, 5, pp. 487–502.

20   Students' Conceptions of Quantum Physics

Azam Mashhadi

Abstract

Elementary particles seem to be waves on Mondays, Wednesdays and Fridays, and particles on Tuesdays, Thursdays and Saturdays. (Sir William Bragg)

Over the last fifteen years there has been considerable research interest in the student's perceptions of phenomena in such areas as energy, motion, the particulate nature of matter, electricity, and light. However, ninety years after the genesis of Quantum Physics significant research on students' understanding of such revolutionary phenomena is only beginning to emerge.

What are electrons really like? Are they like particles or waves? Are they like both particles and waves, or like neither? These questions illustrate the psychological difficulties with which students are confronted when trying to incorporate the concepts of quantum physics into their overall conceptual framework. They also illustrate the difficulties in using analogies taken from ordinary experience (i.e., essentially classical models) to 'explain' the subatomic world. In its predictive abilities quantum theory is the most successful physical theory that has ever been conceptualized, and yet Einstein once remarked that quantum theory reminded him of 'the system of delusions of an exceedingly intelligent paranoiac, concocted of incoherent elements of thought.' (In Arthur Fine, 1986).

Following a review of previous research, and a critical discussion concerning the learning and teaching of Quantum Physics an interim report is presented on a study to elicit students' conceptions of quantum phenomena.

Introduction

> ... I think I can safely say that no one understands quantum mechanics. ... do not keep saying to yourself, if you possibly can avoid it, 'But how can it be like that?' because you will get 'down the drain', into a blind alley from which nobody has yet escaped. Nobody knows how it can be like that. (Richard Feynman, 1967, p. 129)

Especially over the last fifteen years there has been considerable research interest in the students' perceptions of phenomena in such areas as energy, motion, the particulate nature of matter, electricity, and light. Ninety years after the genesis of quantum physics significant research on students' understanding of such revolutionary phenomena is only beginning to emerge. The aims of this new study, the Students'

Conceptions of Quantum Physics Project (SCQP), are to elicit students' conceptions of quantum phenomena, investigate their ideas of metaphors and analogies in constructing conceptual models, develop a model of cognitive adaptation to a new paradigm, and evaluate the efficacy of the incorporation of quantum physics at the pre-university level. The study should lead to more effective teaching and learning strategies, and inform policy and curriculum decision-making.

The basic ideas of quantum physics are more strange than difficult. In some situations, electrons that are usually referred to as 'particles' may exhibit 'wave-like' behaviour. Electromagnetic radiation, known classically as a wave phenomenon, is explained in terms of particles called photons. Both matter and radiation can be viewed as having a dual (wave-particle) nature. What are electrons *really like*? Are they like particles or waves? Are they like both particles *and* waves, or like neither? These questions illustrate the psychological difficulties which confront students when trying to incorporate the concepts of quantum physics into their overall conceptual framework. They also illustrate the difficulties in using analogies taken from ordinary experience (i.e., essentially classical models) to 'explain' the subatomic world. In its predictive abilities quantum theory is the most successful physical theory that has ever been conceptualized, and yet Einstein once remarked that quantum theory reminded him of 'the system of delusions of an exceedingly intelligent paranoiac, concocted of incoherent elements of thought' (In Arthur Fine, 1986).

At present in the UK, upper secondary school students (ages 16–18) wishing to read for a physical science degree at university will follow the two-year Advanced Level Physics course. The quantum physics section of the course syllabus will typically not include the Heisenberg Uncertainty Principle, the Schrödinger wave equation, and there is no explicit mention of introducing students to conceptions of the 'nature of science'.

Previous Research Findings

The most systematic and extensive research to date has been carried out by research groups based in the University of Bremen, and the Free University of Berlin. The research group headed by Professor Niedderer, based at the University of Bremen in Germany, has implemented a teaching approach for grade 13 students (age 18–19) in upper secondary school based on the following principle (from Niedderer, Bethge and Cassens, 1990, p. 67)

> From Bohr to Schrödinger: whereas most teachers at the moment teach Atomic Physics on the basis of Bohr's model, the Schrödinger model, within our more qualitative approach based on the notion of standing waves, allows for more and better explanations, especially in relation to chemistry, and is nearer to what scientists of today believe.

Fischler and Lichtfeldt (1992, p. 183), based at the Free University of Berlin, advocate an approach to teaching quantum physics in which:

1 Reference to classical physics should be avoided.
2 The teaching unit should begin with electrons (not with photons when introducing the photoelectric effect).
3 The statistical interpretation of observed phenomena should be used and dualistic descriptions should be avoided.
4 The uncertainty relation of Heisenberg should be introduced at an early stage (formulated for ensembles of quantum objects).
5 In the treatment of the hydrogen atom, the model of Bohr should be avoided.

Niedderer (1987, p. 345) reported on Bormann's (1987) work on students' attempts to reconcile the wave-particle duality of electrons:

1 The 'strict' particle view
  Students looked at electrons as particles moving along straight lines. The observations of electron distributions were explained by collisions.
2 The particle moving along a wave
  The electron is a particle (mass, velocity, orbit). This particle moves along a wave-orbit. The electron is the oscillator of the wave.
3 The formal wave conception
  The diffraction pattern is explained by an electron wave. Either the electron is a wave itself or there is a new kind of wave (which is influenced by a magnetic field).

In addition Bormann works on the following hypotheses:

• The particle view is easier for students to understand than the wave view.
• The electron is a 'real' particle, the photon is a sort of 'energy particle'.
• Photons and electrons are primarily particles which should have some wave properties to explain special sophisticated experiments.

Niedderer, Bethge and Cassens (1990, p. 77) provide a summary of some of Bethge's (1988) investigation of grade 13 (age 18–19) students:

Characteristics of Students' Own Reasoning

1 Students have a concrete picture of the atom, in terms of mechanics and the everyday life-world.
2 Students tend to use the concepts of movement and trajectory in their own explanations of properties of the atom (even if they deny them!)
3 Students tend to use the concept of energy and mass conservation in their own explanations.
4 On the other hand, students do not spontaneously request further explanations of the existence of discrete energy levels, but tend to use them as a basis for other explanations.

A Second Level of Description Is More Related to Students'
Preconceptions

1   Movement (and trajectory) are continuous; for every two points of the movement, the points between also belong to the movement, even if they are not observed. At the beginning and at the end we have the same body, even if we have not watched it in between.
2   A trajectory is a definite and ordinary path, such as a circle or an ellipse, but not some strange zig-zag-movement.
3   The stability of an atom is the result of a balance between an attractive electric force and the activity (= force or energy!) of the movement of the electron. The electrodynamical problem of stability is not present in students' views.
4   Energy is seen as some activity or general cause which is specified in special situations (sometimes as a force, or as energy in a physical meaning or even as a kind of matter).
5   Probability is seen as some kind of inaccuracy. If you do not know something exactly, you talk about probability.

Fischler and Lichtfeldt (1992, p. 187) found that the following conceptions of the 'atom-electron' were found most often in their study of 240 A-level students (*Leistungskurse* course in the upper Gymnasium or grammar school):

Circle (circular orbit): conceptions of electrons which fly round the nucleus with (high) velocity in fixed, prescribed orbits. In this conception the centrifugal force and the Coulomb (electric) force are brought into equilibrium. The students use their experience with roundabouts first to explain the movement of the planet, and then second to explain the process in atomic shells, without regard to reference systems (63 per cent of 240 students in both groups).
Charge: students have a fixed conception of the repulsion between charges. They often explain the properties of charges incorrectly. The charges of both the proton and the electron cause a distance between the two particles (similar to a bipolar dumbbell). The students assemble a suitable conception from single elements of knowledge (23 per cent of 240 students in both groups).
Shell: conception of a firm casing (shell, ball) on which the electrons are fixed or move (8 per cent of 240 students in both groups).
[After the unit was taught another 'conceptual pattern' was constructed from students' responses:]
Loc. (localization energy): the stability of atoms was regarded by the students as connected with the Heisenberg uncertainty principle. According to this conception, the mere restriction of space results in a rise of the kinetic energy of the electrons, the loci of which are subjected to a statistical distribution. At the same time the students dispensed with statements about single electrons which they thought of as inconceivable.

The research by the Bremen group indicates that for students mechanical thinking in terms of orbits of classical particles is dominant. Fischler and Lichtfeldt (1991, p. 257), in Berlin, interpreted their study as finding that the:

> . . . results of the control group meanwhile pointed to an incorporation of the 'new' phenomenon into the 'old' mechanistic ideas. Here, the different ideas in quantum physics were merely acquired verbally in the science language level and forgotten again afterwards. The conscious top–down process of reconstruction which had to be done by the students in the everyday language was not possible for them.

The German educational system is different from that of the UK and, since the 18–19 age group was being considered, the research findings may not be directly applicable. The quantum physics section of A-level physics syllabuses in terms of both the extent and depth of their coverage of the topic is different to the German syllabus. This current study was, therefore, initially concerned with seeing if A-level students in England hold similar conceptions.

Interim Report on the SCQP Project

This preliminary study of the Students' Conceptions of Quantum Physics Project (SCQP) consisted of a semi-structured questionnaire completed by A-level physics students (N = 57) in three Oxfordshire secondary schools in May 1993. The questionnaire utilized open and closed questions, drawings of particular situations, and attitude scales.

How Do Students View the Atom?

Following an interpretative analysis of responses to questions concerning 'the atom' (see the Appendix, and Questions C2, C5, and C10), the following broad conceptions of the atom were constructed:

- mechanistic picture;
- probabilistic picture;
- 'random' motion picture;
- 'smeared charge cloud'; and
- no visualization possible.

1 **The mechanistic conception** (held by about 25 per cent of the students) consisted primarily of (many) fast-moving electrons in definite orbits, similar in some ways to the planetary model of the atom:

> Because electrons orbit so fast that we can't tell where one is at any time — therefore it is inaccurate to draw them at one place. (22/C10)

The planetary model is not necessarily the same as the Bohr model, not only was there no mention of Bohr's postulates but the term 'Bohr model or atom' was not explicitly mentioned by the students. Elements of language from the Bohr atom were used (e.g., electron orbits, energy levels etc.), but it is doubtful if the students actually had the Bohr model in mind. There was an acknowledgment by many students that the planetary model of the atom is a useful picture but with recognized limitations:

> The analogy has certain likeness but is also dissimilar to the structure of an atom. In a solar system planets are held in orbit by a gravitational force and in an atom electrons are held by an electrostatic force of attraction. However the nucleus of an atom is massive and many times larger than the electrons. Whereas this size discrepancy is not evident in the solar system. Electrons move between orbital whereas planets don't. (10/C2(b))

The orbit is regarded as the result of a 'balance' (as several students expressed it) between the electron's speed and the electrostatic force of attraction between electron and nucleus:

> The electron has a negative charge and is travelling at a certain speed. The nucleus has a positive charge and so attracts the electron. This keeps the electron in place and everything is balanced. (3/C5(a))

2   A significant percentage (about 25 per cent) regarded electron clouds as providing a **probabilistic picture**, but they still thought in terms of 'the electrons', i.e., as particles:

> You can't say where you will find an electron, only draw in areas or more correctly volumes where there is a greater than 95 per cent chance of finding an electron. (43/C10)

The Heisenberg Uncertainty Principle does not form part of the syllabus, and the 'standing electron-wave' model, if it is taught at all, is only briefly touched upon so it is unclear whether this probability view stems from a recognition of the wave nature of the electron or is viewed as the result of imprecision in measurement or randomness in movement. Further study needs to be undertaken of their conceptions of this, as well as their perception of the nature of 'probability'. One student made a specific reference to Heisenberg:

> ... I think this is what physicists argue in accordance with Heisenberg's Uncertainty Principle. Although the notion of fundamental uncertainty makes me dubious as to whether quantum mechanics is a complete model of reality. (40/C7(e))

3   **The 'random' motion picture** (about 23 per cent) consisted of combinations of the mechanistic and probability/random viewpoints involving random movement *within* a bounded region or *at* different energy orbits (a 'shell'):

Electrons do not move in a circle around the nucleus, like a planet does around the sun, instead it moves randomly but in the shape of a certain shell, therefore we can predict that at one instant the electron may be at that point but we can never be sure, therefore they draw a cloud. (46/C10)

4 A smaller number (about 10 per cent) talked in terms of a '**smeared charge cloud**':

Electrons have no shape they are charge clouds and so could not be individual but all together. (26/C10)

When orbiting an atom, the electron does not occupy only one space at any one time but instead is 'spread out' all around its orbit. (32/C10)

5 In addition a few students (about 5 per cent) argued that **visualization was not possible**:

. . . I believe it is very difficult if not impossible to conceive what is actually going on. Our visual models are derived from experience through evolution of the environment we are in the world of miniature particles is totally alien to us. (40/C7(g))

How Do Students View Electrons?

How do students view electrons or the behaviour of electrons when faced with a diffraction effect? Two of the questions focused on the 'electron diffraction tube', and a situation in which electrons encounter a single slit (see the Appendix, Questions C7 and C8). Students' conceptions of the electron when faced with phenomena that illustrates their 'wave behaviour' are quite tangled. Certain broad conceptions do, however, emerge with electrons regarded as:

- 'classical' particles;
- waves;
- linked to 'probability waves';
- 'smeared charge'; and
- cannot be visualized.

1 Many students, just under a third, still adhered strongly to the **classical** particle or 'electron-as-particle' viewpoint, with electrons having a definite trajectory. Comments included:

This implies that electrons are waves, and so must be nonsense because electrons behave like particles, therefore cannot interfere either constructively or destructively. (18/C7(c))

Students with this classical viewpoint adopted a straight line path (in response to Question C8), with the electrons hitting the screen at one point. Typical comments included:

As the slit is so large compared to an electron, I think that they will be unaffected by it and all hit the screen in the same place. (15/C8)

2   In their responses to the diffraction tube roughly two-thirds of the students associated electrons with **waves**, and talked in various ways of 'electron diffraction/ interference'. However this is quite a broad conception, and it is unclear whether they are thinking in terms of electrons as particles with wave properties, particles that turn into waves, or electrons as waves that interfere. Typical comments included:

The electrons are behaving like waves, however the nuclei of the graphite atoms are acting on the electric charge of the electron and diffracting them, the electron waves then meet in certain places and interfere. (32/C7(d))

One student made explicit reference to the 'standing wave' model of the electron-atom:

The energy of the electron. The electron forms a standing wave around the nucleus. If it were to approach closer, the standing wave would be disrupted. (42/C5(a))

3   Only a few of the students (about 4 per cent) talked explicitly in terms of a **'probability wave'**:

The path of a particle is undetermined. There are an infinite number of paths, with paths of destructive interference having the least probability, and vice versa. The path that the electron takes is governed by this probability, and can only be determined when it strikes the screen, i.e., its wave properties are 'removed'. (42/C7(c))

4   Another minority viewpoint (about 4 per cent) regarded electrons as consisting of **'smeared charge'**:

They consist of smeared charge at different distances from the nucleus. (20/C7(a))

5   A similarly small number of students (about 4 per cent) argued that **visualization is neither possible nor desirable**:

... unfortunately all that is known about electrons is just theory because no one can ever see an electron because these are smaller than the wavelength of visible light. So really, it is just a case of whichever theory makes the most correct predictions. (32/C7(b))

Conclusion

The preliminary results are generally consistent with previous research in other countries, and indicate that for students mechanical thinking in terms of orbits of classical particles is dominant. The students, largely, are not conscious of their own conceptions and consequently do not begin to question them. The preliminary results of the study indicate that students have incorporated the 'new' quantum phenomena into the 'older' mechanistic conceptions. The current data implies that most students are not epistemologically aware that quantum physics constitutes a new 'paradigm'.

Further work needs to be carried out using a larger population sample to establish the generalizability of any findings, more focused research on students' conceptions of figurative language, their perceptions of the nature of theoretical entities, and to investigate the interrelationships between conceptions.

References

BETHGE, T. (1988) 'Aspekte des Schülervorverstandnisses zu grundlegenden Begriffen der Atomphysik (Aspects of student's matrices of understanding related to basic concepts of atomic physics)', PhD thesis (in German), University of Bremen, Germany.

BORMANN, M. (1987) 'Das Schülervorverständnisses zum Themenbereich "Modell-vorstellungen zu Licht und Elektronen" (Students' Alternative Framework in the Field of Particle and Wave Models of Light and Electrons)', in KUHN, W. (Ed) *Didaktik der Physik*, Vorträge, Physikertagung 1987, Berlin, Gießen.

FEYNMAN, R. (1967) *The Character of Physical Law*, Cambridge, MIT Press.

FINE, A. (1986) *The Shaky Game*, University of Chicago Press.

FISCHLER, H. and LICHTFELDT, M. (1991) 'Learning quantum mechanics' in DUIT, R., GOLDBERG, F. and NIEDDERER, H. (Eds) *Research in Physics Learning: Theoretical Issues and Empirical Studies*, Proceedings of the International Workshop, Bremen 1991, IPN, Kiel.

FISCHLER, H. and LICHTFELDT, M. (1992) 'Modern physics and students' conceptions', *International Journal of Science Education*, **14**, 2, pp. 181–90.

NIEDDERER, H. (1987) 'Alternative framework of students in mechanics and atomic physics; Methods of research and results', in NOVAK, J. (Ed) *Proceedings of 2nd International Seminar on 'Misconceptions and Educational Strategies in Science and Mathematics*, July 26–9 Cornell University.

NIEDDERER, H., BETHGE, TH. and CASSENS, H. (1990) 'A simplified quantum model: A teaching approach and evaluation of understanding', in LIJNSE, P.L., LICHT, P., DE VOS, W. and WAARZO, A.J. (Eds) *Relating Macroscopic Phenomena to Microscopic Particles: A Central Problem in Secondary Science Education*, Utrecht, CD-ß Press, pp. 67–80.

Appendix: Questionnaire

C2 People sometimes say that the structure of the atom is similar to the structure of the solar system (i.e., the planets in orbit around the sun).
(a) Do you agree with this?   **(b)** Explain your answer.

C5 (a) In many textbooks there is a diagram like the one below, in which an electron is said to be in orbit around the nucleus of the atom. Explain how the electron stays in orbit.

(b) What do you think lies between the nucleus of an atom and its electrons?
(c) Is this sort of diagram useful, or is it misleading? Does it give people the wrong idea about atoms?

C6 In one of the physics textbooks it says that J.J. Thomson *discovered* the electron in 1895. A student on reading this remarked that J.J. Thomson *invented* the electron.

What do you think? Why should the student have felt that the electron was invented, and not discovered?

C7 The diagram below shows an apparatus in which a beam of electrons is accelerated in an electron gun to a potential of between 3500 and 5000 V and then allowed to fall onto a very thin sheet of graphite. Graphite consists of regularly spaced carbon atoms. As you can see a pattern of concentric rings is produced on the fluorescent screen.

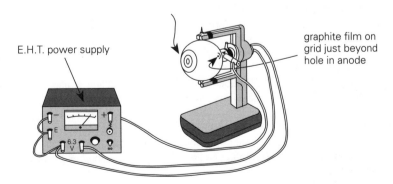

Student A says, 'The pattern isn't being produced by electrons, but by light given off from the hot cathode.' He argues that he can show this to be the case by holding a magnet next to the pattern. Light is not affected by a magnetic field, and so he argues the pattern will stay unchanged. However, to his surprise, when he carries out the experiment, the *pattern is deflected.*

(a) Student B then says, 'These rings are a diffraction pattern. The sheet of graphite is acting just like a diffraction grating.'

If this were the case what would it indicate about the nature of electrons?

(b) At this point student C says, 'That's nonsense, electrons are particles and also negatively charged. Electrons are always repelling each other, and even if tow electrons were to collide they would just bounce off each other. There shouldn't be any pattern at all with electrons. Something else is happening.'

Do you agree or disagree with this? Explain your choice.

(c) Student D forcefully points out, 'Electrons are being shot out of the electron gun. The pattern was deflected by a magnet, so whatever it is must have an electrical charge. That means it isn't due to light being diffracted. That only leaves the electrons. That must mean that *the electrons* are constructively and destructively interfering with each other.'

What do you think? Does this sound reasonable or 'nonsense'?

(d) Student B then says, 'The chemical on the detector screen is glowing brightly whenever an electron hits it and transfers its kinetic energy. So there are places where there are electrons striking the screen, and places where electrons are not striking the screen. The brighter the ring, the greater the number of electrons hitting that area.'

The teacher, at this point, asks the class, 'If this is the case then how come there are areas where the electrons are going to and areas where electrons are not going to?'

What answer would you give?

(e) Having thought about the situation very carefully, student A says, 'If we want to find out where electrons are then they are most likely to be where there are bright rings, glowing on the chemical coating at the end of the tube. In other words the rings are telling us the likelihood or the probability of where the electrons are most likely to strike the detector.'

Does this sound reasonable? Do you agree or disagree (and why) with his argument?

(f) Student C remarks, 'The pattern does look very like the diffraction patterns we were getting when we looked at the diffraction of light. But this must be just a coincidence, as light and electrons are very different things.'

Why should he say this? Do you agree with him?

(g) Student B then says she is very confused by this experiment, and that she is going to adopt the attitude that there is no point in thinking about what electrons are really like or about what they are doing once they leave the electron gun. She is just going to look up in the textbook the formula which will tell her at what points on the end of the tube the electrons will most likely be at (i.e., the formula which will predict the shape of the pattern), and then just use that formula if she is asked to do any calculations.

What do you think of her attitude or approach? Do you agree with it, or not? Explain your answer as fully as possible.

(h) Student A says that they don't know enough about the situation or about electrons. If they knew more they could explain everything perfectly. What do you think?

C8 The apparatus below acts as a source of electrons. It is, however, a very special piece of apparatus. Electrons can only come out of it one at a time. Draw on the diagram below what you think happens to the electrons. Add any words of explanation on the diagram and/ or in the space below.

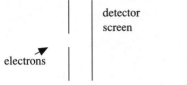

single slit (size about the size of an atom)

C10 In some science textbooks, especially chemistry textbooks, when diagrams of atoms or molecules are drawn they do not show individual electrons in orbit but describe *electron orbitals* or *electron clouds*. Why is this?

Part IV
Students' and Teachers' Perceptions
of the Nature of Science

Introduction
Geoff Welford

This section comprises just two papers, but should not be underestimated in its ability to provoke thought. Leach asks us to consider the fundamental question about the nature and purposes of the science curriculum setting his response in the context of an empirical student-centred study. His findings take us into the twin policy objectives of public understanding and the training of future scientists. Nott and Wellington on the other hand probe teachers' views of the nature of science, suggesting that much teaching about this occurs embedded in most everyday science teaching transactions.

21 Students' Understanding of the Nature of Science

John Leach

Abstract

This paper addresses fundamental questions about the nature and purposes of the science curriculum, in the context of findings from a study of young people's understanding of the nature of science in the 9–16 age range.[1] Three aspects of understanding of the nature of science were addressed in the study, namely the perceived purposes of scientific activity and the demarcation of science as a domain, the nature and status of scientific knowledge and its relationship to empirical evidence, and social and institutional aspects of the scientific enterprise. The representations of science drawn upon by students in a variety of contexts are described. These findings are considered in terms of two commonly stated policy objectives for the science curriculum, namely promoting the public understanding of science and training future scientists.

Epistemology in the Curriculum — An Inevitable Place

Science enjoys a prominent place in the school curricula of most countries in the world, and arguments as to why societies might want their young people to study science have been widely articulated by policy makers. In both the USA and the UK, a case has been made for the place of science in the school curriculum in terms of promoting economic well-being by training the scientists of the future (e.g., Royal Society, 1985; AAAS, 1989; White Paper, 1993). A further rationale is presented which involves promoting 'scientific literacy' or 'the public understanding of science' through school science education for all future citizens: it is intended that individuals will be better equipped to interpret issues with a science dimension as encountered in their personal and professional lives, and thus judgments about the activities of scientists will be made from a more informed viewpoint.

Academics with an interest in science education have also addressed the issues of adult scientific literacy and the public understanding of science. For example, Thomas and Durant (1987) suggest three main reasons for promoting the public understanding of science. The utilitarian case relates to equipping the general public with basic scientific knowledge usefully to inform decisions relating to their everyday lives, the democratic case relates to equipping individuals with appropriate scientific knowledge for their role as citizens in a scientific and technological democracy, and the cultural case relates to citizens knowing about science as an important cultural product, and its role in contemporary scientific and technological societies.

The notions of 'scientific literacy' and 'public understanding of science' are recognized as problematic (e.g., Millar, 1993), and academics appear less sanguine than policy makers about the feasibility of promoting scientific literacy and the public understanding of science through the school science curriculum for reasons that will be discussed later. For whatever reason science is placed in the school curriculum, however, questions emerge as to appropriate curricular content. The content of science curricula has been characterized into three broad areas (e.g., Miller, 1983; Wynne, 1990), which I will refer to as the 'contents' of science (the laws, theories and models of science), the 'methods' of science (the activities through which reliable and valid knowledge claims are arrived at) and the 'social relations' of science (the social processes, both internal and external to science, inherent in scientific activity).

Science curricula designed primarily for the training of future scientists, or for promoting the public understanding of science for utilitarian, democratic or cultural reasons, will include the contents of science at some level. For example, a case could be made for including the germ theory of disease as part of the training of future scientists to work in the life sciences. This content could also be included as part of the general education of individuals because it is useful knowledge in preventing disease (utilitarian), because it is important in understanding issues of broad social relevance such as the cause, transmission and treatment of HIV/AIDS (democratic), or as a cultural product with broad social implications (cultural). The treatment of the germ theory of disease would inevitably be different for each of these purposes: different content would be relevant to young people training as scientists than those encountering the theory as part of a health education programme.

Although it may be possible for science educators to reach some consensus about curricular content appropriate for the training of future scientists, it is more difficult to imagine how such consensus would arise about appropriate content for utilitarian reasons. This would involve making decisions about the scientific knowledge likely to be required by all individuals in the various personal and professional situations in which they find themselves during their adult lives. For this reason, promoting the public understanding of science through the school science curriculum for utilitarian reasons alone seems an overly ambitious aim. Providing adults with all the scientific knowledge likely to be required for democratic reasons during their future lives seems equally implausible. If scientific knowledge were treated as a cultural product in the science curriculum, then questions arise as to what aspects of that cultural product should be included for all students. Although different perspectives would certainly arise, it is not hard to imagine that some degree of consensus could be reached.

Understanding the contents of science inevitably involves some understanding of the epistemological basis of scientific knowledge claims (Driver *et al.*, 1994a). For example, a scientific understanding of the germ theory of disease goes beyond merely knowing descriptions of the actions of particular pathogenic micro-organisms. It also involves knowing about acceptable modes of explanation in science (so that teleological explanations of disease are recognized as unacceptable), and appreciating that scientific knowledge is intended to be parsimonious, logically and empirically consistent and generalizable. Scientific understanding also involves understanding the nature of scientific knowledge and its limitations in addressing real-world problems. For example, the application of abstract scientific models involving entities such as point masses and friction free planes in order to make predictions about motion in real world contexts is inherently problematic.

A case can also be made for including the methods and social relations of science in curricula aimed at training future scientists, or promoting the public understanding of science. A compelling case has been made that in order to understand 'the actual influence and potential power that scientists may exert on society' it is important that scientists themselves develop sophisticated understandings of the institutional nature of science and its political control (Husén *et al.*, 1992). Both scientists and other members of the public are likely to encounter scientific information in their personal and professional lives outside their fields of specialism, however. In order for individuals to interact with such information, judgments about the reliability and validity of that information, and its relevance in a particular context, must be made. Such judgments may well involve drawing upon images of the ways in which scientific knowledge is generated and validated within scientific communities, and the institutions through which scientific communities and other social groups communicate and reach decisions in actual contexts.

Understanding the methods of science and the social relations of science clearly has an epistemological dimension. Individuals living in the UK are exposed to stories in the media involving disputes between scientists relating to the interpretation of scientific data (such as the possibility of a disease such as BSE being transmissible *via* meat from infected cattle to humans), or how scientific knowledge might be drawn upon in making decisions about issues with social, political, economic and ethical dimensions (such as the acceptable uses of DNA technology). The way in which such disputes will be understood depends critically on the epistemological understanding brought to bear on the issue — different interpretations of data are presented, different opinions about acceptable risks are discussed and scientific information is drawn upon in economic, ethical and legal contexts. Such debates may also be presented in an institutional setting: particular opinions are presented as emerging from government departments or scientific institutions.

It appears that whether the science curriculum involves promoting understanding of the contents of science, the methods of science or the social relations of science, there is inevitably an epistemological dimension. As young people encounter science, tacit epistemological understandings will develop. Indeed, empirical work has suggested that teachers' images of the nature of science are very important in influencing the epistemological understandings of their students (Brickhouse, 1989; Lederman, 1986). It is not a simple matter, however, to specify the epistemological understanding that it is hoped that pupils will learn from their science education nor how this should be taught. Recent work in the history, philosophy and sociology of science is portraying science as a multifaceted activity, characterized as much by diversity as by unity, and as such attempts to portray '*the* nature of science' as it should be understood by students appear unhelpfully naive (see, for example, the portrayals of science used by Songer and Linn, 1991 and Kuhn *et al.*, 1988). Furthermore, there is no reason to assume that students' espoused views about the nature of science correspond to the implicit epistemological understandings drawn upon in making decisions and informing actions in particular contexts.

It does not seem possible to specify in a precise way the epistemological understanding that should be learnt by students through school science. By contrast, it does seem feasible to suggest that certain epistemological understandings are more appropriate than others in particular contexts. For example, interpretations of disputes between scientists as inevitably involving incompetence or bias do not recognize the legitimate place of debate within the scientific community. Similarly, Duschl (1990) makes a compelling case for helping school students to judge scientific knowledge claims in the context of their status within the scientific community: if disputes about established theories of science and more peripheral knowledge claims are judged as being equal in status, the role of the scientific community in validating public knowledge is not being recognized. A plausible role for school science might thus be to extend the range of epistemological understandings available to young people to be drawn upon in particular situations, and to offer them insights into how this knowledge might be applied in particular contexts.

The Development of Young People's Understanding of the Nature of Science Between the Ages of 9 and 16

Rationale and Methodology of the Study

We have seen that science curricula of diverse purposes (such as the training of scientists or promoting the public understanding of science) and of varying content (such as the contents, methods and social relations of science) have an inescapable epistemological dimension. When students encounter information about science, whether this is through the media, through school science or through particular professional or personal circumstances in adult life, representations of the knowledge produced by scientific communities may be constructed (Layton, *et al.*, 1993). If developing students' understanding of the nature of science is to be an explicit purpose of science education, rather than an implicit consequence of it, then it is necessary to characterize the nature of the understandings that it is hoped to promote. Decisions about curriculum content, sequence and pedagogy may then be usefully informed by information about the epistemological understanding characteristic of students of given ages as they embark on courses of study (Driver *et al.*, 1994b).

In this part of the paper, findings from a cross-sectional study of young people's understanding of the nature of science between the ages of 9 and 16 are referred to. These findings have been reported in detail elsewhere (Driver *et al.*, in press; Driver *et al.*, 1993a). Aspects of the nature of science which were probed and which are referred to include:

- students' views of the domains and purposes of scientific activity;
- the nature of the basis on which scientific knowledge claims are thought to be made; and
- the nature of the personal and social processes through which scientists are thought to establish knowledge claims, and the social processes governing the application of such knowledge to real world problems.

In probing young people's images of scientific knowledge, a number of methodological questions were addressed.

Nomothetic or Ideographic?

Students' understanding of the nature of science could be elicited and analysed in terms of a normative view of 'the nature of science'. However, philosophical views of the nature of science are characterized by multiple perspectives and dispute. An alternative approach involves characterizing students' epistemological understanding in its own terms, without reference to any particular normative position.

The approach used for data collection and analysis involved identifying key questions around which opinions about the nature of science are addressed (such as the relationship of empirical testing and scientific theory), collecting data about students' views, and characterizing this data in ideographic terms.

Espoused Knowledge or Implicit Knowledge Informing Actions?

A further distinction involves the methods through which students' knowledge of the nature of science is elicited. One approach involves asking explicit general questions with the result that data represents espoused knowledge. For example, Carey *et al.* (1989) asked students questions such as 'What is an hypothesis?'. At the other extreme, inferences can be made about the implicit knowledge informing students' actions as they undertake tasks. For example, Rowell and Dawson (1983) made inferences about the epistemological reasoning underpinning students' investigations of flotation, and compared this with espoused views about the nature of science made by the same students.

For a variety of reasons, asking explicit questions of students about the nature of science was discounted. Students' espoused views of the nature of science may not be reflected in the way that they approach scientific investigations (Rowell and Dawson, 1983). Furthermore, when students answer questions such as 'Why do scientists do experiments?' (Solomon *et al.*, 1992) it is difficult to know what scientists and what experiments students have in mind when answering. A number of tasks were therefore designed around which students' performance would be observed, and their reasons for taking particular actions would be probed.

Scientific Knowledge in What Context?

There is no reason to assume that the epistemological assumptions made by students about science as they encounter it in their own science learning, and their assumptions about the nature of the work of scientists, should be the same. One might speculate that students interpret scientific activity through a variety of common cultural representations of science drawn from the media and from school science teaching; it is certainly unlikely that school age students have any direct experience of the workings of the scientific community to draw upon. It was therefore decided to probe students' representations of science in contexts that would be familiar to them from their science learning in school. We felt it reasonable to assume that a knowledge of how students represent scientific activity as encountered in school science contexts would be insightful in hypothesizing about the issues facing them in coming to understand science when encountered in particular social situations. In order to investigate this issue, the images of scientific knowledge and empirical testing drawn upon by 16-year-old students in interpreting information about controversial issues with a science dimension were also probed.

In investigating students' epistemological understanding, a key question involves the perceived relationship between theory and evidence. Some studies have addressed this issue in domains relatively distant from the scientific knowledge likely to be encountered by students in their studies of science, or through the media (such as the effect of the colour of tennis balls on the quality of service; Kuhn et al., 1988). Other studies have probed students' understanding in domains likely to be familiar to students from science lessons (such as the causes of various astronomical phenomena; Samarapungavan, 1992). As we were interested in students' epistemological understanding in the context of their science learning, tasks were constructed which were embedded in scientific knowledge likely to be familiar to students from their science learning.

Five diagnostic tasks (termed 'probes') were designed and administered to pairs of students aged 9, 12 and 16 (n = approximately thirty pairs for each probe at each age). Interviews were audio-taped and transcribed in full. The sample was drawn from primary, middle and secondary schools in urban and suburban areas in the north of England, and was designed to span the full ability.

A further probe was designed in order to collect data about the understandings brought to bear by young people to interpret situations involving uncertainty in science. The contexts used were the Wegener controversy (a debater broadly located within the scientific community), and the issue of whether food irradiation should be legalized in the UK (a debate of broader social relevance). The mode of administration of the probe involved the presentation of audio-tape material with a supporting booklet to whole classes of students. This was followed by small group work in groups of four, which was audio-taped and transcribed in full. An interviewer was allocated to each group of four to probe and clarify meanings. The sample (n = 51) was drawn from four classes in two urban/suburban schools in the north of England.

Young People's Representations of Scientific Knowledge and the Nature of Empirical Enquiry

The following sections briefly summarize findings from two probes in the study where students' epistemological reasoning was investigated, though both probes were designed with wider purposes in mind. The Scientific Questions probe was designed primarily to elicit information about students' views of the domains and purposes of science, though information about their representations of empirical testing was also elicited. By contrast, the Closure probe was designed to elicit information about students' views of the social processes involved in scientific activity, though information about their representations of scientific knowledge and empirical testing was also elicited.

The 'Scientific Questions' Probe

The Scientific Questions probe was administered to pairs of students in the 9–16 age range (for detailed reporting of findings see Leach *et al.*, 1993 and Driver *et al.*, in press). Our initial conceptualization suggested that issues such as the empirical basis of knowledge claims, the type of phenomena on which knowledge claims focused and the institutional location of activity involved in working with knowledge claims would be important in demarcating science from other domains of knowledge. In order to find out how young people perceived the domains and purposes of science, we devised a probe where various questions had to be classified as 'scientific questions'. Eleven questions were therefore written which varied across these dimensions:

Which kind of fabric is waterproof? Which is the best programme on TV? Is it wrong to keep dolphins in captivity? How do birds find their way over long distances? What diet is best to keep babies healthy? Is it cheaper to buy a large or a small packet of washing powder? How was the Earth made? Is the Earth's atmosphere heating up? Do ghosts haunt old houses at night? What kind of bacteria are in the water supply? Can any metal be made into a magnet?

Students were asked to classify each question as 'a scientific question', 'not a scientific question', or 'don't know', and to justify their classification. They tended to refer to three features in justifying their classifications, which were the nature of the phenomenon at the subject of the question, the perceived personal and institutional features of individuals involved in addressing the question and the amenability of the question to empirical testing, and the nature of testing involved.

The most frequently stated justification for classifying the questions was their amenability to empirical testing, and the nature of the testing involved. A number of different images of empirical testing and its relationship to scientific knowledge were noted in students' explanations (for further details see Driver *et al.*, in press).

- Phenomenon-based reasoning
Description of particular phenomena and possible explanations of the cause were sometimes not distinguished. For example, students suggested that scientists might address the question 'Is the Earth's atmosphere heating up?' by a simple process of observation to see if the Earth is heating up. For such students, testing appears to involve observing the behaviour of phenomena, and scientific knowledge is a description of such phenomena. In effect, no clear separation of theories, explanations, and descriptions was apparent in such responses.
- Relation-based reasoning
Controlled intervention in phenomena, involving the manipulation of key variables, was sometimes seen as leading to knowledge about the cause of particular phenomena. For example, some students suggested that scientists would answer questions such as 'Which is the best diet to keep babies healthy?' by carrying out a 'fair test'. Although theory/explanation and data/evidence were separated in such responses, an answer to the question was thought to emerge in a straightforward way from the data. In addition, explanations of causation were constituted in the same terms as descriptions of behaviour: theories involving new, unobservable entities were not posited.
- Model-based reasoning
Some students' responses suggested an awareness that theories are conjectural, and that enquiry involves the evaluation of theories or conjectures in the light of evidence. For example, scientists would address questions such as 'How was the Earth made?' by evaluating competing theories against evidence from the fossil record. Posited theories and data which could be collected and used in evaluation of theories were separated in such responses.

There is no intention to suggest that these images of empirical testing are hierarchical in any sense. Indeed, each image may be appropriate to use in interpreting the work of scientists in particular contexts: detailed descriptions of the behaviour of phenomena, generalizations about the relationships between variables and theory generation and evaluation all have a place in scientific practice. It is therefore unlikely that young people will use any one image of scientific knowledge and enquiry across a diverse range of contexts. It has already been argued, however, that certain images of empirical testing (and the scientific knowledge that such testing relates to) are more appropriate for making interpretations of particular contexts, than others.

Young peoples' epistemological reasoning in contexts familiar from their science learning was investigated using the Scientific Questions probe and four others. Similar representations of the nature of scientific explanation, the relationship between explanation and description and the nature of scientific enquiry were noted across these probes, and a more generalized characterization of phenomenon-based reasoning, relation-based reasoning and model-based reasoning was generated with the aim of characterizing forms of reasoning likely to be used by students at the population level. It was not intended to characterize individual students' reasoning in a coherent or predictable manner.

The characterization is informed both by data on the epistemological reasoning actually used by young people in particular contexts, and by a view of the relationships between explanation and description. In particular, model-based reasoning is informed by a particular view of the nature of science: although a small number of students showed clear evidence of acknowledging the place of conjecture and theory evaluation in scientific activity, statements tended to be piecemeal and implicit.

The Closure Probe

The forms of epistemological reasoning so far described relate to students' thinking in contexts familiar from their school science learning. To what extent is similar reasoning drawn upon in their interpretations of the work of scientists involved in the generation of scientific knowledge, or its application in complex real-world contexts? The Closure probe was designed in order to address this question. Groups of 16-year-old students were presented with information relating to disputes between scientists about the generation and application of scientific knowledge, and how those disputes might be resolved (see Driver *et al.*, 1993b). For example, in the case of food irradiation students were presented with background information about food irradiation as a method of food preservation, and arguments for and against legalization of food irradiation in the UK. They were asked to discuss a presented dialogue between two scientists, one from a university department and the other from a food industries research department. In the dialogue, evidence from a range of research studies was presented and discussed by the scientists, contrary views about legislation justified in terms of differing interpretations of the research data.

A significant proportion of the 16-year-olds in the sample suggested that the reason for this dispute between the scientists was an absence of facts, and that closure of the debate would be reached by access to more facts (Driver *et al.*, 1993b). The reasoning used by such students seems very close to relation-based reasoning: answers to problems were thought to emerge directly from observable data in an unproblematic way. Many students suggested that quantitative data, or 'hard facts' obtained by means of modern technology, would be most useful in reaching.

A significant minority of students recognized that balance of evidence and judgment were involved in reaching decisions. A number of students articulated a position closer to model-based reasoning than relation-based reasoning:

- 'They are always going to get a slightly different version but they've still got too much sort of either way.'
- 'If you did bigger surveys that wouldn't be conclusive but it will be a step closer to the truth.'
- 'There's going to be no way of proving it. It won't be crystal clear, but it will be a lot easier to see.'

No groups in the study made explicit statements about the need to evaluate risks and benefits in reaching judgments in such situations, and the epistemological reasoning that appeared to underpin the majority of responses portrayed scientific knowledge as emerging in an unproblematic way from data about the natural world. Disputes between scientists were thus portrayed as arising due to an absence of relevant data, or occasionally the personal bias or incompetence of individual scientists.

The Educational Significance of Young Peoples' Epistemological Reasoning

By the end of compulsory science education in the UK, it appears that the epistemological reasoning drawn upon by young people in contexts familiar from their science education is often naive in that the relationships between explanation, data and forms of enquiry are poorly articulated. For example, the predominant portrayal of empirical testing involves a simple process through which reliable knowledge is generated in an unproblematic way. Similar epistemological reasoning seems to underpin interpretations of issues with a science dimension set in contexts of broader social relevance. To this end, although it could be argued that the curriculum is doing an acceptable job of providing a pool of well-qualified young people for future study in science, it is less plausible to argue that the public understanding of science, whether for utilitarian, democratic or cultural reasons, is being promoted. It is, of course, likely that young people would be able to reach more sophisticated understandings of issues with a science dimension following different teaching, and that the epistemological understandings drawn upon at age 16 are likely to develop with future education (in science or otherwise) and life experience.

The feasibility of promoting scientific literacy through the school science curriculum has been questioned (Jenkins, 1992). However, the espoused purpose of teaching science to *all* young people in England and Wales in the 5–16 age range relates to promoting scientific literacy (NCC, 1993). The National Curriculum for Science (NCC, 1992) focuses mainly upon the contents of science. An attempt is made to portray the methods of science, though this portrayal has been criticized as naively empiricist (e.g., Laws, in press) and narrow in focus (Tytler and Swatton, 1992). It seems hard to equate such a curriculum with the aim of promoting scientific literacy.

The social relations of science are afforded a very low profile in the National Curriculum. During the history of the National Curriculum 'the nature of science' has been included explicitly, all but removed, and reinstated in a different form (NCC, 1988; 1992; SCAA, 1994).

The curriculum as written by policy makers is not the same as the curriculum as taught by practitioners, and neither are the same as the curriculum as experienced by learners. The way in which the majority of science teachers in England and Wales portray epistemology through their teaching of the contents, methods and social relations of science is open to question, though attempts have been made to provide strategies and materials for making their knowledge of the nature of science more explicit, and drawing upon this in their portrayals of the epistemology of science to students (e.g., CLIS, 1989; Solomon, 1992; 1993; Nott and Wellington, 1993). Such efforts are likely to be critical to the success of school science education at promoting less naive images of science amongst students.

If promoting the public understanding of science is to be a serious policy aim of school science curricula, critical questions emerge relating to both policy and practice. At a policy level, clarity is needed about the role of school science in the training of scientists and the criteria by which young people are likely to opt into science as a career, and be selected to do so. Decisions must be made about the quantity and nature of science it is necessary for young people to study in order to make career choices relating to science. The situation becomes more complex if supporting the public understanding of science is also to be an aim of school science. Further decisions must then be made about the extent to which it is feasible to support the public understanding of science for utilitarian, democratic and cultural reasons through the school science curriculum, the relative balance of contents, methods and social relations of science thought likely to achieve this aim, and the amount of time to be spent on science.

It is hard to imagine how the school science curriculum could promote the public understanding of science without focusing on the methods and social relations of science, and a case has also been made for addressing understanding of these features in a curriculum for future scientists. The issue of appropriate pedagogical approaches for promoting epistemological understanding arises. The extent to which current undergraduate courses in science equip future scientists with appropriate epistemological understanding is open to question (Sheppard and Gilbert, 1991). This question becomes particularly pertinent when the training of future teachers of science is considered — it is difficult to make a case that current undergraduate science courses or programmes of teacher education in the UK are designed to promote epistemological understanding relating to the contents, methods and social relations of science.

Perhaps the most pertinent question of all relates to the nature of 'authentic activity' in the science curriculum (Seeley Brown *et al.*, 1989). The contents of science are often presented to learners with no emphasis upon how the knowledge arose, its domains and limitations, and our warrants for believing in it. In a curriculum aiming to promote epistemological understanding, a more appropriate approach might involve addressing the development of scientific ideas in a particular domain in historical contexts, focusing upon warrants for belief and the application of knowledge in particular contexts. Recent portrayals of the work of scientist in the UK curriculum have focused upon a personal, empirical process involving skills such as hypothesizing, inferring and observing (e.g., Screen, 1986). In order to learn about the work of scientists, students have been made to undertake empirical investigations based largely upon demonstrating the relationships between variables (NCC, 1992). If an aim of the science curriculum is to promote epistemological understanding as it relates to the methods and social relations of science, the focus is likely to be upon the work performed by various scientists, with epistemological features of the context being addressed explicitly.

Note

1 This paper draws upon findings from the study 'The development of children's understanding of the nature of science' carried out by Rosalind Driver, John Leach, Robin Millar and Philip Scott, which was funded by the Economic and Social Research Council (grant R000233186). The author acknowledges the contributions of Rosalind Driver, Robin Millar and Phil Scott to his thinking in preparing this paper, while accepting responsibility for the arguments expressed.

References

AAAS (1989) *Science for All Americans*, Washington, DC, AAAS.

BRICKHOUSE, N.W. (1989) 'The teaching of the philosophy of science in secondary classrooms: Case studies of teachers' personal theories', *International Journal of Science Education*, **11**, 4, pp. 437–49.

CAREY, S., EVANS, R., HONDA, M., JAY, E. and UNGER, C. (1989) 'An experiment is when you try it and see if it works: A study of junior high school students' understanding of the construction of scientific knowledge', *International Journal of Science Education*, **11**, 5, pp. 514–29.

CLIS (1989) *Interactive Teaching in Science: Workshops for Training Courses*, Hatfield, Association for Science Education.

DRIVER, R., LEACH, J., MILLAR, R. and SCOTT, P. (1993a) *Students' Understanding of the Nature of Science: Working Papers 1–11*, Leeds, Centre for Studies in Science and Mathematics Education.

DRIVER, R., LEACH, J., MILLAR, R. and SCOTT, P. (1993b) *Students' Understanding of the Nature of Science: Working Paper 9: Students' Awareness of Science as a Social Enterprise*, Leeds, Centre for Studies in Science and Mathematics Education.

DRIVER, R., LEACH, J., MILLAR, R. and SCOTT, P. (in press) *Young Peoples' Images of Science*, London, Routledge.

DRIVER, R., LEACH, J., SCOTT, P. and WOOD-ROBINSON, C. (1994a) 'Young people's understanding of science concepts: Implications of cross-age studies for curriculum planning', *Studies in Science Education*, **24**, pp. 75–100.

DRIVER, R., LEACH, J., ASOKO, H., MORTIMER, E. and SCOTT, P. (1994b) 'Constructing scientific knowledge in the classroom', *Educational Researcher*, **23**, 7, pp. 5–12.

DUSCHL, R. (1990) *Restructuring Science Education: The Importance of Theories and their Development*, New York, Teachers College Press.

HUSÉN, T., TUIJNMAN, A. and HALLS, W.D. (1992) *Schooling in Modern European Society: A Report of the Academia Europeae*, London, Pergamon.

JENKINS, E.W. (1992) 'History and philosophy of science and school science education: Remediation or reconstruction?', in HILLS, S. (Ed) *Proceedings of the Second International Conference on the History and Philosophy of Science and Science Teaching*, Kingston, Queens University.

KUHN, D., AMSEL, E. and O'LOUGHLIN, M. (1988) *The Development of Scientific Thinking Skills*, Orlando, Academic Press.

LAWS, P. (in press) 'Scientific investigation: Ideology in the national curriculum for England and Wales', *International Journal of Science Education*.

LAYTON, D., JENKINS, E., MACGILL, S. and DAVEY, A. (1993) *Inarticulate Science? Perspectives on the Public Understanding of Science and some Implications for Science Education*, Driffield, Studies in Education Ltd.

LEACH, J., DRIVER, R., MILLAR, R. and SCOTT, P. (1993) *Students' Understanding of the Nature of Science: Working Paper 4: Students' Characterisations of what Constitutes Scientific Questions*, Leeds, Centre for Studies in Science and Mathematics Education.

LEDERMAN, N.G. (1986) 'Relating teaching behaviour and classroom climate to changes in students' conceptions of the nature of science', *Science Education*, **70**, 1, pp. 3–19.

MILLAR, R. (1993) 'Science education and public understanding of science', in HULL, R. (Ed) *ASE Science Teachers' Handbook, 2nd edition*, Hatfield, Association for Science Education.

MILLER, J. (1983) 'Scientific literacy: A conceptual and empirical review', *Daedalus*, **112**, 2, pp. 29–48.

NATIONAL CURRICULUM COUNCIL (1988) *Science in the National Curriculum*, London, HMSO.

NATIONAL CURRICULUM COUNCIL (1992) *Science in the National Curriculum*, London, HMSO.

NATIONAL CURRICULUM COUNCIL (1993) *Teaching Science at Key Stages 3 and 4*, York, NCC.

NOTT, M. and WELLINGTON, J. (1993) 'Your nature of science! An activity for science teachers', *School Science Review*, **75**, 270, pp. 109–112.

ROWELL, J.A. and DAWSON, C.J. (1983) 'Laboratory counter examples and the growth of understanding in science', *European Journal of Science Education*, **5**, 2, pp. 203–15.

ROYAL SOCIETY (1985) *The Public Understanding of Science*, London, The Royal Society.

SAMARAPUNGAVAN, A. (1992) 'Children's judgements in theory choice tasks: Scientific rationality in childhood', *Cognition*, **45**, 1, pp. 1–32.

SCHOOLS CURRICULUM and ASSESSMENT AUTHORITY (1994) *Science in the National Curriculum*, London, HMSO.

SCREEN, P. (1986) 'The Warwick Process Science Project', *School Science Review*, **68**, 232, pp. 12–16.

SEELEY BROWN, J., COLLINS, A. and DUGUID, P. (1989) 'Situated cognition and the culture of learning', *Educational Researcher*, **18**, 1, pp. 32–42.

SHEPPARD, C. and GILBERT, J. (1991) 'Course design, teaching method and student epistemology', *Higher Education*, **22**, pp. 229–49.

SOLOMON, J. (1992) *Exploring the Nature of Science at Key Stage 3*, London, Blackie.

SOLOMON, J. (1993) *Exploring the Nature of Science at Key Stage 4*, Hatfield, Association for Science Education.

SOLOMON, J., DUVEEN, J., SCOT, L. and MCCARTHY, S. (1992) 'Teaching about the nature of science through history: Action research in the classroom', *Journal of Research in Science Teaching*, **29**, 4, pp. 409–21.

SONGER, N.B. and LINN, M.C. (1991) 'How do students' views of science influence knowledge integration?', *Journal of Research in Science Teaching*, **28**, 9, pp. 761–84.

THOMAS, G. and DURANT, J. (1987) *Why Should we Promote the Public Understanding of Science?*, Scientific Literacy Papers, no. 1, University of Oxford, Department of External Studies, pp. 1–14.

TYTLER, R. and SWATTON, P. (1992) 'A critique of Attainment Target 1 based on case studies of students' investigations', *School Science Review*, **74**, 266, pp. 21–35.

WHITE PAPER (1993) *Realising our Potential: A Strategy for Science, Engineering and Technology*, London, HMSO.

WYNNE, B. (1990) 'The blind and the blissful', in *The Guardian*, 13 April, p. 28.

22 Probing Teachers' Views of the Nature of Science: How Should We Do It and Where Should We Be Looking?

Mick Nott and Jerry Wellington

Introduction

At the European Conference on Research in Science Education held at the University of Leeds in April 1995 we gave a paper on 'Critical incidents in the science classroom and the nature of science' (Nott and Wellington, 1995b). The questions and responses to that paper have made us explicate our methodology for probing teachers' understandings of the nature of science. We are grateful to the discussants and speakers from the audience for the questions which made us do this and to Robin Millar for his comments on a first draft. This paper is our considered response to the points made.

We begin this paper with a review of ways of probing teachers' views of the nature of science and then outline models of teacher knowledge, teachers and science which inform and underpin a probe we have developed. This probe consists of what we call 'critical incidents'. The value and validity of these is discussed and we finish by highlighting what we see as some of the key issues in research into teachers' views of the nature of science.

Past Probes into Teachers' Views of Science

A wide range of probes and instruments for exploring teachers' views of the nature of science have been used in the last thirty years. We do not have the space here to summarize them, but we can say that some have involved written tests and questionnaires, some have involved lesson observation, others have been based on interviews with teachers, others have involved a mixture of classroom observation and interview.

Hewson and Hewson, for example (1989), used 'interviews about instances' to probe pre-service science teachers' conceptions of science and science teaching. Brickhouse (1989, 1990) used a combination of observation and interviewing in an intensive study of three science teachers to examine the connection between their understandings about science and their teaching.

The most widely used probes have been written tests, frequently involving multiple choice items. Each one seems to coin its own acronym leaving a legacy including NOST, TOUS, WISP, NOSS and VOSTS (see Lederman, 1992). Probes into teachers' views of science, especially written tests of this kind, do raise a number of problems about the status and source of teachers' understandings of science.

Firstly, status: authors reporting the results of tests given to teachers in the past have written of 'inadequate conceptions' and the need for 'improvement'. Our view is that it is highly problematic to make value judgments about teachers' views of the nature of science using terms such as 'correct', 'improved' or 'good'. The nature of science is a construct which itself changes and shifts and it is presumptuous of researchers to propose that teachers should measure up to conceptions of the nature of science which themselves may be the subject of academic criticism. This point has been forcibly stated by Lucas (1975) and reiterated by Hodson (1993).

Secondly, the source of teachers' conceptions: in the literature reviewed by Lederman it has been assumed that these spring from teachers' subject knowledge. Thus Lederman writes that 'no relationship was found between secondary teachers' conceptions of science and any of the academic background variables'. As a result of his review of past probes, Lederman notes that 'academic background variables are not significantly related to teachers' conceptions of the nature of science' (Lederman, 1992, p. 345). He concludes that 'there is little reason to continue our reliance on paper and pencil assessments of . . . teachers' conceptions' and he notes that interviews yield richer data than questionnaires.

However, in a study of the way that high school chemistry teachers used curriculum materials to teach chemistry, Lantz and Kass (1987) identified the teachers' academic histories as one of the factors which determine a teacher's 'functional paradigm'. (The functional paradigm is the framework of pedagogical practices that a teacher has.) The academic history of a teacher is made up of the courses that the teacher has studied as a student and the range of classes, students and syllabuses that the teacher has taught in their professional career.

> Apparently a teacher's academic history i.e., training in chemistry and experience of teaching chemistry, *has a considerable influence on shaping his perceptions of the Nature of Science*. (Lantz and Kass, 1987, p. 126 [our emphasis])

The indication is that teachers' knowledge of the nature of science may be as much formed by their teaching of science as informing their teaching of science.

A later study (Lakin and Wellington, 1991, 1994) indicated that teachers found statements from the National Curriculum (DES/WO, 1991) about the nature of science difficult to interpret and showed diffidence with the organization of appropriate teaching activities. The second part of Lakin and Wellington's work used repertory grid methods to determine teachers' personal constructs about science. The science teachers identified science with the 'science as a process' model of science which may be a reflection of the National Curriculum at that time. These processes of science were not perceived as involving questions of emotions, imagination and intuition. The impression is that the small sample of science teachers saw science as an empiricist, positivist activity.

However, and we think this is important, this was the first time these teachers had done anything of this kind — both in methodology and content. It was the first time that they had been asked questions about their views on the nature of science. 'They commented that the method used made them challenge their assumptions and views' and that they were 'thinking on their feet'. This survey confirmed the finding in earlier surveys (Solomon, 1990, 1991) that science teachers could give little account of having reflected on the nature of their subject. It also highlights that teachers' understandings of the nature of science will not be singular things, but will change with time and context as has already been found in another study (Kouladis and Ogborn, 1989).

Armed with the findings from Lakin and Wellington and using items from Kouladis and Ogborn we wrote our own twenty-four item questionnaire which teachers could use to construct a personal profile of their understandings of the nature of science (Nott and Wellington, 1993). The aim was to help teachers elicit and examine their understanding of the nature of science and hence evaluate whether the responses given did have validity with respect to their understanding of the nature of science,

> The activity ... should be treated as a way of getting teachers to think, learn and reflect (about the nature of science) rather than as a valid measurement on some sort of objective scale. (Nott and Wellington, 1993, p. 109)

Many participants, who helped in the development of the profile, thought the results challenged their ideas about the nature of science and, after the discussion work, most recognized that having done the exercise once and discussed it they would not answer in the same way the next time.

This confirmed our suspicions not to place too much faith in paper and pencil tests of teachers' understandings of the nature of science. Lederman's (1992) review rightly recognized that research methods that are more qualitative and phenomenological may provide a better understanding of the interaction between teachers' understandings and their classroom actions.

Three Models

We want to draw upon three models.

A Model of Knowledge

Teachers have a set of knowledge which they bring to the classroom and a set of knowledge which is developed and learned from their classroom experience. These two ways of knowing one's subject have been called, respectively, subject-content knowledge and pedagogical-content knowledge (Shulman, 1986, 1987). The two sets of knowledge interact and inform each other. We would argue that as teachers progress through their careers their pedagogical-content knowledge increases in size and importance relative to the subject-content knowledge that they brought to their career. Beginning teachers recognize the acquisition of this new knowledge (Wilson, Shulman and Richert, 1987).

Science teachers' knowledge of the nature of science is interesting to examine with respect to this model. The evidence is that few science teachers formally study the nature of science in their pre-service education either in subject studies or teacher training (Lederman, 1992). Yet in science classrooms they continuously cope with events like practicals going wrong or ethics and morals to do with the nature of scientific knowledge and scientists. The knowledge to cope with critical incidents will come from their subject-content knowledge and their pedagogical-content knowledge. Knowledge of the nature of science will be brought to the classroom *and* developed through classroom experience. If the former is very thin in any kind of formal sense then the latter may be significant in teachers' learning about the nature of science.

A Model of Teaching

Good teachers continuously make decisions and change their plans. The book which gave us the idea of using critical incidents (Bishop and Whitfield, 1972) proposes that the good teacher is an experienced decision-maker. Bishop and Whitfield proposed that experienced teachers, through experience, classify classroom events into types which allow choice about the appropriate teacher response. They distinguish between long-term choices and on-the-spot choices. Student teachers can be trained in long-term decision making involving planning, theories of learning, discipline matters etc. However, it is difficult to train for on-the-spot decision making because the making of those decisions is not clear. Bishop and Whitfield note,

> Most experienced teachers talk in terms of 'intuition' or 'I just know its one of those . . .' (Bishop and Whitfield, 1972, p. 7)

In summary, we offer a model of a good teacher as an experienced decision maker who makes long-term decisions and who has also learned a set of routines, or on-the-spot decisions, to cope with the conditions of teaching. We believe that the incidents are critical because they occur despite the best of long-term planning and necessitate difficult on-the-spot decisions by the teachers.

A Model of Science

We view the canon of knowledge in school science as a 'black box' (Latour, 1987). The black box model of science states that scientists strive to make their knowledge and procedures acceptable to other scientists and the public at large by making them unproblematic. The knowledge and the procedures are 'just there' and their stability is ensured by them being accepted unquestioningly by those who use them. Theories like germ theories or photosynthesis are like the 'black boxes' of the television set, video recorder etc., used everyday as part of common culture with no thought for how they were created and what kind of infrastructure is needed to maintain them. Scientific knowledge is created like black boxes are created. Once created it has to be maintained. Unless it is maintained the black box breaks down and has to be opened up and repaired or a new black box built.

Science teachers are involved with the continual maintenance of scientific knowledge. This knowledge and the procedures by which they are produced are set out in the National Curriculum (DfE/WO, 1995). Science classrooms in England and Wales are places where 'matters of fact' (Shapin and Schaffer, 1985) are required to be frequently and regularly produced in order to maintain the black boxes of knowledge and procedures. All classroom participants are expected to produce these matters of fact, yet every secondary science teacher will tell you that the black box springs open all the time in that nature appears obdurate and will not behave as it should — practicals continuously go wrong.

Using Critical Incidents As Probes

We have argued that teachers' knowledge of the nature of science will be grounded in pedagogical content knowledge as well as subject content knowledge. By using critical incidents which are qualitative and rooted in classroom experience we believe that the twin problems of source and status, to which many probes are subject, can be avoided.

We define a critical incident as an event which makes a teacher decide on a course of action which involves some kind of explanation of the scientific enterprise. It may be an event like some practical work going wrong or it may be an event which raises moral and ethical issues about scientific knowledge or the conduct of scientists. These events are often stimulated by pupils saying and doing things but they may also arise through the action of the teacher, particularly when a demonstration goes wrong. A wide range of these incidents can be found in Nott and Wellington (1995a). Two such examples are included here to illustrate the type of events we have defined as critical.

- You have set up a demonstration of the production of oxygen by photosynthesis with Canadian pond weed. Just before the lesson when the class are to look at the apparatus again, you notice that there is a small amount of gas in the test tube but not enough with which to do the oxygen test. List the kinds of things you could say and do when the lesson starts.
- You are six weeks into Term 1 with a year 7 group. The unit you are doing is on 'Life and living processes'. One of the pupils states impatiently at the start of a lesson, 'When are we going to start cutting up rats then?' List the kind of things you could say and do at this point.

We are not talking about deliberately arranged events where a teacher may plan for and engineer a debate or argument amongst children. Part of the incidents' criticality is that they evoke responses from the teacher which provide an insight into the teacher's view of science as well as matters to do with teaching and learning. We have argued elsewhere (Nott and Wellington, 1995b) that these responses provide access to teachers' understanding of the nature of science.

We believe that the incidents we provide make teachers draw on their understanding of the nature of science because:

- they relate closely to classroom practice;
- they promote reflection and discussion when used;
- they explore teachers' implicit understandings and help to make them explicit; and
- they 'get at' teachers' knowledge-in-action (Schon, 1983) and their practical wisdom rather than 'academic knowledge'.

This last point is perhaps the most important. The critical incidents can probe understandings from the domain of pedagogical content knowledge as well as subject content knowledge — our argument is that this is the domain where teachers' conceptions of science really lie. We believe that we are now beginning to probe *in the right way, in the right place*.

How the Incidents Have Been Used

The incidents have been used in two ways. First, as an interactive, social, group activity; others have used them with students individually and then shared individuals' responses. We have given them to small groups of teachers and classroom-experienced student teachers to discuss and then to offer their responses. Each response was thrown open to the larger group for discussion. Responses were noted from the lists and the discussion. Second, we have used them as a stimulus in one-to-one interviews. The interviewee has been asked to make responses and then they are asked, 'What do you think children may learn about science and scientists from your response?' and subsequent to their answers the interviewee is asked, 'Is that what you believe about science?'

Summary of the Responses

A fuller discussion and interpretation of the responses can be found in Nott and Wellington (1995a, 1995b).

The incidents where practicals 'go wrong' appear to elicit three categories of response. These are 'talking your way out of it', 'rigging' and 'conjuring' (Nott and

Smith, 1995). The majority of responses are in the first category of 'talking your way out of it' or as we would like to say more positively 'talking your way through it'. When science teachers talk their way through practicals going wrong they often engage the children in a critical evaluation of practical work. In doing so the teachers are conveying the following messages about science. Science is an activity where:

- practicals need to be evaluated and this may involve repeating experiments;
- the null result is as important as the positive result;
- doing science involves sharing results and collectively criticizing, negotiating and deciding interpretations and procedures;
- sometimes, to learn from and do practicals, you need to have an idea in your head before you start;
- there are reasons why experiments go wrong, there is a rational explanation; and
- results need to match to previously accepted knowledge.

'Rigging' is the use of strategies that teachers have learned over the years to ensure (as best they can!) that the apparatus or procedure works. We are still unsure as to whether teachers explain rigging to their pupils. The last category is 'conjuring'. This is where the teacher fraudulently produces the correct 'matter of fact' by sleight of hand. The alleged procedure has *not* produced the result. We have found that student teachers can start to 'conjure' spontaneously or are induced into it by science staff and technicians.

Non-practical incidents are important because the nature of science in school science includes a consideration of '. . . ways in which scientific ideas may be affected by the social and historical contexts in which they develop, and how these contexts may affect whether or not the ideas are accepted' (DfE/WO, 1995). How did the teachers in our sample respond to the incidents which were not connected with practical work? The teachers saw the incidents as opportunities:

- to promote discussion about the ethics of animal experimentation;
- to talk about science as a way, but not the only way, of explaining and knowing;
- to stress the importance of empirical evidence for scientific theories;
- to explain that scientific theories can change over time;
- to discuss the relationship between science and other cultural universals;
- to hold discussions on the benefits and drawbacks of scientific knowledge; and
- to talk about the social responsibility and accountability of scientists.

All these responses tell us something about the teachers' knowledge of the nature of science and their attitudes to the connections, and differences, between science and other ways of knowing the world. The use of the critical incidents in this way provides insights into teachers' subject-content knowledge and pedagogical-content knowledge which may not be revealed by a test or questionnaire.

Conclusion

Teachers are able to express views about science but not in direct response to abstract, context-free questions of the sort, 'What is science?' Classroom events create and confront teachers' knowledge about the nature of science. Responses to these events make teachers express views about their own understanding of the nature of science which are embedded within talk about their professional practice. These answers will be framed with reference to teaching or classrooms and may also be views of science as learned from their own science education and training. Such views of science are grounded in experience and academic and personal history. When teachers talk about their understanding of the nature of science the talk is connected with their pedagogical content knowledge. In some cases this knowledge is nearly totally embedded in their pedagogical knowledge; in other cases pedagogical knowledge and subject knowledge are linked, but one is not necessarily congruent with the other. We believe that much teaching about the nature of science occurs in science classrooms as an everyday occurrence.

A corollary of this view is that teachers do not necessarily have 'inadequate' views of the nature of science. They have teachers' views of the nature of science which are determined by their academic and professional histories. Teachers do not talk about science as science- or education- or science education-researchers. Nor do they talk about science as professional philosophers, historians or sociologists. They talk about science as teachers.

When teachers talk about school science as opposed to 'other' science then the 'other' science is science as done by science researchers — the people who create and discover scientific knowledge or apply scientific knowledge to particular technological problems. Teachers' knowledge about science is illustrated by examples from their own professional practice as much as scientific researchers' knowledge about the nature of science is illustrated by examples from their professional practice (e.g., Wolpert and Richards, 1989; Perutz, 1991; Medawar, 1991). Researcher-scientists know science through their professional practice and, we would argue, so do teachers.

The academic models of the nature of science which emerge from the recent historiography and sociology of science are grounded in the study of the actions of researcher scientists and their explanations of their professional practice (e.g., Gilbert and Mulkay, 1984; Collins, 1985; Latour and Woolgar, 1979). The philosopher-scientists, historian-scientists and sociologist-scientists create an understanding of the nature of science from these theoretical and empirical studies of science and scientists. Researcher-scientists often fail to recognize themselves or their practice in the work of the philosopher-scientists, historian-scientists and sociologist-scientists (see Wolpert, 1993). Different professional groups have different perceptions of the nature of science. These perceptions are grounded in their professional practice. However the expectation of educational researchers has hitherto been that science teachers' understandings of science will match the views of the other groups of scientists identified above.

Our hypothesis is that as much as researcher-scientists, philosopher-scientists, historian-scientists and sociologist-scientists know about the nature of science from their work then teacher-scientists know about the nature of science from their work. It is proposed that teachers of science can be viewed as teacher-scientists. Through their professional practice teacher-scientists have a different way of knowing the nature of science that is not necessarily accessible to other groups of scientists. This way of knowing is rooted in the pedagogical content knowledge of the nature of science.

Teacher-scientists work, by and large, with groups (i.e., classes of children) who do not necessarily share the same norms, assumptions and values. In addition the groups who teacher-scientists work with by definition do not share the same rationality about the production and reliability of knowledge claims. Teachers have to persuade children that the scientific explanations and empirical evidence are reasonable. Frequently children's attempts to produce empirical evidence that supports teachers' science fail to work (Atkinson and Delamont, 1976; Wellington, 1981). Therefore teacher-scientists often encounter classroom episodes where they have to make explicit how scientific knowledge is constructed and made stable — the black box springs open. This form of explication about the nature of science to this type of audience is not necessarily encountered by researcher-scientists etc. Teacher-scientists will build explanations about the nature of science from their work based on their academic history and their classroom experiences. These explanations of the nature of science may be different but as valid as those of other groups of scientists.

It is important to study these explanations as they are another way of knowing about science which has had some study so far (French, 1989; Selley, 1979; Wellington, 1989) but mainly from the viewpoint about appropriate ways of teaching science rather than from a perspective on teachers' understandings of the nature of science. A question which may elicit more about teachers' understandings is to ask them what they think children *ought* to be taught about the nature of science. This may identify a set of moral positions and cognitive values that teachers have about the nature of science that their professional practice has not yet let them express. This is the avenue we hope to explore next.

The last point we wish to make is that we are not arguing that teachers' understandings of the nature of science are sufficient in themselves or immune from criticism. We believe that we will be left with extreme relativism if the nature of science is that 'anything goes' and that 'anything goes' for understanding the nature of science. There would be no point in having such a nature of science in the curriculum! We recognize that teachers need to know about and understand the nature of science as expressed by researchers, historians etc. and that those understandings should be reciprocated. We believe there is a need to forge a broad consensus about the nature of science in school science. In other words what ought to be taught and how.

References

ATKINSON, P. and DELAMONT, S. (1976) 'Mock ups and cock ups: The stage management of guided discovery instruction', in HAMMERSLEY, M. and WOOD, P. (Eds) *The Process of Schooling*, London, RKP.

BISHOP, A. and WHITFIELD, R.C. (1972) *Situations in Teaching*, Maidenhead, McGraw-Hill.

BRICKHOUSE, N. (1989) 'The teaching of the philosophy of science in secondary classrooms: Case studies of teachers' personal theories', in *International Journal of Science Education*, **11**, 4, pp. 437–49.

BRICKHOUSE, N. (1990) 'Teachers' beliefs about the nature of science and their relationship to classroom practice', *Journal of Teacher Education*, **41**, 3, pp. 53–62.

COLLINS, H. (1985) *Changing Order: Replication and Induction in Scientific Practice*, London, Sage.

DES/WO (1991) *Science in the National Curriculum*, London, HMSO, p. 22.

DfE/WO (1995) *Science in the National Curriculum*, London, HMSO, p. 24.

FRENCH, J. (1989) 'Accomplishing scientific instruction', in MILLAR, R. (Ed) *Doing Science: Images of Science in Science Education*, Lewes, Falmer Press.

GILBERT, G.N. and MULKAY, M. (1984) *Opening Pandora's Box*, Cambridge, Cambridge University Press.

HEWSON, P. and HEWSON, M. (1989) 'Analysis and use of a task for identifying conceptions of teaching science', in *Journal of Education for Teaching*, **15**, 13, pp. 191–209.

HODSON, D. (1993) 'Towards a more critical approach to practical work in school science', in *Studies in Science Education*, **22**, pp. 85–142.

KOULADIS, V. and OGBORN, J. (1989) 'Philosophy of science: An empirical study of teachers' views', in *International Journal of Science Education*, **11**, 2, pp. 173–84.

LAKIN, S. and WELLINGTON, J. (1991) *Teaching the Nature of Science: A Study of Teachers' Views of Science and their Implications for Science Education*, Division of Education, University of Sheffield.

LAKIN, S. and WELLINGTON, J. (1994) 'Who will teach the nature of science?: Teachers' views of the nature of science and their implications for science education', in *International Journal of Science Education*, **16**, 2, pp. 175–90.

LANTZ, O. and KASS, H. (1987) 'Chemistry teachers' functional paradigms', in *Science Education*, **71**, pp. 117–34.

LATOUR, B. (1987) *Science in Action*, Milton Keynes, Open University Press.

LATOUR, B. and WOOLGAR, S. (1979) *Laboratory Life: The Construction of Scientific Facts*, Princeton, Princeton University Press.

LEDERMAN, N. (1992) 'Teachers' and students' conceptions of the nature of science: A review of the research', in *Journal of Research in Science Teaching*, **29**, 4, pp. 331–59.

LUCAS, A. (1975) 'Hidden assumptions in measures of 'knowledge about science and scientists', in *Science Education*, **59**, 4, pp. 481–5.

MEDAWAR, P. (1991) *The Threat and the Glory: Reflections on Science and Scientists*, Oxford, Oxford University Press.

MULKAY, M. and GILBERT, N. (1984) 'Theory choice', in Mulkay, M. (1991) *The Sociology of Science*, Milton Keynes, Open University Press.

NOTT, M. and SMITH, R. (1995) 'Talking your way out of it', 'rigging' and 'conjuring': What science teachers do when practicals 'go wrong', in *International Journal of Science Education*, **17**, 3, pp. 399–410.

Nott, M. and Wellington, J. (1993) 'Your nature of science: An activity for science teachers', in *School Science Review*, **75**, 270, pp. 109–12.

Nott, M. and Wellington, J. (1995a) 'Critical incidents in the science classroom and the nature of science', in *School Science Review*, **76**, 276, pp. 41–6.

Nott, M. and Wellington, J. (1995b) *Critical Incidents in Science*, Milton Keynes, Open University (Science Document 14 in the Teaching Science in Secondary Schools series).

Perutz, M. (1991) *Is Science Necessary?: Essays on Science and Scientists*, Oxford, Oxford University Press.

Schon, D. (1983) *The Reflective Practitioner*, New York, Holt Rinehart.

Selley, N. (1979) 'Scientific models and theories: Case studies of the practice of science teachers', Unpublished PhD Thesis, University of London.

Shapin, S. and Schaffer, S. (1985) *Leviathan and the Air Pump*, Princeton, Princeton University Press.

Shulman, L. (1986) 'Those who understand: Knowledge growth in teaching', in *Educational Researcher*, **15**, 2, pp. 4–14.

Shulman, L. (1987) 'Knowledge and teaching: Foundations of the new reforms', in *Harvard Educational Review*, **57**, 1, pp. 1–22.

Solomon, J. (1990) *A Report on the Implementation of Attainment Target 17*, Oxford University, Department of Educational Studies.

Solomon, J. (1991) *Teachers' Perceptions of the Strategies to be used for Teaching the Nature of Science*, Oxford University, Department of Educational Studies.

Wellington, J. (1981) 'What's supposed to happen, sir?: Some problems with discovery learning', in *School Science Review*, **63**, 222, pp. 167–73.

Wellington, J. (1989) *Skills and Processes in Science Education*, London, Routledge.

Wolpert, L. (1993) *The Unnatural Nature of Science*, London, Faber and Faber.

Wolpert, L. and Richards, S. (1989) *A Passion for Science*, Oxford, Oxford University Press.

Wilson, S., Shulman, L. and Richert, A. (1987) '150 different ways of knowing: Representations of knowledge in teaching', in Calderhead, J. (Ed) *Exploring Teachers' Thinking*, London, Cassell.

Part V
Social Interactions in Science Classrooms

Introduction
Geoff Welford

The first paper in this section by Whitelegg is a collection of excerpts from a symposium on gender effects in science classrooms with contributors from Sweden, Denmark and England. A variety of research methods reveal interesting similarities in attitudes among children from all three countries. Bezzi takes us into the geosciences to show us how students represent the disciplines and view the quality of their teachers and teaching in this area. Scott draws upon a Vygotskian perspective to advance a theoretical framework of the way in which social processes affect personal meaning-making in high-school pupils. Spiliotopoulou and Ioannidis are concerned with Greek primary school-teachers' models of the Universe, basing their study of teachers' knowledge on drawings and accompanying explanations. Campbell and Ramsden round off the section with an examination of student teachers' perceptions of the value of classroom research in their learning to become teachers.

23   Gender Effects in Science Classrooms

Elizabeth Whitelegg

Abstract

This paper provides excerpts from five contributions to a symposium on gender effects in science classrooms. The symposium contributors from England, Denmark and Sweden came together initially because of a shared concern about the potential influence of gender on learning. The Northern European Network for Gender and Science Research formed in Copenhagen in 1993 as a result. The contributions reflect a range of perspectives on gender and its influence on science learning at primary and secondary level and describe some aspects of current research. The research includes detailed observational studies looking at collaboration amongst small groups of children and larger scale interview and questionnaire surveys focusing on attitudes to science and science education amongst girls and boys. The contributions reveal interesting similarities in gender attitudes, coming as they do from different educational systems in Northern Europe.

Introduction[1]

The first three contributions share a common theme — examining primary children's attitudes to group work in science, the nature of their collaborations and the subsequent implications for learning. The first, from the Collaborative Learning and Primary Science (CLAPS) project examines girls' and boys' attitudes to group work and group composition. Ros Smith, in her contribution, reports her research into patterns of discourse in collaborative learning situations and considers the significance of gender differences in relation to teachers' goals for collaborative learning. The contribution from Denmark describes the ELIN project. In this contribution, the authors describe the results of a study of girls' and boys' ways of working together.

In the next contribution, Gaynor Sharp describes her research examining girls' and boys' attitudes to science learning and subject preferences in secondary science classrooms in England. The research aims to establish whether the CASE (Cognitive Acceleration in Science Education) interventions differentially affect girls and boys.

The final contribution by Else-Marie Staberg describes the gender differences that girls and boys bring to science in Sweden and how these are responded to by teachers. She brings a sociological perspective to her analysis and so relates the need for a more inclusive science curriculum to the needs of a changing society.

The symposium contributions reveal some common findings about the effects of gender in science classrooms. All the studies reveal that gender plays a subtle but significant role in children's thinking and learning and in teachers' responses to children. These gender effects are multifarious and can be additive. To promote effective learning for all children, it is crucial that gender differences and similarities are recognized and responded to appropriately by teachers.

Gender and Attitudes to Group Work

Introduction

This paper reports on some of the initial results of the Collaborative Learning and Primary Science (CLAPS) project (Murphy *et al.*, (in press); Scanlon *et al.*, 1995). The central aim of the project is to better understand how science learning in groups evolves in the context of primary classrooms. The questions we are concerned to address are.

- What interactions between children, children and teachers, and children and tasks are productive in terms of learning outcomes?
- What fosters such interactions?
- What limits them?

One dimension of interest in the research is the potential influence that gender might have on children's collaborative learning. In particular, we are concerned to establish whether children's attitudes to groups varies between boys and girls and if so, how they impact on their learning. This aspect of the research is focused on briefly here.

Background

Research into learning has provided evidence that children benefit from working in groups. However the benefits appear to depend on the children who collaborate and the tasks selected as vehicles for learning. What also remains uncertain is how this enhanced learning in groups occurs.

Much of the research evidence collected has been from pairs of children working in clinical settings. Very little work has been done in curriculum areas like science, looking at groups selected by teachers working in classrooms as part of the normal school day. Nor has much attention been paid to teachers' and children's views of group work and the consequences of these for learning in classroom contexts. The link between attitudes and learning has been well established by research particularly in the area of gender differences. It is important therefore in researching collaboration to consider how children's beliefs about it might influence their approach to it. Whilst many studies have examined the influence of task structure on collaboration very few of these studies have considered teachers' choices of tasks and children's perceptions of these.

Approach

For the reasons outlined the project conducts naturalistic case studies of ongoing classroom work. We have observed children from year 3 to year 6 in a number of different schools in England. We video the science work in a class over a period of five to eight weeks. We focus on target groups of children recommended by the teacher as collaborating well. Each target child is radio-miked. We interview teachers and children at various points and monitor the children's learning. A questionnaire is administered to all children in a class to establish their attitudes both to group work and group compositions. Using this and video snippets we probe target children's attitudes to group work and their effects on their learning.

Some Findings

The questionnaire we have developed provides children with photographs of a variety of typical primary classroom working situations. The questionnaire uses a four point 'smiley face' scale with associated descriptions of alternative views. The results presented here are from one case study of year 6 children. There were thirty-one children in the class, fourteen boys and seventeen girls. The results for this class have been repeated in other schools and for different aged children.

Children were asked in the questionnaire their preferred way of working. The only significant difference that emerged was that girls rate working on their own higher than boys. In terms of who they worked with, boys and girls were equally positive about working with their friends. Their attitude to working with children chosen by the teacher did vary significantly. Girls were more tolerant of this than boys. This may be influenced by the number of girls and boys in a class. The clear message emerging was that both boys and girls much prefer working in groups of the same gender. They actively dislike working with groups made up of the opposite gender. Mixed gender groups are seen as preferable to this by both boys and girls. Two-thirds of the class expressed a very positive preference for working with children of the same sex and the remaining third were positive. Three-quarters of the children, on the other hand, claimed 'not to like at all' or not to 'enjoy' working with children of the opposite sex.

There was a significant difference between girls' and boys' enjoyment of working in mixed sex groups with girls rating it more highly. This accords with other findings that it is more difficult for boys to cross, or be seen to cross, gender boundaries.

Children's views were probed in interviews to find out why they preferred particular groupings to others. It was clear that both boys and girls believe that girls' and boys' approaches to group work differ. The children's beliefs about the nature of the difference were consistent between girls about boys, and between boys about girls. For many children the problem was that girls' and boys' ideas were different. In a group situation where a single idea had to be agreed upon this was made difficult if one gender was outnumbered hence their preference for groups to have equal numbers of girls and boys.

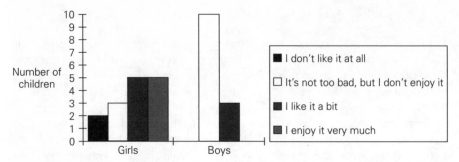

Figure 23.1: Girls' and boys' attitudes to working in mixed gender groups

Boy 1   I don't mind girls, but boys and girls like different things.

Boy 2   Say we were making a puppet theatre like we did last term the girls want pretty flowers on the side and the boys were saying that they wanted army tanks.

Girl 1   Well I just find that I am working with Peter, say if Peter had an idea then Sean would go with Peter because Peter is a boy.

Girls and boys also interpreted each others roles in groups differently. Boy 1 talking about Girl 1:

Boy 1   Cheryl is explaining what to do and I'm not listening . . . My dad does it all the time.

Interviewer   Why don't you listen?

Boy 1   Because I'm fed up of her telling me what to do.

However from the girl's point of view:

Girl 1   I get lumbered with the writing and then they [the boys] get mad with me because I make the decisions but then they won't listen to me if I'm asking them if the decision is ok.

Girls tended to agree that boys left them to do the work in mixed groups.

Girl 2   They just like to sit there chatting all the time and leaving us to do all the work. When you are with girls people work with you and other people.

Girl 1   The boys are always going on about different things and not their work. I like to stick to the work but they are so picky and choosy about what I do.

The girls' view that boys do not listen and the boys' view that girls are bossy was a common theme in the children's interviews and strongly influenced their views about groups.

Boy 3   Girls always have to have their own choice. They know that they have got more power so they can boss us around.

Boy 4   If there were three girls and only one boy they wouldn't listen to you but if there were two boys they would listen.

Girl 2   All the boys work together and you are like sitting in the corner working on your own.

Concluding Remarks

The children's responses revealed significant differences in the groups they preferred to work in. The interviews show that these differences reflect strong beliefs about typical girls' behaviour and typical boys' behaviour. These beliefs indicate that girls and boys interpret the same behaviour in different ways. Of particular interest is the impact of these different attitudes and beliefs on the roles children play in groups and the consequences of these for children's learning. Evidence emerging from the research already indicates that gendered roles have a significant impact on the level of children's cognitive engagement with tasks.

Gender Differences in Patterns of Discourse: Implications for Collaboration?

Introduction

Research has shown that boys and girls vary in their preferred styles of discourse (Reay, 1991). It is also suggested that girls' attempts to achieve consensus through discussion affect the progress of their scientific conceptual development.

Current research particularly that of the SLANT (Spoken Language and New Technology) project has examined the relationship between the discourse interactions of primary aged children and their learning. The research suggests that *exploratory* talk — where hypotheses are proposed, objections are made and justified and relevant information is introduced — helps to develop children's ability to reason through talk (Mercer, 1994).

This paper reports on a research study that is related to the 'Collaborative Learning and Primary Science Project'. A major element of the project research is to examine the potential impact of gender differences on children's learning in science.

A key reason for advocating group work is that it enables collaboration. Effective discourse is an essential means of achieving this.

The questions considered by the study are:

- do gendered patterns of discourse occur?
- does the discourse affect children's learning?
- does the discourse required to achieve success in science disadvantage girls and alienate them?

It is the first question that is considered briefly here in light of a small-scale pilot study.

Approach

Small groups of children aged 9–11 years were videoed working on science invest-igations and their dialogue transcribed and analysed. The pilot results are from four groups. The analysis looked at the number of turns taken by each child, i.e., the number of interjections and their length. These were then categorized using the SLANT categories of 'disputational' talk, 'cumulative' talk and 'exploratory' talk.

Outcomes of the Initial Analysis

The analysis showed that girls were more likely to take 'turns' than boys. Boys' talk also contained more exploratory talk than girls. Boys more than girls appeared to hold strong views and to be unwilling to listen or to acknowledge others' sug-gestions. They also interrupted more in order to create opportunities to express their views. Girls on the other hand were more likely to accept the status quo and were anxious to follow what they understood to be teachers' instructions.

Concluding Comments

There are interesting overlaps between the findings of this pilot and those of the CLAPS project. Teachers' goals related to social interaction may, quite uninten-tionally, actively discourage exploratory talk. As girls are more likely than boys to respond to teachers' directions to 'work together' they may be disadvantaged by this. The next stage for the study will be to consider a wider sample and to look at patterns of discourse over time and importantly in relation to the progress and learning outcomes of the investigative activity.

Action Research and Teacher Development in Denmark: Towards a Gender Inclusive Classroom

The implementation of 'Nature/technique', a new integrated science subject in Danish primary schools, is a challenge for all involved, teachers, educators and researchers. How to support all pupils' learning is an important research question.

The Project ELIN

The intention of the project ELIN (a Danish acronym for Pupils Learning in Nature/technique) is to investigate and reinforce classroom situations that allow pupils ownership of their own learning situations.

The starting points for the project are:

- the new science subject at primary level from grade 1 to 6;
- the aims of the Folkeskole which emphasizes pupils' participation in deciding content and working methods; and
- a wish to support a gender-inclusive science environment.

This action research follows up results from school-based innovations in place before the introduction of the new education act. Science projects in the Danish curriculum from 1988 to 1991 differed in many ways *vis-à-vis* aim, subject matter and teaching strategies. Findings reported by Andersen, Lütken and Veje (1992) point out several problems concerning the pupils' conditions of learning. Science was also included in the curriculum under the heading 'Equal worth and multiplicity' (Pedersen and Reisby, 1992). This report emphasizes that working with physics and chemistry as part of primary science does not necessarily result in equality in education.

Danish research on girls and physics and chemistry has shown that girls evaluate whether they as people have an interest in the subject and the content as well as the ways of working in each specific teaching episode (Sørensen, 1991, 1993). Thus at any grade level both factors influence whether girls continue to participate actively in the teaching and learning of physics and chemistry, a finding supported by Craig (1986). As girls prefer to have influence on their own working situation, teachers are key persons in enabling gender fair science education (Sørensen 1991, 1993; Alting and Wagemans, 1991).

In the ELIN project we work with teachers and their pupils. These teachers have worked in primary science long before the subject became compulsory and this year they attended a primary science in-service course at the Royal Danish School of Educational Studies. On the course, ideas for changing science teaching are introduced through discussion and sharing of experiences. As educators and researchers we do not want to impose our ideas of how to teach science on the teachers — the initiatives have to come from the teachers as well as the pupils. We give feedback and discuss ideas about different strategies and offer advice on the planning of teaching sequences.

Snapshots from Classroom Observation

Up until now we have been focusing on a year 4 and a year 6 class. Both classes have worked with gender inclusive strategies. Observations illustrate that focusing on learning means more than looking for the mastery of concepts and skills.

In an episode video-taped in the year 4 class, the pupils worked in groups with Lego. The groups were selected by their teacher. These were girls-only, boys-only and mixed groups. In the girls-only groups, the girls shared the work — all hands literally working together. They all took part in the planning and construction of the Lego model. In boys-only groups, the boys either took turns with the construction or one of the boys had the leading role, with the working partner(s) just finding the bits and pieces.

In most mixed groups the boys had the leading role and did the construction work with the girls as 'helpers'. In a following lesson, the pupils, the teacher and the observer watched the video-tape and discussed the cooperation in the groups. The class is used to evaluating both product and process. Each child was asked about the experience of cooperation in their group.

The discussion confirmed that girls and boys had different understandings of 'working together'. For the girls this meant sharing the work equally. For many of the boys it meant allocating different tasks to individuals so that the task could be completed. It became evident that in some groups social status determined who was going to do the most demanding and interesting work. No discussion preceded this division of labour. In most of the mixed groups the boys' understanding of 'working together' set the scene. The girls complained, they wanted to share the work more equally. The class also had a general discussion of the advantages and disadvantages in the different choices for sharing the work.

In a subsequent lesson the pattern where one boy took the lead occurred again. When the observer focused the video camera on the group of boys, they started to share the work more equally without being told and without a word to each other. To recognize the situation is the first step towards change.

Some Results of a Questionnaire

The two classes have answered a translated and adapted version of the 'smiley faces' questionnaire developed by Patricia Murphy as part of the CLAPS project, discussed earlier in this symposium collection.

The answers mirror the pupils' experiences of working in groups as they show the same pattern of gender differences in the two classes. The different interpretation of 'working together' expressed by the girls and the boys may be one reason for the children's preference for working in single sex groups and in pairs. Boys more than girls enjoy working in larger groups and as a class. To change the ways of working from classroom settings to small groups may be a simple strategy towards developing more inclusive settings for girls. But then it becomes important to qualify the pupils' skills in working together by explicitly making them aware of the roles they take in the collaboration, and by supporting their sharing of ideas and negotiation of working conditions.

Concluding Comment

In Danish schools gender differences in approaches to group work and the preference for work in single sex settings are most evident in years 4 to 7. In the last years of schooling the pupils claim that the personality of group members is more important for creating good collaborative working conditions than biological sex (Sørensen, 1990).

Improving Science Learning: A Case of Gender

This contribution describes research that is concerned to examine the way interventions affect learning in science, particularly in respect to gender. The interventions focused on are derived from the CASE intervention project (Adey and Shayer, 1993). The CASE project was developed in the early 1980s and was designed to enhance thinking and reasoning skills in science for 11–13-year-olds. The intervention lasts for two years and includes thirty-two activities, to develop skills in areas such as variable manipulations, proportionality, probability and classification. The intervention is designed to replace 25 per cent of the science curriculum for 11–13-year-olds in schools. When public examination results at 16 plus were analysed, enhanced achievements in English, maths and science were noticed two or three years after the intervention for some groups of pupils. There were gender differences in the effects observed. In science, boys' groups starting the intervention when 13 years of age and girls' groups starting the intervention at 12 years of age showed significant improvements in their science GCSE grades, after individual pre-test results were taken into account. However, 13-year-old girls' and 12-year-old boys' performance did not improve. The CASE project is exerting considerable influence currently in science education so further examination of these differences is warranted.

In the contribution the preliminary results on attitudes to science learning, confidence in science and subject preferences are reported. The research is expected to span a period of five years, following four groups of pupils in different schools through their secondary school science programme. The initial stage was to establish pupils' aptitudes and preferences in science and to later relate these to the CASE intervention. In each of the four schools 30 per cent of all 11-year-olds completed a questionnaire in their first term of secondary school. Of these pupils one-third were randomly selected to provide a sample for interview.

Attitudes to Science Learning

Piburn and Baker (1993) suggest that there is a consistent decline in attitudes to science between the ages of 9 and 13 years. They report that the majority of 13-year-olds felt unsuccessful in science. This negative attitude was related to the growing abstraction and complexity of science. The research reported here indicates that pupils come to secondary schools with exciting images of science lessons and these are thwarted in the first few months of secondary school. Many pupils complain about the lack of time in science lessons — time for reflection on completed work is rare. Fensham (1985) mentions 'bad habits' in science teaching, for example — superficiality and premature closure. Many of the pupils interviewed would prefer to study subjects over longer periods of time, in more depth and with less teacher direction and more teacher support. All these factors have an effect on learning outcomes.

Confidence and Topic Preferences

A questionnaire was administered to establish pupils' confidence and preference for particular science topics. The questionnaire results showed that boys were more confident than girls, particularly when assessing their own performance in science, explaining what they had found out and about practical work. Girls, in the single sex school were least confident about their success.

The data from the 12-year-olds showed that overall boys preferred more science topics than girls. Sixteen topics normally covered at this stage in the science curriculum were listed and the pupils were asked to tick those topics they most enjoyed. On average boys chose one more topic area than girls. In the individual topic areas there was a substantial difference between the boys' and girls' interest and enjoyment:

- boys' preferences: acids and alkalis, microbes, forces, energy, space, electricity and magnetism;
- girls' preferences: environment, food, health, animals and human body.

With regard to school subjects music and French were seen as the most popular choices for 'girls' subjects' and maths, technology and PE were the most popular choices for 'boys' subjects'.

These differences in girls' and boys' preferences are very similar to those established by the Assessment of Performance Unit Science Project (DES, 1989). They may arise because boys have greater confidence in their own ability than girls and greater interest in scientific subjects due to exposure to a range of out of school experiences that have reinforced subject stereotyping, for example parental attitudes, media representations etc.

The CASE Intervention: The Impact of Differential Attitudes

The next stage in the research is to see if these differences influence pupils' views of interventions in science designed to promote learning. Initial results show that girls in the single sex school had a distinct dislike for the CASE intervention, although a few pupils did mention certain experiments as being interesting. The girls commented that the work was too hard or did not fit in with what they were doing.

At school 2 the pupils had a more positive approach to the intervention, but did find some of the activities hard. A number of girls found the maths-based activity on ratio particularly hard — this was not the case with the boys.

It the next stage of the research questionnaires will also be given to CASE teachers to ascertain their levels of interest and enthusiasm for the intervention and their views of the training they received prior to and during the intervention period.

Concluding Remarks

These initial findings have revealed that even a decade on and after may initiatives in science, gender differences in pupils' views of science topics remain in the schools under study. Initial results from these schools using the CASE interventions do show some gender differences which may have consequences for the effectiveness of the intervention.

Gendered Voices from Swedish Secondary Schools

Introduction

The concept of 'gender' describes the social construction of the meaning of girls and boys, women and men, in contrast to the concept of 'sex', which stresses the biological distinction between women and men — the sexes. In this contribution the sex–gender dichotomy is challenged. Many feminist researchers want to avoid this dichotomy preferring to examine the relationship between sex and gender. They also consider gender to be a way of structuring society (Harding, 1991). Even so, in discussing education it can be fruitful to concentrate on gender as a social construction.

Focusing on gender may obscure other differences, such as social class, race and ethnicity. But, as the philosopher Jane Martin says, differences do not exclude similarities. The important things to consider are the similarities between girls and women that are relevant (Martin, 1994) and exist in the area of education in science and technology, where there are gender differences. Researchers look for explanations for gender differences in three overlapping categories:

- at the individual level, drawing on psychoanalytic theories;
- at a structural level, focusing on the gendered division of labour with science and technology as male domains; and
- in discourse analyses where beliefs and truths about girls/women and science/technology are deconstructed.

Perspective from Swedish Schools

In a study of science and technology education in Swedish compulsory schools gendered preferences and learning styles were found (Staberg, 1994). Girls for example prefer topics connected with their own and others' lives, while boys are mainly interested in using apparatus and in making things. Girls enjoy working together and rely on books in their work and on reading and writing. Boys however tend to play both with apparatus and with each other. They compete more than they collaborate and study less than the girls.

Girls' learning can be characterized by work and boys' by play and this is particularly obvious in laboratory work. This difference is interesting considering the analysis of Valerie Walkerdine and her co-workers (1989). They showed that in the context of mathematics, boys' play is considered the appropriate approach to learning whilst girls' hard work is interpreted as indicating a lack of real understanding.

Boys' interests and learning style are favoured in classrooms often because the subject content is masculinized and because boys influence the teaching methods. On the other hand, girls' capacity to collaborate is not being used to their best advantage. Girls' need to use language is not really taken into account. Girls' theoretical way of approaching the subjects, partly owing to their unfamiliarity with tinkering (Sjøberg and Imsen, 1991) and partly due to their learned diligence, is one of the reasons of their need to understand and to achieve overall comprehension.

The social construction of chemistry, physics and technology as masculine domains continues in the classroom alongside the construction of gender (Cockburn, 1991; Kelly, 1985). At the individual level the notions of physics and chemistry vary, depending partly on family background, but the majority of the girls have, over the years, come to construct femininity — and maturity — as not fitting in with an enjoyment of science. Boys also perceive these subjects as masculine. This is even more obvious in technology, with the consequence that the majority of girls reject technology.

All girls, even the successful ones, tend to question their own understanding in science. This can be interpreted as knowledge (conscious or unconscious) of the relationship between gender and science at the symbolic level. Girls are aware of the views in society of women as illogical, irrational, soft and subjective, and thus unfit for science. Girls therefore doubt their own competence in science.

A study currently being conducted in upper secondary school students in Sweden indicates that girls see themselves as less clever than boys but there are signs of resistance towards this belief. In chemistry and physics girls find that both the teaching and the subject content is more appropriate for boys. They also point out that in order to be engaged with these subjects they have to be interested in the things that are largely of interest to the boys. They also comment that boys answer more questions than girls and when teachers treat boys preferentially, the girls notice.

Due to recent reorganization in upper secondary schools in Sweden, technology is studied by all pupils in the science programme in the first year. The technology curriculum includes history and the social consequences of technology as well as more traditional areas. None of the interviewees had anything to say about the discussions of social consequences of technology. Is this new curriculum emphasis possible? In the technology curriculum, the mutual social constructions of gender and technology are manifest — the gender differentiation is obvious.

Learning Styles

The stories girls tell in interviews demonstrate the importance of the affective factors in learning which have been well studied (Beyer, 1992; Murphy, 1993). As far as can be seen from these interviews and pupils' past experience there is very little that could be recognized as the development of metacognition amongst these pupils. The girls report that they study and do not have time for reflection — they seem to think that discussion and reflection are not appropriate in science and technology. It is not solely a female virtue to reflect on what is learned and what use can be made of what is learnt, but in both studies it was found that it is mostly girls who raise these questions. What is more, the notions of the science curriculum held by many authors and teachers seem to be symbolically male. It is believed that science should be effective, rational and hard. Is it possible to change this image in a society where male dominance over science and technology is accepted as the norm?

Concluding Remarks

In a world aiming for sustainable economic and social development we have to place much more emphasis on values in order to foster socially responsible and gender-inclusive science and technology.

Note

1   With thanks to the following contributors:
    Patricia Murphy, Kim Issroff with Barbara Hodgson, Eileen Scanlon and Elizabeth Whitelegg, The Open University, Milton Keynes, England.
    Ros Smith, Homerton College, Cambridge, England.
    Annemarie Møller Andersen and Helene Sørensen, The Royal Danish School of Educational Studies.
    Gaynor Sharp, Coventry, England.
    Else-Marie Staberg, Umeå University, Sweden.

References

ADEY, P. and SHAYER, M. (1993) 'An exploration of long term far-transfer effects following an extended intervention programme in the high school science curriculum', *Cognition and Instruction*, **11**, 1, pp. 1–29.

ALTING, A. and WAGEMANS, C. (1991) 'The influence of the physics teacher: Towards a research model and a formulation of the 'ideal' physics teaching for girls', *Contributions on the Sixth international GASAT Conference*, Melbourne, Australia.

ANDERSEN, A.M., LÜTKEN, H. and VEJE C.J. (1992) 'Experiences from science development projects: Summary of evaluation', *Report for the Innovation Council of the Folkeskole*, Copenhagen, Royal Danish School of Educational Studies.

BEYER, K. (1992) 'Det er ikke taenkning det hele', NIELSEN I.H. and PAULSEN A.C. (Eds) *Undervisning i Fysik — Den Konstruktivistiske*, København, Gyldendal.

COCKBURN, C. (1991) *Brothers: Male Dominance and Technological Change*, London, Pluto Press.

CRAIG, J. (1986) *A Longitudinal Study of the Effect of Differing Primary Science Experience on the Level of Interest and Achievement of Girls and Boys in Their First Year of Secondary Science*, UK, Loughborough University of Technology.

DEPARTMENT OF EDUCATION AND SCIENCE (1989) *Science at Age 13:A Review of APU Survey Findings 1980–84*.

FENSHAM, P.J. (1985) 'Theory in practice: How to assist science teachers to teach constructively', in ADEY, P., BHOS, J., HEAD, J. and SHAYER, M. (Eds) *Adolescent Development and School Science*, Lewes, Falmer Press.

HARDING, S. (1991) *Whose Science? Whose Knowledge?*, Milton Keynes, Open University Press.

KELLY, A. (1985) 'The construction of masculine science', *British Journal of Sociology of Education*, **6**, 2, pp. 133–54.

MARTIN, J.R. (1994) 'Methodological essentialism, false difference, and other dangerous traps', *Signs*, **19**, 31, pp. 630–57.

MERCER, N. (1994) 'The quality of talk in children's joint activity at the computer', *Journal of Computer Assisted Learning*, **10**, pp. 24–32.

MINISTRY OF EDUCATION (1994) *Act on the Folkeskole: The Danish Primary and Lower Secondary School*, Consolidation Act No. 311 of 25 April 1994, Copenhagen, Denmark.

MINISTRY OF EDUCATION (1994) *Formål & Centrale Kundskabs og Færdighedsområder Folkeskolens Fag* (Aims and central areas of knowledge and skills), Copenhagen, Denmark.

MURPHY, P. (1993) 'Assessment — A constructivist perspective', Paper presented at the *Nordic Symposium on Science Education*, Gilleleje, Denmark.

MURPHY, P., ISSROFF, K., SCANLON, E., HODGSON, B. and WHITELEGG, E. (in press) 'Developing investigative learning in science — The role of collaboration', in AKKER VAN DEN, J., KUIPER, W. and HAMEYER, U. (Eds) *Issues in European Curriculum Research*, (Paper presented at ECUNET, University of Twente, 31 August–2 September 1994)

PEDERSEN, G. and REISBY, K. (Eds) (1992) *Equal Worth and Multiplicity — On Equal Opportunities in School*, Copenhagen, The Royal Danish School of Educational Studies.

PIBURN, M. and BAKER, D. (1993) 'If I were the teacher . . . a qualitative study of attitude toward science?', *Science Education*, **77**, 4, pp. 393–406.

REAY, D. (1991) 'Intersections of gender, race and class in the primary school', *British Journal of Sociology of Education*, **12**, pp. 163–83.

SCANLON, E., MURPHY, P., ISSROFF, K., HODGSON, B. and WHITELEGG, E. (1995) 'Collaboration in primary science classrooms: Implications for teaching and learning', Paper presented at Science Education Research in Europe, Leeds Conference, 7–11 April.

SJØBERG, S. and IMSEN, G. (1991) 'Gender and science education: 1', in FENSHAM, P. (Ed) *Developments and Dilemmas in Science Education*, Lewes, Falmer Press.

SØRENSEN, H. (1990) 'When girls do physics', in *Contributions to the European and Third World GASAT Conference*, Sweden, Jönköping.

SØRENSEN, H. (1991) 'Physics and chemistry in the Danish primary school — Seen from girls' perspective', in *Contributions to the Sixth International GASAT Conference*, Australia, Melbourne.

SØRENSEN, H. (1993) 'You will have to do more than just tell them', in *Contributions to the Seventh International GASAT Conference*, Canada, Waterloo.

STABERG, E-M. (1994) 'Gender and science in the Swedish compulsory school', *Gender and Education*, **6**, 1, pp. 630–57.

WALKERDINE, V. (1989) *Counting Girls Out*, London, Virago.

24 Geology: A Science, a Teacher, or a Course?: How Students Construct the Image of Geological Disciplines and That of Their Teachers

Alfredo Bezzi

Abstract

The repertory grid technique, developed out of personal construct psychology (Kelly, 1955), has been used to elicit information about students' outlook on some geosciences and relevant instructors. Constructs were elicited from a sample of six undergraduate students. Six subject areas of geosciences (geography, geology, stratigraphy, mineralogy, palaeontology, petrology) constituted the range of convenience of the grid elements. Examples are given of students' representations of the geosciences and their teachers as well as the cluster analysis of the grid elements in terms of geological disciplines and relevant instructors. This study demonstrates that this technique is useful to understand better the nature and range of students' ideas about the image of geosciences and the quality of teaching.

Introduction[1]

Some years ago I read a sentence that struck me for the acuteness of the observation that was made: 'Ask a student to tell you about a discipline and he/she will answer telling you about his/her teacher.' This assertion grasps the essence of the role of teachers in communicating the image of their subject with the relevant implications, both in more general cultural terms and in terms of motivation and learning.

In international literature, there are plenty of 'traditional' studies, regarding the image of science and scientists: see, for instance, the articles of Acevedo Diaz (1993), Àlvarez *et al.* (1993), Carrascosa *et al.* (1993), Guasch *et al.* (1993), Lederman (1992), Lumb and Strube (1993), Nesbitt (1992), Newton and Newton (1992), Schibeci (1990). In these works, investigations were carried out directly by interviewing people (laypeople, students, teachers), or indirectly by means of analysis of textbooks, newspapers, lessons and television programmes, or written drafts and drawings produced by people who were interviewed. In the direct investigations, a variety of elicitation methods were used: mainly, the more conventional semantic differential methods and various types of Likert scales, and the more innovative 'Draw-a-Scientist Tests' by Chambers (1983) and 'Views on Science-Technology-Society' (VOSTS) by Aikenhead and Ryan (1992). It is interesting to note that there is a general agreement in the results reached by the various authors in the various contexts; contexts which were very dissimilar both with respect to geographic location, school and cultural level, and socio-economical situation. The images described are usually rather naive, stereotyped, and almost never reflect the real practice of science (usually seen as positivist and objective) or scientists (generally perceived as unusual, negative people who, more often than not, are busy pursuing their own purposes rather than the 'good of mankind').

A different type of methodological approach, used to explore both students' and teachers' ideas about certain scientific issues and their attitudes towards the aspects of science, was used by Stead (1983), Happs and Stead (1989) and Lakin and Wellington (1994). These authors used the repertory grid technique, whose theoretical framework resides in the Personal Construct Psychology (PCP) of Kelly (1955). This theory is based on the assumption that individuals psychologically work in accord with their attempts to give their surrounding world a meaning, through a continuous process of 'anticipation' of future events. This anticipation basically consists of hypotheses which are either refuted or verified as a result of subsequent experiences: hence, Kelly's emblematic metaphor about 'Man-the-Scientist' to identify the type of activity in which every person is involved in an attempt to construct the surrounding world. The basic units of these hypotheses are the 'personal constructs' which, as a whole, offer an idiosyncratic model of reality for each individual. Each personal construct is defined as the representation of a linear, bipolar and discriminating dimension (e.g., *beautiful/ugly*, or *interesting/boring*), in which it is possible to place events, objects, people, etc. called 'elements'. A particular set of elements, evaluated (for example with a numeric scale from 1 to 5) on the basis of a series of constructs in a rectangular matrix, defines a 'repertory grid'. The lack of space prevents all the aspects of Kelly's theoretical framework regarding both methodologies and techniques of grids' analysis from being described in detail. These details, as well as the various fields where PCP has been applied up until now (psychology, psychotherapy, industry, education), can be found, in addition to the above mentioned papers, in Bannister and Fransella (1989), Dalton and Dunnett (1989), Fransella and Bannister (1977), Pope and Keen (1981), Shaw (1980, 1981), Shaw and McKnight (1981).

In general, the research on the image of science was mainly oriented towards its general aspects rather than to single disciplines; for geology, in particular, there is a lack of specific works, with the exception of those by Arthur (1993) and Jordan (1989) which are articles of opinion rather than actual research.

The present study, planned according to the theory of PCP, tries to fill this gap in terms of this discipline, taking into account that, unlike research of a 'motivational' character, there was no specific aim to assess any learning achievement. In fact, the main purposes are to outline the image of some geological disciplines, constructed by university students, and to verify whether their teachers affect the construction of such an image.

The six subjects who took part in this research were university students at the end of their course of study who are referred to, in this paper, by pseudonyms. Some of the most significant geological disciplines (geography, geology, mineralogy, palaeontology, petrology, stratigraphy) were chosen as elements of the repertory grid. The elicitation of constructs took place at two different stages, with two different areas of exploration. In the first case, the students were asked to consider the elements of the grid exclusively as scientific disciplines, while in the second case the elements of the grid were constituted by the relevant teachers. Constructs were elicited by students using the triadic fashion, i.e., comparing groups of three elements indicated by a dot on the grid supplied. Students had to think of some way in which two of the three elements were alike, yet different from the third. Then they had to write the way the elements were alike on the left hand and the opposite on the right hand of the grid's rows. The two similar elements were rated 1 and the different element 5. The remaining three elements of the row were rated accordingly to the elicited constructs and the 1 to 5 scale. Students repeated this process filling each of the twenty rows of the provided grid.

Analysis of the Content: The Disciplines

In an initial analysis of the repertory grids in which the elements were the disciplines, the 120 elicited constructs (20 for each grid) were grouped together into common categories based on the author's judgment of the meaning of the construct. These categories can group either 'internal' or 'external' type constructs: in the first case the discipline was identified as a science with its objects (A1) and techniques of investigation (A2) (with constructs such as *magmatic/sedimentary processes; use of fossils/minerals*). In the second case the discipline was described in terms of the features of the course and the type of study (B3; with constructs such as *vague/clear programme*; *rote learning is/is not important*) or in terms of the professional aspects or the type of work (B4; with constructs such as *work in an office/in the field*). Other categories can be added concerning the epistemological aspects (C5; with constructs such as *reality/fantasy*), the affective aspects (C6; with constructs such as *negative/positive impression*), and the general cultural aspects (C7; with constructs such as *specific for geology/in other university courses*). Unimportant or incomprehensible constructs are included in the last category (D8) (Table 24.1).

As Table 24.1 shows, the teaching aspect (B3 = 42.50 per cent) is the prevailing one, exceeding, even though only slightly, the more strictly 'scientific' aspect (A1 + A2 = 39.17 per cent). The affective (C6 = 7.50 per cent) and the professional aspects (B4 = 4.17 per cent) are left significantly behind. It is particularly interesting to note that the methodological and the epistemological aspects, as well as the aspects relating to the philosophy of science, are basically absent (C5 = 2.50 per cent).

However, in a Kellyan qualitative approach, which tends to point out the idiosyncratic system of the individual constructs, statistical considerations are undoubtedly less significant (also because they are intrinsically prevented by the small number of subjects), while the personal 'configurations' of the subjects are much more important. In Table 24.1, even though there is a still prevailing presence of

Table 24.1: Analysis of categories of the elicited personal constructs concerning the geological disciplines

Categories	Bob	Tom	Liz	Students Meg	Emy	Sam	Total	Percentage
A1	7	1	0	8	3	1	20	16.67
A2	8	4	7	4	3	1	27	22.50
B3	1	11	11	5	12	11	51	42.50
B4	3	1	0	1	0	0	5	4.17
C5	1	1	0	0	1	0	3	2.50
C6	0	2	0	1	0	6	9	7.50
C7	0	0	2	0	1	0	3	2.50
D8	0	0	0	1	0	1	2	1.67

Notes
(A1 = objects of investigation; A2 = techniques of investigation; B3 = features of the courses and the type of study; B4 = features of the type of work and the profession; C5 = epistemological aspects; C6 = affective aspects; C7 = general cultural aspects; D8 = others)
The figures represent the number of constructs for each category (row) and for each student (columns). The total and the relevant percentage refer to each category.

teaching aspects, there is a rather varied range of images of the disciplines: starting from the more strictly 'scientific' image (Bob) to that where the discipline's aspects basically disappear and the teaching and the affective aspects prevail (Sam) to the image where there seems to be some balance between the components (Meg).

The analysis of the clusters, obtained with the FOCUS programme (Shaw, 1980), gives further interesting indications: the process groups the elements which, thanks to a high *similarity match* between one and the next, appear to have common features in the eyes of people from whom the constructs have been elicited. As can be seen in Figure 24.1, for the students who preferably used or gave some prominence to 'internal' constructs with respect to the disciplines (Bob, Meg and Emy), there is a logical and tight link between the sectors of disciplines of similar geosciences, as far as objects and methods of investigation are concerned. In fact, these students present a cluster, consisting of geology, stratigraphy and palaeontology, where the first two are similar to one another at 75.0 per cent, 73.8 per cent and 71.2 per cent respectively for Bob, Meg and Emy and where the third is linked to the previous two at 60.6 per cent, 68.1 per cent and 60.0 per cent respectively. Another cluster is prominent to the three students, that of mineralogy and petrology in percentages of similarity match at 55.6 per cent, 75.6 per cent, 65.0 per cent respectively. The situation is instead very different and somewhat incongruous for those students who paid more attention to the 'didactic' side of the subjects: a basic overall similarity in all the disciplines (Liz), or some rather unconventional clusters (palaeontology/mineralogy/petrology — Tom), and some rather unusual matches (such as geography/mineralogy — Sam) can be found. Moreover, it is interesting to note that for almost all students (except for Sam), the position of geography is essentially detached from the general context, with a minimum similarity match of 43.1 per cent (Meg) and a maximum of 55.0 per cent (Emy).

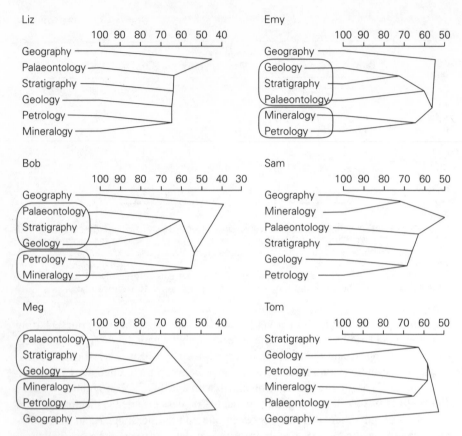

Figure 24.1: Clusters of the elements (geological disciplines), on the basis of personal constructs which were elicited in each individual and analysed by means of the FOCUS programme
Note: The horizontal scale shows the percentage of the similarity match.

In order to explain better these types of constructs (which will be useful for a subsequent comparative analysis) and the idiosyncratic images resulting from the grids, the following paragraphs report some of the 'profiles' of individual disciplines which can be obtained through a logical reconstruction of the constructs supplied by some students. The phrases in square brackets have been added by the author to the original constructs to facilitate comprehension.

(Emy, 20 constructs) geology: it is a discipline that fairly requires reasoning; it taught me the 3-D vision, it gave me an idea of petrogenetic processes; it gave me a fair idea of geological times and Ligurian geology (though it left me unsatisfied about the answers in the geological context) and it taught me sufficiently about a careful observation. I did not learn how to read topographic maps, or how to use a microscope and the teaching of the procedural technique was insufficient. Geology starts from unverifiable dogmatic assumptions, the evidence of which can be found in the rocks and, by means of a fair use of the macroscopic identification of rocks (the pressure/temperature genetic conditions of which can be more or less determined). It detects the geological structures of an area and reconstructs the past life, more or less defining the dates. This discipline is connected to the other subjects and sufficiently stimulated me to further investigations; it is usually connected to practical applications.

(Liz, 20 constructs) palaeontology: [from the point of view of the techniques of investigation] it is based on field evidence and requires equipped laboratories; it can require, or otherwise, the identification of the rocks, but not the identification of minerals; the microscope, also, may or may not be necessary. It is a discipline with a lot of theory, with independent topics and one rather specific topic. This discipline can be a prerequisite to other disciplines, or otherwise; it does not require a knowledge of physics and chemistry, or the use of formulas; it requires a knowledge of geology, fossils and geological eras; [in palaeontology] there may be, or otherwise, a handling of 3-D objects; palaeontology is known only by the geologist, but you can also find it in other university courses; it is useless for reading geological maps. The learning of this discipline requires a mnemonic study.

(Bob, 20 constructs) petrology: it deals with magmatic and magmatic-metamorphic processes, tectonic deformations and it studies surface and subsurface structures; it mainly studies the paleoenvironments with respect to the present rock basements and investigates the tectonic weathering using non sedimentary geochemistry. It does not study geological or topographic maps, but it studies and uses minerals, though only the metamorphic ones; the identification of the textures, of the minerals and of the microscopic textures of minerals is carried out using samples. The problem of geological dating is not important. Petrology involves either a wide or limited fieldwork, with surveys and work in the laboratory: therefore, it is a job that can be carried out both in the office and in the field. It is interesting for the mining sector. Its results reflect reality rather accurately and its study usually involves the memorization of only a few names.

Analysis of the Contents: The Teachers

Similarly to the procedure used for the disciplines, in the analysis of the repertory grids in which the elements were the instructors of the geological disciplines, the 120 elicited constructs (20 for each grid) were grouped by the author into two main categories relevant to the teaching and to the personality of the teachers. These main classes have been in turn subdivided respectively into teaching professional aspects (A1) and features of the courses (A2), and into personal aspects of his or her character (B3) and his or her rapport with the students (B4). Examples of students' constructs pertaining to these categories are the following.

Table 24.2: Analysis of the categories of the elicited personal constructs concerning the teachers of geological disciplines

	Bob	Tom	Liz	Students Meg	Emy	Sam		
Categories							Total	Percentage
A1	7	5	3	4	12	9	40	33.33
A2	0	6	11	1	2	0	20	16.67
B3	8	3	4	13	4	1	33	27.50
B4	4	5	2	1	2	10	24	20.00
C5	1	1	0	1	0	0	3	2.50

Notes
(A1 = professional teaching aspects; A2 = features of the courses; B3 = personal character; B4 = rapport with the students; C5 = others)
The figures represent the number of constructs for each category (row) and for each student (columns). The total and the relevant percentage refer to each category.

(A1) he/she makes the course interesting/unbearable; he/she involves/does not involve; he/she wants/does not want rote studying; competent/incompetent; he/she requires/does not require precision of terms);

(A2) lessons with/without logic consequence; use/non-use of samples; tough/reasonable timetable);

(B3) he/she gets/does not get angry at the exams; liberal/conservative; pleasant/unpleasant; on time/late for the lesson; schematic/imaginative);

(B4) he/she is approachable/not approachable; he/she is friendly/unfriendly; he/she is prepared to discuss/always right; he/she offends/does not offend)

(Table 24.2)

A mature critique by the students emerges from the overall analysis of these constructs, in that, quantitatively, the aspects of the teaching and the human qualities of the teachers are almost totally balanced (A1 + A2 = 50.00 per cent against B3 + B4 = 47.50 per cent) and, moreover, the aspect that appears more important to them is undoubtedly the professional teaching one (A1 = 33.33 per cent). It is worth pointing out the significant weight (B4 = 20.00 per cent) given to the personal rapport between teacher and learner.

With the teachers as elements of the grid, the analysis of the constructs allows the determination of how students perceive the quality of the received teaching. In fact, 82 per cent of the elicited constructs contained a judgment expressed by the rating of the elements on the scale from 1 (positive evaluation) to 5 (negative evaluation). Examples of these constructs containing an evaluation of the teaching situation related to the identified categories are:

(A1) *he/she can/cannot explain;*

(A2) *boring/not boring lessons;*

(B3) *kind/rude;*

(B4) *available/not available*

Table 24.3: Statistical analysis of the personal constructs in which the students expressed a judgment about the teaching situation (Constructs elicited with the teachers of the geological disciplines as elements of the grid)

	Categories			
	A1	A2	B3	B4
Number of constructs	35	7	32	24
Mean	2.98	2.82	2.76	2.99
Std deviation	0.5650	0.3772	0.4610	0.5949
Std error	0.0955	0.1426	0.0815	0.1214
Minimum*	1.67	2.33	1.50	1.67
Maximum*	4.33	3.33	3.50	4.17

* minimum and maximum refer to the means of the ratings for each construct
Notes
(positive evaluation = 1; negative evaluation = 5; codes identify the same categories as in Table 24.2; the figures of the first row represent the number of constructs for each category in which a judgment is expressed)

Table 24.3 shows that in all the categories the mean tends towards neutral and clearly the students do not perceive the offered teaching performance as satisfactory.

The analysis of the individual students (Table 24.2) obviously presents a more complex situation, from which the 'constructive' individual features emerge. In fact, it appears that, as far as the construction of the teachers' image is concerned, some students devote much more attention to aspects of character (Meg), while others principally to the teaching professional aspects (Emy), or to the features of the courses (Liz), or even to the rapport with the student (including the teaching aspects) (Sam). The most 'balanced' construction seems to be Tom's, where the various categories coexist in a relatively homogeneous mix. Bob's profile is similar to this, but, for him, all the aspects of teaching concentrate on the personality of the teacher, neglecting the features of the courses.

The comparative analysis of Table 24.1 and Table 24.2 shows us that, at least in one case, there is a clear interaction between the image of disciplines and the image of teachers. We can see, in both elicitations, the strong 'didactic' characterization that has left its mark in Liz's and Emy's constructs. For the other students it is more difficult to establish meaningful connections with such analysis.

Similarly to the disciplines, we report some examples of the 'profiles' of the teachers. For reasons of discretion we have not mentioned the names of the disciplines.

(Sam, 20 constructs) *Discipline X*: [This teacher] is not prepared to discuss and he/she imparts a feeling of detachment between him/herself and the students; usually he/she is not available to go on fieldwork, dedicate extra hours, or give further explanations. He/she is not available out of the classroom, though seldom postpones the lesson; he/she is indifferent to exams being moved, but will not add extra exams and does not allow the extra use of laboratories. His/her explanations are insufficient, he/she is reluctant to explain and nothing of his/her explanation sinks in; he/she does not provide links with other subjects and he/she does not show samples or geological maps; he/she does not require rote studying. He/she arrives on time to lessons and wants his/her course to be attended, since he/she believes that his/her subject is important.

(Meg, 20 constructs) *Discipline Y*: [This teacher] has personality, he/she is friendly, smiling, imaginative, dynamic, strict, disorganized, he/she smokes moderately, he/she is fairly authoritarian, [inspires] joy and sympathy; he/she is not detached, though he/she does not act too friendly; he/she does not agree or disagree [with colleagues]. He/she is competent, you can disturb him/her, he/she involves the students, he/she teaches fairly enough and his/her lessons are not boring. He/she is a naturalist, [and he/she teaches in] other courses.

(Tom, 20 constructs) *Discipline Z*: [This teacher] sufficiently accepts comments, he/she answers questions, he/she gets rather angry over exams, he/she offends the students, he/she does not quite understand the problems and is rather subjective in his/her judgments. He/she is a good teacher, but he/she makes the course unbearable; he/she explains everything, very well and clearly. The use of maps is not needed [in the course] and maximum commitment is required; the exam, which is subdivided into several parts, is one of the most difficult in the course and requires the use of samples and a microscope. He/she treated me badly, but he/she did not fail me, and I would allow him/her to continue the professorship.

The analysis of the clusters in Figure 24.2, compared with those in Figure 24.1, allows us to make some rather interesting comments, taking into account that the constructs used by the students in the two different grids were stated with very different verbal meanings. The geology/stratigraphy/palaeontology cluster is not only found in Bob, Meg and Emy, who had already 'created' it for the disciplines (and in this case with even higher percentages of similarity match), but also in Tom who, as far as disciplines were concerned, displayed rather uneven matches, and in Liz, for whom the similarity match is not so close. In four cases out of six the mineralogy/petrology cluster occurs (Liz, Emy, Sam, Tom). The same cluster occurred only three times (Bob, Meg, Emy) when the elements of the grids were the geological disciplines. Geography, in a way, returns to have some form of relevance, presenting here higher percentages of similarity match, though 'inconsistent' (if we think of it as a scientific discipline) when placed in the mineralogy/petrology sector with three very clear-cut clusters (Emy, Sam, Meg) and two less correlated ones (Bob, Tom). In one case only (Liz), geography is related to the geological sector, with which it should have a greater affinity. Therefore, in conclusion, this overall picture is, in a way (and surprisingly enough), internally more consistent than that of the disciplines. Consequently it allows us to perceive the influence that personal and professional features can have on the construction of an image of science (or sciences).

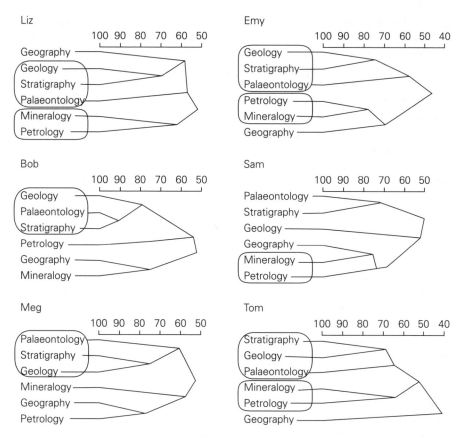

Figure 24.2: Clusters of the elements (teachers of the geological disciplines) on the basis of personal constructs which were elicited in each individual and analysed by means of the FOCUS programme
Note: The horizontal scale shows the percentage of the similarity match.

Conclusion

Similarly to the cited works, and particularly for the geological domain to Happs and Stead (1989), this paper also seems to confirm the usefulness and validity of the application of PCP and repertory grids to elicit both quantitative and qualitative data. These data focus on the image of science (and of sciences), developed by individuals, and likewise with the various kinds of clinical interviews mentioned in the introduction. It is clear that the size of the sample can allow neither conclusions nor universal extrapolations. Nevertheless, the data obtained are interesting both for their intrinsic qualitative aspects and because they could constitute a starting point for further considerations and a reference point for other similar research.

Moreover, it is worth reaffirming that, in the 'classical' methods of elicitation, there is always a constraint, to a greater or lesser extent, inferred by the interviewer. The quality and quantity, the formulation and the structure of spoken or written questions which obviously depend upon the cultural assumptions and purposes of the researcher, imply an unavoidable implicit bias of method. Aikenhead and Ryan (1992) underline the fact that the instruments of investigation, commonly used in these types of research, are applied with the assumption that students perceive and interpret the statements of the tests with the same meaning as given by the researchers. However, Aikenhead and Ryan proved that this fact was controversial and that great ambiguity was undermining the validity of the interpretation of the results. So much so that their VOSTS instrument is not the result of theoretical statements, based on the researchers' point of view, but it was instead developed empirically, through a complex and difficult process, starting from opinions by students, just to avoid this implicit and inevitable difficulty. The VOSTS instrument is undoubtedly more efficient than others, both for its reliability and the variety of topics which can be investigated. Nevertheless, the elicitation of results seems to take place through a rather laborious process.

On the contrary, as also pointed out by Lakin and Wellington (1994), when using the repertory grid there are no previously established dimensions, except those of the system of constructs, used by the subject to give meaning to his or her own experience, to anticipate events or to build, subjectively and idiosyncratically, his or her surrounding reality. Therefore, before other subsequent interactions can take place between the interviewer and the subject, the original elicitation of constructs reveals strictly personal (cognitive, value-related, affective) dimensions and meanings, free from external influences, which are true indicators of the uniqueness of the individual terms of reference. This is the assumption that allows us to state that the images, outlined by the students' elicitation of constructs, are as significant as they are illuminating as far as the real dimensions of personal constructions are concerned. We could say (paradoxically) that the more they reflect the subjectivity of individuals, the more 'objective' they are. Therefore, the overall picture is not an average expression resulting from the statistical analysis of constructs prearranged by others. On the contrary, it is the totality of single and individual images allowing a global vision without, nevertheless, sacrificing the singularities.

In conclusion, we can say that the present findings show that teachers have a great responsibility for conveying (implicitly and explicitly) not only the content but also the image of a science. These results confirm the statement of Eichinger (1992), who analysed the perceptions of some college science majors regarding their formal science experience in school, that teachers' personalities have greater influence than classes and are likely to be the most significant factors affecting student perception of science classes.

In the geological disciplines, teachers should not leave out, both for the general formative aims and for the more specific professional aims, the aspects regarding the nature of science in general (and of geology in particular) which allow appreciation and meaningful use of the scientific knowledge, neither should he or she omit to provide a continuous reference to the overall vision of both the education project and the geosciences. Failing this, it is quite likely that the image of geology, in its broadest sense, would be constructed not as the image of a science, built on a consistent series of principles, methods and paradigms, but rather as the image of a series of lectures, linked or otherwise to one another. On account of the present level of discomfort shown by students as far as the quality of teaching services and the modalities of this conveyance are concerned, it is hoped that in the future it will be possible to state about teachers what Robert C. Whisonant's wife told her husband when he received the Neil Miner Award (the prize awarded by the National Association of Geology Teachers to teachers who distinguished themselves for the excellence of their teaching) 'The thing that really sets you apart from a lot of teachers is that you actually treat your students like they're human beings' (Whisonant, 1994).

Note

1   The present work was supported by a contribution of the Italian National Research Council (CNR). The author is grateful to the students who took part in this research and to Bianca De Bernardi, Psychology Institute of the University of Verona, for the critical reading of the manuscript. Thanks go to Beatrice Bezzi and Anthony Westwood for the translation in English of this paper.

References

ACEVEDO DIAZ, J.A. (1993) '¿Qué piensan los estudiantes sobre la ciencia?: Un enfoque CTS', *Enseñanza de las Ciencias*, Número extra (IV Congreso Internacional sobre Investigación en la Didáctica de las Ciencias y de las Matemáticas), pp. 11–12.

AIKENHEAD, G.S. and RYAN, A.G. (1992) 'The development of a new instrument: 'Views on Science–Technology–Society' (VOSTS), *Science Education*, **76**, 5, pp. 477–91.

ÀLVAREZ, M., SONEIRA, G. and PIZARRO, I. (1993) 'Cómo percibe el alumnado algunas interacciones entre Ciencia–Tecnología–Género–Sociedad', *Enseñanza de las Ciencias*, Número extra (IV Congreso Internacional sobre Investigación en la Didáctica de las Ciencias y de las Matemáticas), pp. 19–20.

ARTHUR, R. (1993, April) 'Public understanding of geology: Do we mean "What is geology?"', Paper presented at the First International Conference on Geoscience Education and Training, Southampton, UK.

BANNISTER, D. and FRANSELLA, F. (1989) *Inquiring Man: The Psychology of Personal Constructs*, London, Routledge.

CARRASCOSA, J., FERNANDEZ, I., GIL, D. and OROZCO, A. (1993) 'Análisis de algunas visiones deformadas sobra la naturaleza de la Ciencia y las características del trabajo científico', *Enseñanza de las Ciencias*, Número extra (IV Congreso Internacional sobre Investigación en la Didáctica de las Ciencias y de las Matemáticas), pp. 43–4.

CHAMBERS, D.W. (1983) 'Stereotypic images of the scientist: The draw-a-scientist test', *Science Education*, **67**, 2, pp. 255–65.

DALTON, P. and DUNNETT, G. (1989) *A Psychology for Living; Personal Construct Psychology for Professional and Clients*, London, Dunton Publishing.

EICHINGER, J. (1992) 'College science majors' perceptions of secondary school science: An exploratory investigation', *Journal of Research in Science Teaching*, **29**, 6, pp. 601–10.

FRANSELLA, F. and BANNISTER, D. (1977) *A Manual for Repertory Grid Technique*, London, Academic Press.

GUASCH, E., DE MANUEL, J. and GRAU, R. (1993) 'La imagen de la ciencia en alumnos y profesores: La influencia de la ciencia escolar y de los medios de comunicación', *Enseñanza de las Ciencias*, Número extra (IV Congreso Internacional sobre Investigación en la Didáctica de las Ciencias y de las Matemáticas), pp. 77–8.

HAPPS, J.C. and STEAD, K. (1989) 'Using the repertory grid as a complementary probe in eliciting student understanding and attitudes toward science', *Research in Science and Technological Education*, **7**, 2, pp. 207–20.

JORDAN, R.R. (1989) 'Do we teach geology?', *Journal of Geological Education*, **37**, 3, pp. 168–70.

KELLY, G.A. (1955) *The Psychology of Personal Constructs, Vols I and 2*, New York, Norton.

LAKIN, S. and WELLINGTON, J. (1994) 'Who will teach the "nature of science"?: Teachers' views of science and their implications for science education', *International Journal of Science Education*, **16**, 2, pp. 175–90.

LEDERMAN, N.G. (1992) 'Students' and teachers' conceptions of the nature of science', *Journal of Research in Science Teaching*, **29**, 4, pp. 331–59.

LUMB, P. and STRUBE, P. (1993) 'Disturbing the boundaries: The science/literature membrane', *Science Education International*, **4**, 1, pp. 5–8.

NESBITT, J.E. (1992) 'A look at two minority families and their beliefs about who can become a scientist', *Science Education International*, **3**, 4, pp. 23–5.

NEWTON, D.P. and NEWTON, L.D. (1992) 'Young children's perceptions of science and the scientist', *International Journal of Science Education*, **14**, 3, pp. 331–48.

POPE, M.L. and KEEN, T.R. (1981) *Personal Construct Psychology and Education*, London, Academic Press.

SCHIBECI, R.A. (1990) 'Public knowledge and perceptions of science and technology', *Bulletin of Science, Technology and Society*, **10**, 2, pp. 86–92.

SHAW, M.L.G. (1980) *On Becoming a Personal Scientist: Interactive Computer Elicitation of Personal Models of the World*, London, Academic Press.

SHAW, M.L.G. (Ed) (1981) *Recent Advances in Personal Construct Technology*, London, Academic Press.

SHAW, M.L.G. and McKNIGHT, C. (1981) *Think Again: Personal Decision-making and Problem-solving*, Englewood Cliffs, Prentice-Hall Inc.

STEAD, K. (1983) 'Insights into students' outlooks on science with personal constructs', *Research in Science Education*, **13**, pp. 163–76.

WHISONANT, R.C. (1994) 'Robert C. Whisonant: 1993 Neil Miner Awardee. Acceptance remarks', *Journal of Geological Education*, **42**, 1, pp. 7–9.

25 Social Interactions and Personal Meaning Making in Secondary Science Classrooms

Phil Scott

Abstract

This paper draws upon a Vygotskian perspective in developing a theoretical framework to characterize the way in which social processes support personal meaning making in high school classrooms. Case study data taken from lessons focusing on the teaching and learning of specific science concepts are presented with particular attention being given to the ways in which the teacher frames discourse to support learning. Analysis of data will draw upon Vygotskian theory in: characterizing learning of scientific concepts in terms of movement between social and psychological planes; stressing the fundamental importance of language in the teaching and learning process; viewing teaching as assisted performance and in acknowledging the interweaving of different ways of knowing (spontaneous and scientific) during learning. The notion of a 'narrative of introduction' is used to characterize the way in which the teacher supports learning over an extended period of time.

Introduction

Classrooms are peculiarly interesting places! It is in the classroom that young people are introduced to the publicly available knowledge which societies regard as being sufficiently important to pass on from generation to generation. It is in the classroom that the teacher is charged with the responsibility of establishing shared understandings of that knowledge with groups of thirty or more students at a time.

In this paper, I shall draw upon Vygotsky's socio-cultural perspective on learning as a basis for analysing and attempting to characterize the ways in which classroom teaching can lead to individual learning. In particular, I shall focus upon the ways in which the social interactions of the classroom can prompt development of personal understandings of science concepts.

Learning Science Concepts

At the start of instruction about a particular science concept it is unlikely that the students will be aware of the science view which is to be introduced, although they may be familiar with the various phenomena to which the science concept relates. Thus, for example, students' everyday experiences are likely to include events and phenomena such as drinking orange through a straw and the action of a rubber sucker; they are, however, unlikely to be familiar with the concept of air pressure and how that can be related to these instances.

Taking this example a little further, the concept of air pressure is not something that the student can discover for themselves through their own empirical enquiry. Whilst current thinking (see for example Chalmers, 1982) emphasizes the point that there is no single, widely accepted view of the nature of scientific knowledge, there is general agreement about the consensual nature of scientific knowledge and the 'imaginative leap' which separates data from scientific theories. That is, the concepts of science are constructs which have been developed in attempts to interpret and explain phenomena. The objects of science are not the phenomena of nature, but the constructs advanced by the scientific community to interpret nature (Driver *et al.*, 1994). Scientific knowledge is constructed and transmitted through the culture and social institutions of science. Learning science therefore involves being introduced to the ways of knowing of the scientific community.

Such a perspective on learning science has strong resonances with the views expressed by Vygotsky in his account of learning and development. Central to the Vygotskian account is the notion that the learner can only have access to higher mental functions, such as the science concepts which are of interest to us here, through processes which are initially external to the learner. Thus, in his frequently cited general genetic law of cultural development, Vygotsky argues that:

> any higher mental function was external and social before it was internal. It was once a social relationship between two people . . . We can formulate the general genetic law of cultural development in the following way. Any function appears twice or on two planes . . . It appears first between people as an intermental category, and then within the child as an intramental category. (Vygotsky, 1960, pp. 197–8)

Learning thus involves a process of internalization in which concepts are first rehearsed between people, prior to being developed within the learner as an intramental feature. On the intermental plane, language and other semiotic mechanisms (see Lemke, 1990) are used to develop meanings between individuals. That language then provides the symbols or tools that mediate individual cognition. From this perspective, language is absolutely fundamental to learning, learning is mediated through language.

It is important to bear in mind that the process of internalization does not simply involve transfer of concepts, via language, to the individual. The learner *reorganizes* and *reconstructs* experiences of the social plane, a point which has been made by Leontiev:

> The process of internalisation is not the transferral of an external activity to a pre-existing, internal plane of consciousness: it is the process in which this plane is formed. (Leontiev, 1981, p. 57)

The process of internalization thus encapsulates not only the Vygotskian condition that learning starts on the social plane, but also the Piagetian insight that to understand is to reconstruct (see Fischer and Bullock, 1984). Reconstruction, in this sense, involves the coming together of the learner's existing, everyday notions and the scientific concepts. Vygotsky emphasized that everyday and scientific concepts are interconnected and interdependent; their development is mutually influential. It is through the use of everyday concepts that children make sense of the definitions and explanations of scientific concepts; everyday concepts provide the 'living knowledge' for the development of scientific concepts (Moll, 1990). That is, everyday concepts *mediate* the acquisition of scientific concepts. Additionally, Vygotsky (1987) proposed that everyday concepts also become dependent on, and are mediated and transformed by, scientific concepts; they become the gate through which conscious awareness and control enter the domain of everyday concepts.

How might these theoretical ideas be brought to bear on characterizing teaching and learning science concepts in classroom settings?

Teaching and Learning Science Concepts in Classroom Settings

The author is currently engaged in a programme of research, based on the theoretical perspective just outlined, and which takes the whole activity of teaching and learning science concepts as its focus. The aim of the research is to develop and elaborate Vygotskian theory as a means of characterizing and analysing what is involved in teaching and learning science concepts.

In this research, the author has worked closely with five teachers to develop instructional sequences which focus on particular science concepts. In planning this teaching the literature on children's alternative conceptions was drawn upon to enable the teachers to become aware of students' likely thinking in each domain and to help in developing teaching strategies to respond to that thinking. The strategies used include rather more opportunities for teacher–student and student–student dialogue than would be expected in 'traditional' teaching approaches. The concept domains which have been worked on include air pressure (see Scott, 1994) and chemical reactions (see Scott, Asoko, Driver and Emberton 1994).

This way of working with teachers has allowed researcher and teachers to develop a shared aim for each instructional sequence and clear views about the overall structure of the teaching. In each case these were agreed in advance, ahead of any teaching. What could not be anticipated was how teaching and learning would proceed, through the interactions of individual teachers with their classes, once the lessons were under way.

Each sequence has been taught through by at least two teachers in the group and in each case what happened in the classroom was closely monitored. Data collection was planned both to illuminate the interactions of the intermental plane and to gain insights of any changes in students' intramental functioning. Thus all of the teacher talk, all of the whole class discussion and as much of the student small-group talk as possible was recorded and transcribed. In addition, all of the students completed diagnostic tests prior to and after teaching and target groups of students were interviewed about their developing understandings periodically throughout the teaching.

The remainder of this paper will draw upon the data collected in these studies and will focus upon the social interactions of the inter-mental plane.

Teaching Science: The Social Interactions of the Intermental Plane

According to the Vygotskian perspective, the teacher has a central role to play in introducing students to the science way of knowing about the world. The teacher is the key figure on the intermental plane who prompts science learning in the students. Bruner (1985) has referred to the teacher as being the 'vicar to the new culture'; the principal metaphor used by Vygotsky (1978) in describing teaching is that of 'assisting performance' (see also Newman, Griffin and Cole, 1989).

The notion of 'assisted performance' was originally outlined by Vygotsky in relation to an adult working closely with a single child. The question to be addressed here is how does the teacher assist the performance of a class of students in such a way that they are able to develop personal understandings of the science concepts under consideration? What do the teachers actually do and how might their actions be characterized?

The Narrative of Introduction

The first striking feature to emerge from watching the teachers at work, in the project outlined above, was the way in which they introduced new scientific concepts by means of a public 'performance'. This performance might last for as long as three or four hours, extending over a number of different lessons, on different days, and in different weeks. In many ways this performance resembles telling a story; the presentation of a narrative which gradually unfolds under the direction of the teacher, a 'narrative of introduction' to the science way of knowing.

That teaching should be thought of in terms of presenting a narrative is not unusual when one considers the ubiquitous presence of story-telling in day-to-day interactions. Bruner, for example, (1990, p. 67) suggests that the method of negotiating and renegotiating meanings by the mediation of narrative interpretation is one of the crowning achievements of human development. What then is the nature of this narrative of introduction as it applies to teaching and learning science?

The first point to be made is that performance of the narrative of introduction is dialogic or interactive. There are interesting comparisons which can be drawn here between the interactions of the classroom and performances in the theatre. In the theatre, the narrative is carried principally by dialogue between the players. In Vygotskian terms the intermental plane is laid out, literally on the stage, in front of the audience, and each member of the audience makes sense of (or internalizes) the performance in their own terms.

The situation in the classroom is similar, but different. Here the students both take part in, and are an audience to, the performance. The teacher who is sensitive to students' thinking will respond to their ideas as the narrative progresses; students' comments and questions can alter the way in which the narrative develops. Individual students take their place on the stage from time to time, to contribute to the developing narrative; the skill of the teacher is to control the interactions so that the performance retains its coherence and also leads to the portrayal of the science way of knowing. To achieve this the teacher can draw upon a range of activities (talking, listening, doing practical work, discussing in small groups, writing, reading) all of which are mediated by, and sustained through, language.

From observations of classrooms-in-action it has been possible to identify a number of features which characterize the development of the narrative of introduction. Some of these features are introduced in the following paragraphs.

Features of the Narrative of Introduction

The development of the narrative of introduction is controlled by the teacher and includes a number of features, or lines, each of which has a particular function. Four features have been identified:

- the conceptual line;
- the epistemological line;
- promoting shared meaning; and
- maintaining the form and direction of the narrative.

Central to the narrative is the portrayal or public enactment of the science way of knowing which the teacher aims to introduce to the class. It is this public enactment which prompts personal reflection and internalization and thus promotes development of shared meanings, or common knowledge (Edwards and Mercer, 1987), between teacher and students.

1 The conceptual line

The *conceptual line* of the narrative is a complex one involving all of those events and interactions which contribute to the portrayal of the science way of knowing. The teacher manipulates the events and interactions so that students can develop personal understandings of the science way of knowing, starting from their everyday understandings. Bruner (1990, p. 47) refers to the power of narrative in 'forging links between the exceptional and the ordinary', in 'rendering departures from the norm meaningful in terms of established patterns of belief'. In a similar way, learning science very often involves leaving behind the canonical meanings of everyday discourse for the 'unnatural nature of science' (Wolpert, 1992).

How then can the teacher assist students in coming to understand familiar phenomena in terms of science ways of knowing? From observing the classes of the teachers involved in this study, it has been possible to identify a number of ways in which teachers guide the development of the conceptual line of the narrative.

In interacting with students, the teacher makes various *selection* moves. The teacher might:

- select a student response, or part of a student response;
- select and modify a student response;
- implicitly accept a student response;
- retrospectively elicit a student response;
- overlook a student response; and
- overrule a student response.

Thus teachers select from a range of ideas offered by students. In some cases this selection is made explicitly, at other times what the student suggests is taken to be the case and is thus implicitly accepted. The teacher might refer back to an idea introduced at some point earlier in the lessons and use that idea to enable development of the narrative. In discussion, some students' ideas are simply overlooked, whilst others may be challenged and overruled.

In *shaping* the development of concepts in the narrative, the teacher might:

- make links between ideas;
- demonstrate equivalence of ideas;
- differentiate ideas; and
- sets up polar contrasts between ideas.

Thus the teacher might demonstrate that two ideas have the same meaning or set up polar opposites to allow pupils to appreciate the range of meanings between extreme perspectives. In carrying out these idea-shaping moves, the teachers are sensitive to students' existing understandings and are able to see the differences between these understandings and the science view.

A further issue in the portrayal of the conceptual line concerns the way in which the teacher might act to signal those points in the narrative which are of central importance to its development. In *emphasizing key ideas* in the conceptual line the teacher might:

- repeat an idea;
- repeat and rephrase an idea;
- ask a rhetorical question;
- reformulate and repeat a question;
- ask a student to repeat an idea;
- enact a confirmatory exchange with a student; and
- use a particular intonation of the voice.

Repetition of key ideas was used a great deal by the teachers who were observed. Thus, they might state an idea, repeat the statement, rephrase and repeat the statement, prompt students to repeat the statement and so on. Teachers also used the intonation of their voice to mark, very effectively, the relative importance of particular sections of talk. The teacher might, for example, draw attention to a key idea by slowing down their speech and talking in a slow, clear and deliberate way.

2 The epistemological line

Entry into scientific meaning involves more than developing understandings of science concepts and the relationships between them. In addition there is the issue of coming to appreciate the *nature* of the scientific knowledge which is being taught and learned; alongside the conceptual line of the narrative is a second feature which is referred to as the 'epistemological line'.

Earlier on, attention was drawn to the distinction between everyday and scientific ways of knowing. This distinction can be made both in terms of the conceptual structures used and the ways in which the knowledge is framed and used. Thus, for example, learning about the scientific concept of air pressure involves developing both an appreciation of what is meant by air pressure and also an understanding of the nature of the concept. Learning about the nature of the concept would include coming to understand that its application is not limited to particular contexts (such as the one in which it is first introduced) but can be applied to many situations. Everyday ways of knowing place less value on this property of generalizability which is one characteristic of scientific knowledge.

As might be anticipated, classroom observations have shown that the epistemological line of the narrative, dealing with the nature of the scientific knowledge, is not represented as prominently in the narrative as the conceptual line. Nevertheless it has been possible to identify instances in which teachers:

- make distinctions between different kinds of knowledge: everyday and science views;
- refer to the generalizability of scientific theories;
- consider how new phenomena might be accounted for by an existing theory;
- distinguish between different kinds of variables; for example, identifying the difference between a variable which is essential to a particular phenomenon happening (for example, water is essential for rusting to occur), and a variable which influences that phenomenon but is not essential to it (for example, salt in solution affects the rate of rusting but is not essential for rusting to occur).

3 Promoting shared meaning

A third feature of the narrative concerns the ways in which performance of the narrative acts to promote shared meanings amongst teacher and students. The goal of establishing shared meanings within a whole class of adolescents is, even under the most favourable of conditions, an ambitious one. Experiences of talking with just one other person so often provide evidence of the difficulties involved in establishing shared meanings through discourse. It is one matter to negotiate a shared understanding with one other person, how can teachers hope to achieve this with around thirty other people?

Wertsch (1985, p. 159) uses the notion of 'situation definition' to draw attention to the fundamental problem which is inherent in establishing shared meanings. He defines situation definition as the way in which objects and events in a situation are represented or defined by individuals. Wertsch makes the point that although the same concrete objects and events may be perceptually available to both teacher and students, they may not 'be in the same situation' because they do not define these objects and events in the same way. The extent to which individuals share the same situation definition is a measure of the 'intersubjectivity', which exists between them.

What steps, then, can teachers take to promote intersubjectivity between themselves and their students? It seems that there are two parts to this issue. The first is concerned with making the narrative *available* to all of the students (promoting intersubjectivity) and the second focuses on clarifying meanings and checking the extent to which intersubjectivity exists (monitoring intersubjectivity).

Classroom observations have shown teachers making the narrative available to all of the students (promoting intersubjectivity) in a number of ways. For example, the teacher might:

- present ideas to the whole class;
- share individual student ideas/findings with the whole class;
- share group ideas/findings with the whole class;
- repeat a student idea/response to the whole class; and
- jointly rehearse an idea with a student in front of the whole class.

All of these teacher actions relate to the function of making ideas publicly available in the classroom. This aspect of what happens in the classroom is very much the 'performance' part of the narrative of introduction. The teacher controls the activity or performance and pupils are directly involved 'in performing' to greater or lesser extents. Some pupils like to be actively involved in answering questions and making suggestions; others will sit back listening to, and watching, the action of the unfolding narrative.

In addition, the teacher checks meanings of the students (monitoring intersubjectivity) in a number of ways. For example, the teacher might:

- ask for clarification of student ideas;
- check individual student understanding of particular ideas; and
- check consensus in the class about certain ideas.

Thus the teacher might: ask a student for clarification of an answer which has been offered in class; ask the class to predict the outcome of a demonstration; call for a vote on alternative explanations for a phenomenon; ask students to write out their own answers to questions probing particular understandings.

4 Maintaining the form and direction of the narrative

A final feature of the narrative concerns the way in which the teacher takes action to maintain its form and direction. Here the teacher is involved in talking *about* the narrative, rather than talking the narrative. Talk about the narrative might involve offering an organizing commentary to help students keep track of developing arguments. This kind of commentary is especially important if the narrative is dealing with a difficult concept and lasts over an extended period of time. Thus, for example, in maintaining the form and direction of the narrative, the teacher might:

- review the progress of the narrative to date;
- refocus discussion;
- rehearse possible outcomes; and
- summarize ideas.

These, then, are four features of a narrative of introduction to science ways of knowing. As the narrative is enacted, then the different features thread their way through the performance, appearing and overlapping at different points. In the final part of this paper I should now like to turn to a teaching and learning episode in order to exemplify some of the features outlined above.

Teaching and Learning about 'Air Pressure': A Brief Episode

The following short sequence is taken from a class of 13–14-year-old pupils in a comprehensive school in the north of England. The lessons focus on the concept of 'air pressure' and as we join the class the problem of why a plastic bottle collapses when the air is removed from inside it, is being considered. A pupil called Jamie is talking (the transcript which follows is interspersed with interpretative notes):

Jamie   Well, when all the air's been sucked out, it's er . . . there's nowt in there so you'll have . . . air pressure's pushing the side of the bottle in.

Teacher Which air pressure Jamie?

Jamie   From the outside.

Teacher Say that again so that people can hear.

Jamie has the correct idea, 'air pressure's pushing the side of the bottle in'. The teacher asks for clarification of *which* air pressure is pushing. This question is an important one in that it shows the teacher's awareness of the crucial issue of which air pressure is acting. It is common for pupils of this age to believe that the air inside the bottle 'sucks the sides in'; the teacher is aware of this and makes absolutely sure that Jamie is offering an explanation which has air pressure acting from the outside. In this way the teacher *differentiates* between everyday and scientific ideas.

That the air is pushing from the outside is central to the science explanation, and the teacher asks Jamie to *restate* which air he is referring to. The teacher then asks Jamie to *repeat* his ideas 'so that people can hear'. As it happens, Jamie is not the kind of boy who has problems in making himself heard in class. At most times the converse is true! However, through this brief exchange the teacher has set up a public dialogue to rehearse the new idea in front of the whole class. In this sense Jamie is being used as a (willing) partner in *rehearsing* the new idea publicly and *emphasizing* its key importance.

Jamie   [Jamie repeats his explanation.]
Teacher   Right, so you're saying that when we suck the air out of the bottle, there's less air inside the bottle, so there's less pressure, less air pressure. . . . and why did the sides push in? What did you say again?
Jamie   Cos there's more air pressure outside.
Teacher   Because there's more air pressure on the outside pushing it . . .

Jamie repeats his explanation and the teacher responds by *rephrasing* and *repeating* the explanation, 'Right, so you're saying . . .' The teacher breaks off mid-way through and returns to Jamie with a question, 'why did the sides push in?' The teacher knows that Jamie understands the explanation, the purpose of the teacher's question is not to check Jamie's understanding but to break down the science explanation into its component parts and to *rehearse* it once more before the whole class.

Teacher   [continues] . . . That's what we're going to call the new way of looking at it. The new explanation is that there's two lots of air involved here not one. There's one lot inside the bottle and there's one lot in this room immediately surrounding it. And if we take air out of the bottle, that means there's less air pressure inside the bottle than there was before . . . there's more pressure outside, then, and it pushes it in.

In this passage the teacher offers an 'organizing commentary' in referring to the science explanation as the new way of looking. He is explicit in stating that the new explanation involves two lots of air rather than just one (the air inside sucking) and goes on to once again repeat the formalism which is the new way of explaining. By referring to the science explanation as the new way, the teacher helps the pupils to clarify their thoughts and to *differentiate* two (labelled) sets of ideas. The teacher is also, implicitly, making the *epistemological* point that more than one way of knowing exists.

It is interesting to note that as the teacher starts, 'The new explanation is that there are . . .', then he reduces the loudness of his voice and speaks in a slow and very deliberate manner, thus *emphasizing* the importance of the ideas to the narrative. This is the final performance of the new way of explaining; it is also the fifth time that it has been *repeated* in about the same number of minutes!

Teacher	So why is it then, why is it that if I don't attach an air adder or remover to the bottle it stays with fairly straight sides? Why is it . . . Nick?
Nick	Because the air pressure's the same inside the bottle as the outside of the bottle.
Teacher	The air pressure's the same inside the bottle and outside. Good lad. Now we're going to be using that idea a lot . . . so I want you to make a note of it in your book . . .

Here the teacher draws attention to a 'boundary condition' for the new explanation. The teacher is *checking* to see whether the students are able to apply the new explanation to the 'null' condition of the bottle. Nick provides a good answer, the teacher *repeats* it and signals that it is an acceptable answer (Good lad!). The teacher then provides an *advance warning* that the new idea is going to be used a lot; this particular narrative is not yet finished!

This is a very brief part of one lesson, but some insight is offered into how the teacher introduces the new way of knowing: how the ideas are rehearsed in front of the whole class in a bid to develop shared meanings; how explanations are taken apart and repeated; how the teacher differentiates between different ways of knowing; how the teacher is central to the introduction and development of the science way of knowing.

Conclusion

In this paper, I have drawn upon Vygotsky's sociocultural theory of cognition to characterize how students are able to learn science concepts in classroom settings. The analysis has focused on the interactions between teacher and students, the interpsychological functioning of the classroom which is the necessary precursor to internalization and development of personal understandings. The interactions have been framed and described in terms of a 'narrative of introduction' to a new way of knowing. The general nature of the narrative has been outlined in terms of four features. These features have been elaborated by reference to observations of lessons in which a small group of teachers has been engaged in teaching particular science concepts in a way which is sensitive to students' thinking and recognizes the importance of dialogue in learning.

What is presented here is, of course, only half of the teaching and learning story. The question of what personal understandings students develop as a consequence of the teaching is not addressed. This must remain to be considered at another time, although the problematic nature of internalization (with attendant reconstruction) has been hinted at here.

The performance of the narrative of introduction is witnessed by all of the students in the class and each one of those students makes sense of, or internalizes, the performance in their own way. The teacher can control the development of the narrative on the interpsychological plane, but internalization remains the personal province of each individual child. The question was posed in this paper as to how

Phil Scott

the teachers observed introduce their pupils to the science way of knowing. The answer is here. They control the development of the narrative of introduction within the social setting of their classroom. That is as much as any teacher can do. Individual student understanding cannot be guaranteed.

References

BRUNER, J. (1985) 'Vygotsky: A historical and conceptual perspective', in WERTSCH, J. (Ed) *Culture, Communication and Cognition: Vygotskian Perspectives*, Cambridge, Cambridge University Press, pp. 21–34.

BRUNER, J. (1990) *Acts of Meaning*, Harvard University Press, Cambridge, MA.

CHALMERS, A.F. (1982) *What is This Thing Called Science?*, 2nd edition, Milton Keynes, Open University Press.

DRIVER, R., ASOKO, H., LEACH, J., MORTIMER, E. and SCOTT, P. (1994) 'Constructing scientific knowledge in the classroom', *Educational Researcher*, **23**, 7, pp. 5–12.

EDWARDS, D. and MERCER, N.M. (1987) *Common Knowledge: The Growth of Understanding in the Classroom*, Methuen, London.

FISCHER, K.W. and BULLOCK, D. (1984) 'Cognitive development in school-aged children: Conclusions and new directions', in COLLINS, W.A. (Ed) *Development During Middle Childhood: The Years from 6 to 12*, Washington, DC, National Academy Press, pp. 70–146.

LEMKE, J.L. (1990) *Talking Science: Language, Learning and Values*, Norwood, New Jersey, Ablex Publishing Corporation.

LEONTIEV, A.N. (1981) 'The problem of activity in psychology', in WERTSCH, J.V. (Ed) *The Concept of Activity in Soviet Psychology*, Armonk, NY, Sharpe.

LEONTIEV, A.N. and LURIA, A.R. (1968) 'The psychological ideas of L.S. Vygotsky', in Wolman, B.B. (Ed) *Historical Roots of Contemporary Psychology*, New York, Harper and Row, pp. 338–67.

MOLL, L.C. (1990) *Vygotsky and Education: Instructional Implications and Applications of Sociohistorical Psychology*, Cambridge, Cambridge University Press.

NEWMAN, D., Griffin, P. and COLE, M. (1989) *The Construction Zone: Working for Cognitive Change in School*, Cambridge, Cambridge University Press.

SCOTT, P. (1993) 'Overtures and obstacles: Teaching and learning about air pressure in a high school classroom', in NOVAK, J. (Ed) *Proceedings of the Third International Seminar: Misconceptions and Educational Strategies in Science and Mathematics*, Cornell University, Ithaca, USA.

SCOTT, P., ASOKO, H., DRIVER, R. and EMBERTON, J. (1994) 'Working from children's ideas: An analysis of constructivist teaching in the context of a chemistry topic', in FENSHAM, P.J., GUNSTONE, R. and WHITE, R. (Eds) *The Content of Science: A Constructivist Approach to its Teaching and Learning*, London, Falmer Press.

VYGOTSKY, L.S. (1960) *The Development of Higher Mental Functions*, Moscow, Akad.Ped.Nauk.RSFSR.

VYGOTSKY, L.S. (1962) *Thought and Language*, Cambridge, MA, MIT Press.

VYGOTSKY, L.S. (1978) *Mind in Society: The Development of Higher Psychological Processes*, Cambridge, MA, Harvard University Press.

WERTSCH, J.V. (1985) *Vygotsky and the Social Formation of Mind*, Cambridge, MA, Harvard University Press.

WOLPERT, L. (1992) *The Unnatural Nature of Science*, Faber and Faber, London.

26 Primary Teachers' Cosmologies: The Case of the 'Universe'

Vassiliki Spiliotopoulou and George Ioannidis

Abstract

This study concerns primary teachers' models of the universe. These models are part of the whole system of teachers' knowledge, considered as their cosmology. An individual's cosmology includes not only conceptions and theories, but also beliefs, feelings, intentions and philosophies. These cosmologies not only develop with the age, but they evolve throughout the individual's life. Teachers' models are derived from their drawings of the universe and from explanatory comments on them.

This activity was part of a district-based teacher training course on science education. It aimed to help teachers reflect on their own thinking about the universe with respect to known scientific models and to become more sensitive to children's ideas.

The results are discussed and compared with children's models, using as a base a specially developed systemic network. This network is expected to be a formulation broad enough to cover the different ways of thinking about the universe and is tested through primary teachers' drawings and how well their ideas fit in with it.

Introduction[1]

The topic of teachers' thinking about scientific concepts has been increasingly studied recently. It is of great importance for both the design of pre- and in-service teacher training courses and for what actually takes place in school classrooms. Research on in-service teachers' thinking has demonstrated that teachers' attitudes and beliefs influence their perception and understanding of classroom events and may, therefore, affect their classroom behaviour (Clark, 1989).

Similarly cognitive characteristics in teachers' thinking may affect children's development of scientific concepts.

The purpose of this contribution is to describe a developed course in science education which was realized in a district-based Teacher Training Centre in Patras and to analyse teachers' thinking concerning the universe.

It is widely known that primary school teachers themselves have cognitive inadequacies. Help in this direction can be given by programmes aimed at developing teachers' ability to reflect on aspects of science. In order for reflection to have meaning, Bullough (1989) suggested that such a programme must become part of a coherent conceptual framework, a statement of personal philosophy and values.

Lesh and Kelly (1994) made a general assertion that constructivism is not simply a statement about how children think, but rather it is a statement about the nature of thinking. As a corollary to their premise they suggested that whatever characteristics we ascribe to children's thinking, we should be willing to ascribe both to teachers' thinking and to our own thinking as researchers.

If we claim that children construct internal representations, or models of the world, and that these models are incomplete, flawed, subject to revision and evolve over a long period of time, then we must apply these principles equally to our study of the models of the world held by teachers and researchers. (Lesh and Kelly, 1994, p. 282)

We share this view and we have incorporated it into the teacher training course in which we are involved for two purposes: to develop teachers' knowledge through reflection on their own models and through comparison with the historical ones, and to help them to include into their role as teachers the making sense of children's models and the importance of such models in the learning of science.

The Context of the Research

Three years ago, teacher training courses in Greece were shortened from one year to three months. Thus, in each year there are three groups of teachers in training. The first is made up of the pre-service teachers, who are going to teach after the completion of the course. The other two groups are teachers already in the service. The courses involve a lot of subject matter among which is a component called science education.

The system of selection for universities in Greece has led students with high attainment in language subjects to become primary school teachers and as a result these teachers have a poor background in science and mathematics. In order for a training programme to be effective in such cases, specific strategies have to be employed. Based on Licht, Brinkman and Vonk's report on the aims of pre- and in-service courses for accounting for 'Students Alternative Conceptions in Science in European Institutions' (Licht *et al.*, 1991), we organized a teacher training course the starting point of which was to focus on the importance of 'awareness of personal ideas on science and the learning of science'. After two years' experience of using fragmented units, we unified the parts and organized a teaching and learning model, Figure 26.1, which can be shown to be useful in some science content areas. The model comprises three phases with the order of the second and third phases inter-changeable (Spiliotopoulou and Ioannidis, 1993). The aims of the third phase are related to the objectives 4 and 5 of the report mentioned above, and refer to the application of diagnostic tools and teaching strategies. The aims of the second phase are the objectives 1, 2 and 3 of the report and are related to teachers' awareness and knowledge of children's alternative conceptions.

Objective 6 of the Licht report was the aim of the first phase of our

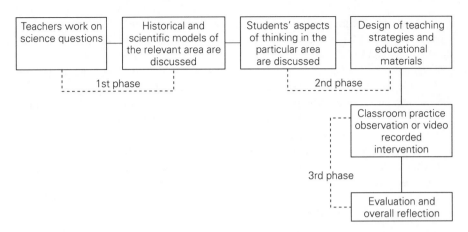

Figure 26.1: The three phase model for teacher training courses

programme. Teachers are caused to think and reflect on their own ideas and beliefs with respect to historical and scientific concepts and processes, the nature and range of science and of the nature of learning processes. This is also considered to be an opportunity for teachers to compare their ideas with the scientific ones and, where necessary, to change their existing beliefs.

We have applied this three phase model to different subject matter areas. The first and main areas, *cosmology* and particularly its component 'Earth and universe', has been chosen for a number of reasons. Firstly, it is a domain of general interest which touches cultural and personal beliefs. It is also an area where scientific knowledge is still evolving and there is no certainty about the unique and 'true' model of the universe. This area we considered to create a feeling of openness for the activity, to make teachers' involvement easier and to facilitate teachers' acceptance of their own personal, possibly alternative, and often incomplete or inaccurate, models. Another reason was that we already had some experience of working with children's cosmological models (Spiliotopoulou, 1994).

Cosmology

In this study we use the term 'cosmology' in two ways. In the scientific world, cosmology is considered to be the science or, for some, 'the natural science', of the universe that attracts and fascinates most people. Harrison (1981) discusses three senses of cosmology.

In one sense it is the science of the large scale structure of the universe, of the realm of extra-galactic nebulae of a distant and receding horizon, of the interplay of cosmic forces and of the dynamic Curvature of Universal Space and Time.

In another sense, he claims that both ancient and modern cosmology are the grand science that seeks forever to assemble all phenomenological knowledge into a world picture. Most sciences tear things apart into smaller and smaller

339

constituents for the purpose of examining the world in progressively greater detail, whereas cosmology is the one science that is devoted to putting the pieces together into a 'mighty frame'.

In yet another sense, cosmology is the history of humankind's quest for understanding of the universe, a quest that began long ago in the Age of Magic and the subsequent Age of Mythology. We cannot study cosmology in the broadest terms unless we take into account the pageantry of world pictures that have shaped the history of the human race.

Metaphorically speaking, we consider cosmology to be the whole system of an individual's knowledge. An individual's cosmology includes not only conceptions and theories, but also beliefs, feelings, intentions and philosophies. It is considered to be a more general way of seeing the world and it does not necessarily have the need for an internal consistency that a scientific-like theory does. Children's cosmology develops naturally when they speculate, imagine and experience parts of the world and in interaction with the cosmology generally accepted at the time they live. In young children's cosmologies, we discern a common-sense type of reasoning based mainly on beliefs and feelings, while scientists' cosmologies are characterized by a more sophisticated type of reasoning, based on a higher order rationale and thinking.

An individual's cosmology is not considered as a structure but rather as a system with internal organization and rules. It tends to acquire a structure, so we can only talk about 'instant structures', although this is something that itself opposes the meaning of structure.

Cosmology involves some ways of thinking that apply in particular areas only, and others that apply to a wide class of aspects of the world. But what cosmology is most interested in is how different personal expressions fit into a general frame of knowing and the rules of this organization. Cosmology also involves the description of the development or the evolution of the specific subsystems of cosmology throughout an individual's life.

In this study we are interested in primary teachers' cosmologies and particularly in Harrison's first sense — primary teachers' thinking about the large-scale structure of the universe.

Buck (1990) described a nesting system with which he tried to depict the material world. The galaxy is described as a system of which our solar system is part, our solar system — the sun together with the planets — the Earth being one element of this system, is thus a system within this system. The Earth itself can be said to comprise a system of mountains, rivers, cities and so on, each progressively smaller, until you end up with elementary particles, probably quarks. According to this 'system–component' approach the qualities and properties of a given component are usually not the same qualities of the whole system of which it is a part.

Teachers' thinking about the widest, most complete, system, that of the universe, seems to be of interest as it includes an appreciation of qualities of both the parts and the whole in a way analogical to the material, natural world. These component parts of the universe are included in the school curriculum and teachers have to help children to form an understanding of them.

Ojala (1992) reports that entrants to the teaching profession in Finland have not understood matters relating to planetary features as a consistent system and that the significance of various phenomena and their interrelationships remain obscure and unstructured. The phenomena are also considered by these beginner-teachers to be separate entities. The researchers are pessimistic about the future capability of these students to teach similar subject content in school.

Teachers' Models of the Universe

How do primary teachers view the structure of the universe? Do they think of it as finite and what point of view do they evidence when they are asked to draw it?

Similar questions have been studied as part of broader research into children's cosmologies. For the universe, in particular, a systemic network (Bliss *et al.*, 1983) has been specially developed to describe children's models. This was considered to be a formulation broad enough to cover different ways of thinking about the universe. This network was tested, here through primary teachers' drawings and how well their ideas fitted in with this formulation. Bliss's network has been extended by adding one more category that expresses an aspect of teachers' thinking about the universe, not met among children. The extended systemic network is presented in Figure 26.2.

Initially drawings were analysed in two dimensions: one concerning the kind of model, the other concerning the limits of the universe. In terms of the limits, the drawing or the comments on it show if the universe is considered by the teachers to be 'finite' or 'infinite'. The category 'indefinable' locates drawings which do not have an indication of the person's beliefs regarding the finite nature of the universe.

The question of finiteness was, and still is giving rise to discussion and debate, and is concerned with religious beliefs and ideologies.

The *Apeiron*, the Greek word for infinity, combines the meaning of 'unbounded' with that of 'unrestricted' and 'unlimited'. The Apeiron is not only devoid of boundaries, it is also capable of taking all possible forms and properties; the usual translation, 'the Infinite', is accordingly quite inadequate (Toulmin and Goodfield, 1961).

A teacher commenting on the universe and its limits said:

> The universe is something that contains in its interior everything. Schematically we could perhaps represent it with a circle in the interior of which, the Apeiron would be involved.

He seemed to consider the universe to be very big, infinite even, but not without limits.

A lot of teachers used the mathematical symbol of infinity, ∞, to represent the infiniteness of the universe. (This symbol has not been met by children.) Others describe infinity by using the word *chaos*.

In terms of the kind of model that the drawing expresses, three categories have

Question: What do you think the 'universe' looks like? Make a drawing.
(The Universe is everything you see, you know or you imagine
exists around us as far as you can possibly think.)

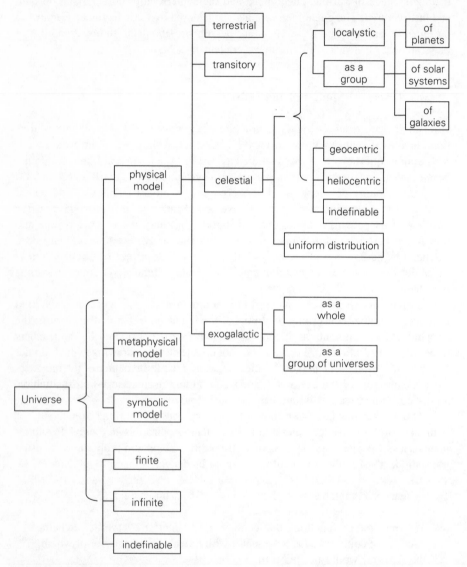

Figure 26.2: Network for models of the universe

been identified: the physical model, the metaphysical model, and the symbolic model. The category of symbolic models was added after careful study of the data collected from teachers. This shows an interesting and important aspect of thinking that has not been met among children and goes back to representations and ideas from the pre-historic years.

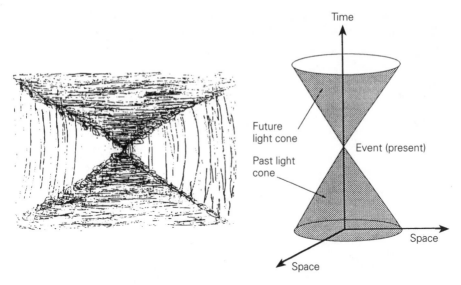

Figure 26.3:   A teacher's symbolic model showing a mathematical sense of the universe and the light cones

Von Glasersfeld (1987) argues that we use the term 'symbolic' when we want to indicate that some item was *arbitrarily* chosen to stand for something else, and does not arise out of ordinary inductive inferences. With regard to Piaget's distinction between the 'figurative' and the 'operative' icons, von Glasersfeld discusses figurative symbols and operative symbols.

Drawings that fall into the symbolic category can perhaps be considered as figurative symbols, as they refer to a figurative item or a sensory-motor situation, such as the universe. Two drawings characteristic of the symbolic category are presented in Figures 26.3 and 26.4.

The symbolic model expressed in the drawing of Figure 26.3 probably shows a mathematical sense of the universe, and can be considered schematically analogical to the future and past light cones (Hawking, 1988).

The drawing in Figure 26.4 exhibits a cognitive and humanistic approach to the universe.

The teacher's comments on this are quite interesting:

The drawing shows . . . the cerebral or cognitive effort to understand the universe. With the information of fundamental elements (centre), we are continuously increasing our intelligence in stages, until we realize that it is impossible to understand or describe it. The perimeter of the paper represents man's mind.

Other teachers approach the universe in symmetric or humanistic ways, while still others symbolize the universe as a continuously moving non-definable situation.

Figure 26.4:   A teacher's symbolic model showing a cognitive approach to the universe

As an example of the *metaphysical* model we can consider a drawing which attempts to show aspects beyond physical entities, entities that belong to a non-physical realm, such as heaven, angels, souls of evil or good spirits, or the eye of God. This expresses the individual's needs to give physical existence to things that are considered to be the most valuable and powerful: souls, minds, emotions, powers, Gods. Figure 26.5 shows a universe of physical existence in the centre of which is located the *Absolute Good Thought* that spreads throughout the universe and probably affects physical entities such as galaxies.

The drawing in Figure 26.6 represents a naive approach to the universe, but it reminds us of Aristotle's doctrine of the four 'elements' of the universe: earth, fire, air and water, each of which possessed two of the four basic properties of the world: hot, cold, wet and dry.

Moreover, it exhibits this person's belief that there is a powerful anthropomorphic figure that controls these four elements.

The category 'physical' model contains drawings which are physical representations of the universe. Modern cosmology studies a physical universe that includes all that is physical and excludes all that is non-physical. In science the definition of physical includes all those things that are observed, that are studied in controlled experiments and are explained by quantitative and predictive theories vulnerable to disproof.

'Terrestrial, transitory, celestial' and 'exogalactic' models are all subcategories of the 'physical' model. Both terrestrial and transitory models were only encountered among the responses given by young children and were not found among teachers.

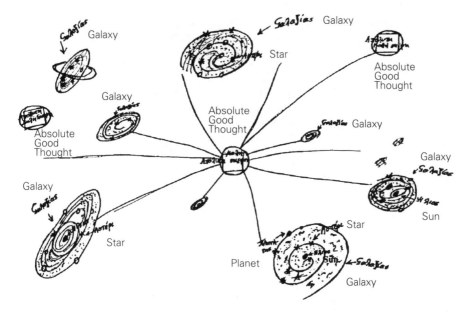

Figure 26.5: A teacher's metaphysical model of the universe The 'Absolute Good Thought' is located in the centre

In the 'terrestrial' models, the universe has a very narrow range. It includes the part of the Earth where we live and familiar entities like buidings, trees, sun, stars, space above us, which we experience in our everyday life.

In 'transitory' models, both terrestrial and celestial entities are exhibited. These models show a phase during which the children are trying to cope with new information about the structure of the universe without being able to overcome their experiential adherence to the sky existing at the top or to the sea at the bottom of the universe.

'Celestial' models can be 'localystic' or can represent 'groups of planets, solar systems', or 'galaxies'. In a localystic model, the universe is again narrow in range, but now the narrowness concerns a part of space. Such drawings show that the person who draws takes a perspective close to a planet, usually the Earth, or close to the sun and represents the nearby area. Such a model is presented in Figure 26.7 and can be found both in children's and teachers' approaches to the universe.

Similarly, drawings with 'groups of planets, solar systems' or 'galaxies' (Figure 26.8) were drawn by both groups.

Localystic models and group models can also be described as 'geocentric' or 'heliocentric'. When there is no clear indication of its geocentricity or heliocentricity, the drawing falls into the category named 'indefinable'.

In the 'uniform distribution' model the point of view taken by the person who represents the universe is such that the cosmic entities are represented as undifferentiated in their structure, usually as dots or small arrows, distributed over the whole area, in a uniform way.

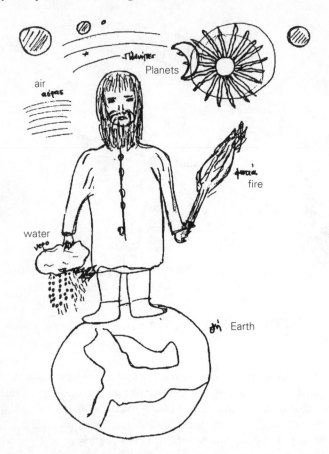

Planets

air
αέρας

fire

water
νερό

Earth

Figure 26.6:   A teacher's metaphysical model representing the four 'elements' of the universe

'Exogalactic models' are considered to be the ones that represent either the universe *as a whole* or *as a group* of universes. These models seem to be closer to a scientific model of the universe.

It seems that what differentiates these models is the person's point of view when representing the universe. A gradual increase in the distance the person takes to view the universe from his or her actual position on Earth, can be identified. The shortest distance is considered to be taken when the person gives a terrestrial model and the farthest when he or she gives a model of a group of universes. This does not necessarily mean that the people who take a further view show an increased understanding of all the characteristics of the universe, or that their model is always closer to a scientific one. An example is a teacher, who despite her knowledge that the universe is probably spherical like a balloon, felt the need to draw a support underneath, in order for the universe to be supported and not to fall! (Figure 26.9).

This identified axis of differentiation can be considered analogous to the evolution of historical models of the universe. We do not claim any exact correspondence, but the history of science shows, and especially with the universe, that every

Figure 26.7: A teacher's localystic model

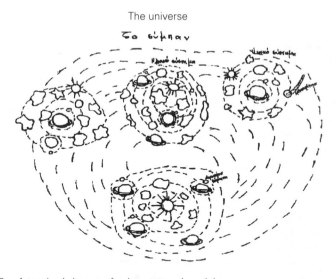

Figure 26.8: A teacher's 'group of solar systems' model

new step man has taken towards scientific thinking was a step further away from the location of his or her existence. Metaphysical and symbolic models fall beyond the axis as they express other aspects of thought.

Mason (1987) developed a metaphor in which a symbol is seen as a door between two worlds; the material world of physical objects, and the virtual

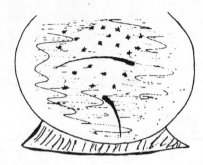

Figure 26.9: A teacher's model showing how the universe stands

reality. The process of symbolizing is then seen as a passage between these two worlds.

Metaphysical models, on the other hand, show the individual's beliefs and their subsequent interference with the physical models. These models are met in both teachers' and children's approaches and express a spiritual dimension to seeing the world. This is not only an inner experience, but also shows human needs, fears and beliefs.

Conclusion

It seems that the network described here was a valuable formulation of the different ways of thinking about the universe. It worked with primary teachers' drawings and was enriched by an interesting category the 'symbolic models'. It was expected that in this case, the two primary categories of 'terrestrial' and 'transitory' models would be unpopulated.

That symbolic models were only given by teachers, may show children's adherence to realism, or it may show teachers' tendency to avoid a reality with which they do not feel very safe, in the cognitive approach. But even if the latter is the case, the construction of a symbolic universe means that the person can use a complex inner experience or belief system, and this can be considered as an extension of usual representative thinking.

Moreover, what this formulation shows, supports our idea of people using a system analogous to a cosmology. The categories which were developed here reflect directions of thinking that are something more than just cognitive about scientific knowledge. Beliefs, symbols, experience and received information become interwoven in a whole system of thought which surely evolves, but not always in an even way, towards absolute scientific thinking. This network was also used with teachers to represent their models, the historical models and the children's models. Teachers seemed fascinated by this approach to probing understanding, knowledge and beliefs. They developed a consciousness of their own models' deficiencies and they were motivated to learn more about the universe. Having done so they went on to organize interesting and effective teaching strategies for their children.

As Potari (1994) claims, teachers learn better about teaching approaches by experiencing the concepts as learners. The historical evolution of science can also support the humanistic view of science and show more clearly its nature as a human and social activity.

Note

1   We wish to thank Jon Ogborn for his valuable comments on an earlier form of the network used in this paper.

References

BLISS, J., MONK, M. and OGBORN, J. (1983) *Qualitative Data Analysis for Educational Research*, Groom Helm, London.

BULLOUGH, R. (1989) 'Teacher education and teacher reflectivity', *Journal of Teacher Education*, **40**, 2, pp. 15–21.

BUCK, P. (1990) 'Jumping to the atoms: Introduction of atoms via nesting systems', in LJINSE, P.H., LICHT, P., DE VOS, W. and WAARZO, A.J. (Eds) *Relating Macroscopic Phenomena to Microscopic Particles*, CD-6, University of Utrecht, pp. 212–19.

CLARK, C.M. (1989) 'Asking the right questions about teacher preparation: Contributions of reaearch on teaching thinking', *Educational Researcher*, **17**, 2, pp. 5–12.

HARRISON, R.E. (1981) *Cosmology, the Science of the Universe*, Cambridge University Press, Cambridge.

HAWKING, W.S. (1988) *A Brief History of Time — From the Big Bang to Black Holes*, Bantam Books, London.

LESH, R. and KELLY, A.E. (1994) 'Action-theoretical and phenomenological approaches to research in mathematics education: Studies of continually developing experts', in BIEHLER, R., SCHOLZ, R.W., STRÄSSER, R. and WINKELMANN, B. (Eds) *Didactics of Mathematics as a Scientific Discipline*, Kluwer Academic Publishers, Dordrecht, Netherlands, pp. 277–86.

LICHT, P., BRINKMAN, F. and VONK, H. (1991) 'Guest Editorial', *European Journal of Teacher Education*, **14**, 1, pp. 5–8.

MASON, J.H. (1987) 'What do symbols represent?', in JANVIER, C. (Ed) *Problems of Representation in the Teaching and Learning of Mathematics*, Lawrence Erlbaum Associates, Hillsdale, NJ, pp. 73–81.

OJALA, J. (1992) 'The third planet', *International Journal of Science Education*, **14**, 2, pp. 191–200.

POTARI, D. (1994) 'Integrating learning mathematics and teaching mathematics in a pre-service elementary education program in Greece', *The Mathematics Educator*, **5**, 2, pp. 42–4.

SPILIOTOPOULOU, V. (1994) 'Children's cosmologies', in *Proceedings of the Second PhD Summerschool, European Research in Science Education* (in preparation).

SPILIOTOPOULOU, V. and IOANNIDIS, G. (1993) 'Earth and universe: A constructivist approach in pre- and in-service primary teachers' training', *Sixth Panhellenic Congress of Hellenic Physicists Association*, Komotini, Greece, March.

TOULMIN, S. and GOODFIELD, J. (1961) *The Fabric of the Heavens*, Hutchinson, London.

VON GLASERSFELD, E. (1987) 'Preliminaries to any theory of representation', in JANVIER, C. (Ed) *Problems of Representation in the Teaching and Learning of Mathematics*, Lawrence Erlbaum Associates, Hillsdale, NJ, pp. 215–25.

27 The Perceived Value of Classroom Research as an Element of an Initial Teacher Education Course for Science Teachers

Bob Campbell and Judith Ramsden

Abstract

Initial teacher education in England is school-based. Many teachers and others who influence the curriculum of initial teacher education place little value on educational research and dismiss this as theory unrelated to practice. There is a danger that the perceived gulf between theory and practice may widen. This has implications for the future viability of the science education research community. What is described here is how a research element was introduced into a postgraduate training course for beginning science teachers. The programme aimed to highlight the value of research in science education, encourage reading of research literature, engage students in classroom research, promote the notion that teachers can be researchers and establish that research can both explain and inform classroom practice. Student reactions were gained by a structured questionnaire. The evidence indicates that students valued the opportunity to do research and benefited from their engagement with it. Most saw research as providing valuable insights into teaching and learning. Several students indicated an intention to repeat or extend their studies when in post and/or to use their research skills to investigate other areas of practice.

Introduction

The relationship between theory and practice, and the extent to which one is influenced by the other, has long generated considerable debate in education. In a climate of considerable educational upheaval, such as has been the case recently in England, there is a serious risk of the perceived gulf between science education research and classroom science teaching becoming wider. Busy teachers, struggling to cope with the classroom reality of policy changes in science education, are unlikely to view educational research very positively unless they feel it has something of direct value to offer to their immediate day-to-day concerns (Rudduck, 1985; Pimenoff, 1995). Furthermore, it becomes increasingly less likely that teachers will wish to engage in research in science education themselves or to view a career in science education research as being worthwhile. Such a trend is particularly worrying as it suggests that one of the important growth areas in educational research over the past two decades, that of practitioner research (see for example Stenhouse, 1975; Ebbutt and Elliot, 1985), is under threat. One effect of the practitioner research movement, or teacher-as-researcher movement, has been to provide a number of valuable insights into educational processes and practices at the classroom level (for example the collection of Webb, 1990) and this provides an important bridge between theory and practice.

The implications of a widening gulf between researchers and classroom practitioners are very serious for the long-term viability of the science education research community. Many of those currently active in science education research are university-based teacher educators and have a background in school science teaching. One could therefore assume that some of the science education researchers of the future will similarly be university teacher educators and former classroom teachers. It therefore becomes particularly important to develop and foster an interest in research in those embarking on science teaching as a career. However, recent changes in legislation in England have resulted in a move towards more school-based initial teacher education, with a correspondingly reduced role for university-based staff. There is, therefore, a very real danger that students emerging from initial teacher education courses will have had few opportunities to engage with aspects of educational research, and are thus unaware of ways in which research might inform their classroom practice. Rudduck and Wellington (1989) argue that student teacher involvement in school-based collaborative enquiry is important because it has the power to change the attitudes of both teachers and students towards educational theory and research so that they are seen as more relevant and accessible. Liston and Zeichner (1990) argue for courses of initial teacher education to produce reflective practitioners (Schon, 1983) and claim that a reflective approach to teaching encourages a reasoned argument for educational actions. They see a need for teacher educators to help students examine their beliefs and values and relate these to central educational traditions. Set against these, reflecting on the move to school-based teacher education in England, McClelland (1993) is not optimistic that educational research will have a high profile in teacher training courses.

The move towards more school-based training does have benefits. At York, one of these has been the increased dialogue which has taken place between university-based staff and school-based staff as they have worked together to plan an integrated programme of learning experiences for students on the Post-Graduate Certificate in Education (PGCE) programme. Course planning meetings have provided a valuable forum for discussion and the opportunity for both university-based and school-based staff to review their contributions to initial teacher education. Such dialogue proved particularly beneficial in helping to identify the unique features each group had to offer to students, and the ways in which these might be drawn together to provide a coherent, cohesive and comprehensive programme for beginning science teachers. Emerging from the discussion was a strong view, expressed by both parties, that one particular contribution university staff were well-placed to make was to provide opportunities for students to access research literature on aspects of science teaching and to help them relate this to classroom experience. In addition, it was recognized that the PGCE course itself generated a research agenda (Fensham and Northfield, 1993). This resulted in the introduction of a research-based element, the 'In-depth focus', into the PGCE course programme. While not a collaborative research activity of the kind developed by Erickson (1991), it was designed to have value to students and to teachers.

The In-depth Focus

The 'In-depth focus' has a number of aims:

- to provide students with a focus for reading research literature in science education;
- to provide students with the knowledge and skills to design, carry out and report on a small-scale research study which links their reading to their classroom practice;
- to encourage students to value the contributions educational research can make to improving classroom practice and pupils' learning;
- to make students aware that teachers are important members of the science education research community.

Two main criteria were used in selecting areas for inclusion as topics for an 'In-depth focus'. Firstly, it was necessary that the range of areas taken as a whole should be sufficiently broad to encompass a range of student interests. Secondly, it was considered important to be able to identify for each individual area, a small number of research papers appropriate for use by PGCE students as entries into the literature. Four topics were finally included on the list:

- Science and society;
- Language and science education;
- Exploring pupils' understanding of scientific concepts; and
- Promoting equal opportunities in science education.

These titles reflect some of the interests of the team of science education tutors, one of whom was responsible for the development of each unit. Additionally, it was decided to offer students the opportunity of negotiating an independent project if they so wished.

The learning materials were developed as supported self study units, of a broadly similar structure. The exact format and presentation differs slightly from unit to unit, reflecting the different natures of the areas covered and possible research projects which might form the basis of an 'In-depth focus'. Typically, the units consist of a briefing sheet, accompanied by introductory tasks and preliminary readings. Briefing sheets normally contain the following:

- information relating to the preliminary readings;
- guidance on the associated preparatory tasks;
- outline suggestions for possible research projects;
- suggestions for further readings, both on the subject matter and on methods of gathering data; and
- guidance on how to structure the research report.

The 'In-depth focus' forms a significant component of the summer term programme on the PGCE course. At this stage in their course, students have completed a full term of teaching but still spend two days each week in school. Students are given two weeks to design their study, and during this time they are encouraged to discuss their ideas with both university and school staff to ensure they have a viable research project. The expectation is that students will collect the data for their research project when they are in school, and that the data will be collected from the groups they are teaching or other groups to which they can easily gain access. Students then have a month to carry out and write up their study. Students are permitted to work jointly on the design of research projects, but each has to submit a separate written report. The report takes the form of a written description and evaluation of the project undertaken, and is expected to be of approximately 2500 words in length. Reports are assessed by university-based staff. In addition, students give an oral presentation of their findings to school staff towards the end of the term and produce summary posters of their projects for display to their fellow students at a research symposium at the university.

Evaluation of the In-depth Focus

Evaluation of the 'In-depth focus' was undertaken to determine the extent to which the aims had been realized. Feedback was collected from both students and school-based staff. Feedback from school-based staff was collected informally at an end-of-year meeting. In general, their feedback was very positive, with staff commenting particularly on the value of the task in stimulating student interest in what can often be a difficult term. Additionally, several staff felt that the presentations had been particularly useful in enabling both staff and students to consider how the data collected had been or would be of use to the science department.

Feedback from students was obtained by means of a detailed questionnaire. It was considered important that students had the opportunity to raise issues which they felt to be significant to them and so a preliminary stage in the design of the questionnaire was a discussion with students on what questions they thought should be asked of them. These questions then formed the basis of the questionnaire put together by the course tutors. In practice, all the areas the tutors wished to include in the questionnaire were also suggested by the students.

The questionnaire sought responses under two main headings, the first relating to practical aspects of the design and execution of the project (for example, reasons for choosing the topic, time considerations, negotiating access to pupils, levels of school and university support) and the second relating to students' perceptions of the benefits and drawbacks of the 'In-depth focus' assignment. In this latter area, students were asked questions relating to the following:

- their views on the most and least valuable aspects of the 'In-depth focus' study;
- the knowledge and skills they felt they had developed in doing research;
- the use they considered the knowledge and skills gained would be to them, both in the short and longer term;
- the appropriateness of the 'In-depth focus' assignment at this stage of their PGCE course;
- how carrying out the assignment had influenced their views on links between educational research and classroom practice; and
- the value of teachers carrying out educational research.

Outcomes

The 1993–4 science PGCE group consisted of thirty-six students. All the focus areas offered were selected for study. Exploring pupils' understanding of particular science concepts (for example, particulate theory, gravity and energy) was the most popular (fourteen students). Eight students opted to study aspects of language in science such as the use of directed activities related to text (DARTS). The seven students studying aspects of science and society worked on topics such as pupils' perceptions of scientists and pupils' perceptions of links between science, technology and society. Four students carried out projects on the relationships between gender, hobbies and interest in science. The other three students carried out projects of their own choice.

Thirty-three students of the thirty-six students in the group completed the questionnaire. Responses were made anonymously. The main findings to emerge from the questionnaire are summarized below. Numbers following quotations refer to the codes allocated to individual questionnaires.

Feedback from the students was generally very positive. Students welcomed the opportunity to engage in a research study of their choice and valued the insights and knowledge that resulted. Almost without exception, students commented that they had enjoyed the task and found it worthwhile, with several saying it had been one of the most useful and informative components of the course. Students welcomed the flexibility offered by the programme:

> . . . being given a chance to look at an aspect of science education that is personally interesting. (Student 15)

Many of the students had structured their 'In-depth focus' studies in such a way that it enabled them to explore in greater depth a problem or area of interest which they had encountered during their main teaching placement the previous term. The opportunity to do this had clearly been welcomed by students, who had found the experience to be of considerable benefit in terms of increasing their understanding of some of the barriers to learning encountered by pupils in science lessons, and ways in which these might be overcome:

I felt I would gain a better understanding of how pupils think about energy, and hoped it would give me ideas on how to approach the topic in my lessons. (Student 17)

It was very useful to find out the difficulties pupils have in this area [states of matter]. (Student 30)

Students also commented on their experiences of using research techniques:

The interviewing technique is a good research method, but it must be done in the right way. (Student 18)

In addition to the development of skills relating to data gathering, students also commented on the way in which carrying out the '*In depth focus*' had contributed to the development of more general abilities:

I have developed the ability to criticize and appraise my own work. (Student 23)

Many of the students commented that their actions in the classroom and/or future resources they produced would be influenced by the findings of their study.

I'd read about it [children's misconceptions] in the autumn term, but doing this really brought it home to me and made me think about what I need to do with *my* groups in *my* lessons. (Student 9)

There was a divergence of opinion as to the timing of the task, with some students feeling they would have preferred the task to be set in the spring term to allow for more time to gather data. This is likely to relate to the fact that the main practical problem encountered by students was in finding the time in school for data collection, a particular problem for those students who missed several days in school to attend job interviews.

In terms of students' views on teachers as researchers, most felt that it was very important for teachers to undertake small-scale research for themselves and that opportunities should be provided for this to take place. Their reasons were clear:

[Research] . . . improves the effectiveness of teachers because of insights they have gained. (Student 1)

For some students, carrying out the 'In-depth focus' made them think in terms of reading about research and in undertaking further research themselves:

. . . at some stage I'd like to take a higher degree in educational research. (Student 15)

Conclusion

In very broad terms, the 'In-depth focus' research task has proved a useful addition to the initial teacher education programme at York. It has been perceived by students to be an appropriate activity which has enabled them to pursue worthwhile professional interests and gain understandings from reading about and carrying out research. Most importantly, students have indicated that their professional actions are likely to be influenced as a result of carrying out their research studies. In the short term, it is intended to follow up the study by contacting each of the students to establish the extent to which their stated intentions have been realized in practice. Longer-term effects will be more difficult to evaluate. However, there is a clear indication in the responses made by students that many of them have gained an appreciation of the value of the findings of educational research to their classroom practice and of the potential contribution that they, as teachers, have to make to the science education research. Rudduck (1985) is clear in her view that attitudes and habits which are supportive of research need to be developed during initial teacher education. Although this study is small scale and based in only one institution, the evidence presented does suggest that there is considerable benefit to be gained from the inclusion in initial teacher education courses of opportunities for students to gain research experience. We contend that it is important that initial teacher education promotes the value of educational research and creates opportunities for students to gain research experience if we are to win respect for and gain new members to the science education research community.

References

EBBUTT, D. and ELLIOT, J. (1985) 'Why should teachers do research?', in EBBUTT, D. and ELLIOT, J. (Eds) *Issues in Teaching for Understanding*, York, Longman.

ERICKSON, G.L. (1991) 'Collaborative enquiry and the professional development of science teachers', *Journal of Educational Thought*, **15**, pp. 228–45.

FENSHAM, P.J. and NORTHFIELD, J.R. (1993) 'Pre-service science teacher education: An obvious but difficult area for research', *Studies in Science Education*, **22**, pp. 67–84.

LISTON, D.P. and ZEICHNER, K.M. (1990) 'Reflective teaching and action research in preservice teacher education', *Journal of Education for Teaching*, **16**, 3, pp. 235–54.

McCLELLAND, V.A. (1993) 'National case studies/Etudes de cas nationaux: England', *European Journal of Teacher Education*, **16**, 1, pp. 13–19.

PIMENOFF, S. (1995) 'Seeking the truth?', in *The Education Guardian*, 11 April, p. 6.

RUDDUCK, J. (1985) 'Teacher research and research-based teacher education', *Journal of Education for Teaching*, **11**, 3, pp. 281–9.

RUDDUCK, J. and WELLINGTON, J. (1989) 'Encouraging the spirit of enquiry in initial teacher training', *Forum*, **31**, 2, pp. 50–1.

SCHON, D.A. (1983) *The Reflective Practitioner*, New York, Basic Books.

STENHOUSE, L. (1975) *An Introduction to Curriculum Research and Development*, London, Heinemann.

WEBB, R. (1990) *Practitioner Research in the Primary School*, Basingstoke, Falmer Press.

Part VI
Language and Imagery

Introduction
Geoff Welford

This two paper section somewhat inevitably has some overlap with Part II on developing and understanding models in science education, since the basis of much modelling is arguably through language. However, these two papers are convened together to highlight the importance of the area in science education research. Merzyn presents a linguistic analysis of fifteen biology textbooks along six linguistic parameters, arguing that his methods provide richer insights than the more usual readability values. Somewhat surprisingly perhaps, his findings show remarkable similarities in presentational style regardless of the age and ability of the target audience. Sumfleth, Körner and Gnoyke looked at the importance of mental representations to the learning process. They ask us to question the assumption that pictures are supportive of learning and go on to explore how pictures have to be constructed if they are to function as an aid to learning.

28 A Comparison of Some Linguistic Variables in Fifteen Science Texts

Gottfried Merzyn

Abstract

The language of fifteen biology textbooks at lower secondary school level is compared. All of the texts conform with the same state-given curriculum. The complete text of the books was scanned onto computer and the technical vocabulary analysed. Among the six linguistic parameters examined are the number of technical terms, their density and their frequency. The results give a better insight into linguistic properties and potential difficulties of a text than usual readability values do. They demonstrate for the fifteen books an unexpected uniformity of the style of presentation. Nearly no adaptation to different ages and abilities can be observed. In addition, specific disadvantages of the long and the short texts are revealed.

Language and Science Education

The development of language and understanding are closely interwoven, as studies by Piaget, Vygotsky, Bruner and many others have shown. Understanding is a constructive process in which the reader interacts with the text. Science educators in many countries, therefore, attach greater importance to the role of written and spoken language in effective science teaching than before. They are more aware of the part language plays in the growth of ideas. Important books have been written on this topic (Prestt, 1980; Vollmer, 1980; Bulman, 1985; White and Welford, 1988).

Nevertheless investigations into the language of science education still show serious problems. A recent survey with German physics teachers, asking them about the physics textbooks in a free answer format, gave language and understandability as the most often mentioned negative aspect of the books (Merzyn, 1994). Free answers of pupils (grades 7 to 10) about their physics texts led to the same result (Bleichroth, Dräger, Merzyn, 1987).

Other researchers have reported that:

- Science textbooks list too many concepts at a high level without dealing with them in sufficient depth (Graf, 1989; DiGisi, 1995).
- The way of talking about science phenomena more than the phenomena themselves often remains mysterious to the pupils (Sutton, 1974).

- Pupils object to the plethora of technical terms used (Kelly and Monger, 1974).
- The linguistic demands in the sciences are considerably higher than in foreign language learning (Merzyn, 1987; Brämer and Clemens, 1980).
- Dialogues in the classroom suffer from a deep linguistic gulf between teacher and pupils (Barnes, 1971).
- Also in the non-specialist parts of their language, teachers use a whole range of forms outside the experience of their pupils (Gardner, 1972).
- In a surprising number of cases pupils take even the opposite meaning to what is intended out of the teacher's language (Cassels and Johnstone, 1985).
- Certain kinds of talking and writing widely employed in science education force pupils into intellectual strategies which are the opposite of those intended by the teacher (Sutton, 1974).

Apparently, too much reliance is still laid on the common-sense and linguistic intuition of science teachers and textbook authors. Too little attention is given to the influences that act on them when speaking or writing; that is influences of the discipline (Borsese, 1994) and the curriculum, of their own education, of their peers. This paper will give further evidence that common-sense and intuition are not sufficient to ensure an easily understandable language, a language appropriate to the audience.

This research concentrates on a small aspect of language, namely on the technical vocabulary of textbooks. This kind of research has a rather long tradition. Extensive contributions were already made by Curtis in the USA in the 1930s (1938). Other well-known contributors were e.g., Evans in the UK (1976) and Gardner in Australia (1972). ('Technical vocabulary' and 'technical terms' in this article mean what Curtis has called 'scientific terms', Evans 'technical terms', Brämer (1980) '*Fachvokabeln*', and Graf (1989) '*biologische Begriffe*'.)

Of course, language is far more than difficult words and science education more than textbooks. An advantage of our concentration on a few aspects is that, in a rather reliable way, findings from a broad textual base can be gathered. A further advantage is that authors can use these methods as a corrective to their intuition. They can review and improve their writing even before printing. Improving the written material will probably also improve the spoken language since it is well-known that many teachers rely heavily on the textbooks as pedagogical guides (Howe, 1990; Merzyn, 1994).

Characteristic Features of Biology Texts

Most features and problems of language are similar in all three sciences. There are, however, some differences, too, as has been pointed out already by Curtis (1938) and recently confirmed by comparison of some German results. To eliminate the variable 'school subject', the following discussion shall concentrate on biology texts. I shall look at the complete written text of fifteen books of the lower secondary school. To make the comparison as fair as possible, all texts regarded here were approved by the same regional Minister of Education, thus in accordance with the same state-given curriculum.

My considerations are based on figures that Graf collected in his dissertation (1989). Parameters I will look at are

- the length of the book;
- the total number of technical terms (that is a term is counted anew every time it occurs), the total number of different technical terms (every term is counted only at its first appearance);
- the density of technical terms (that is, their ratio to all the words in the book);
- the density of new technical terms (a density of new technical terms of 5 per cent means, for instance, that on average every twentieth in the book is a new technical term);
- the average frequency of the technical terms (that is, the ratio of the total number of technical terms over the total number of new technical terms).

Compared to readability studies, where usually one figure is formed to express the syntactical properties, another to express the properties of vocabulary, and a readability value derived from these two, I shall leave aside the syntax. This makes sense for two reasons:

- the vocabulary is the most important feature of the language of every science (Althaus, 1980); and
- the syntax of the natural sciences is usually rather normal in its difficulty; the vocabulary, however, is demanding (Fucks, 1968).

On the other hand, where readability studies usually give one single figure as result, the six variables listed above give more information than does one single readability value. Hence one can learn which aspect of a given text is difficult and thus how to influence its difficulties.

For the fifteen biology books, the length of the books varies considerably, between 40,000 and 97,000 words. There are big differences, too, when looking at the number of technical terms (6900 . . . 17300) and at the number of different technical terms (1480 . . . 3820). These three variables are, however, not independent from each other. On the contrary, the length of book and number of technical terms are proportional (Figure 28.1); that means, the density of technical terms is, within a small margin, the same in all fifteen books.

Certainly the authors do not consciously strive to reach this constant density. Instead, it seems that all of them have a common idea of the way a school text has to be written. They have grown up in the same tradition for decades and they accept it unconsciously. None of the authors tries to vary the density of technical terms to adapt for different abilities or needs of the pupils — a very unexpected and

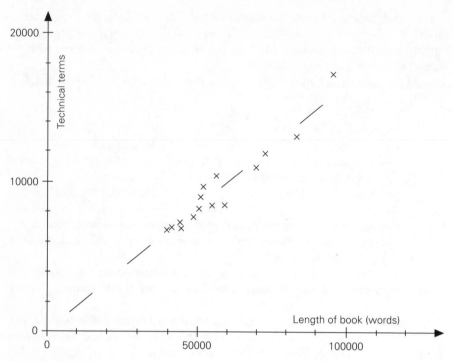

Figure 28.1: The density of technical terms is the same in all fifteen books

unwanted uniformity of style. Apparently neither authors nor teachers have perceived it so far.

A completely different finding was expected: Since the curriculum is given, one could have imagined that the technical vocabulary would be roughly the same in all books, independent of their thickness. The thicker book would use lengthier explanations, give more examples from daily life and have thus a lower density of technical terms.

The next important linguistic parameter is the average frequency of the technical terms. It varies at around a mean of 4.5 (i.e., every technical term appears four or five times in the text), with a minimum of 3.4 and a maximum of 5.7. The longest book of all has an average frequency of 4.53 — precisely the average value of all fifteen books. That means: The longer book does not repeat terms more often. It does not explain the same matter with more words, with more redundancy. Instead it seems to give additional details and elaboration.

The shortest books all have frequencies below the average. That indicates, that these books are shorter because they tackle the single concept more briefly. This property becomes even more apparent if one looks at the density of new technical terms, that is the distance (measured in words) until the next new technical term is used. This distance is forty words for some books (density 2.5 per cent). Short books, however, usually have short distances, close to twenty words (Figure 28.2); that is they have high densities of new technical terms.

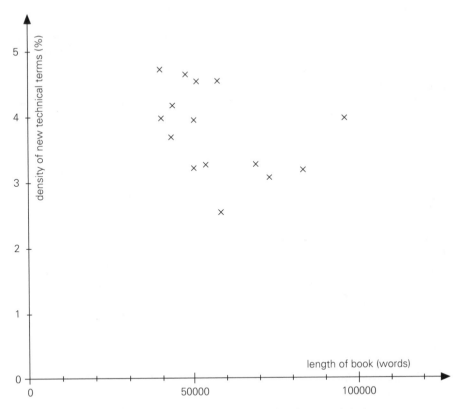

Figure 28.2: The shortest books have the highest density of new technical terms

Therefore, these thin books are by no means easier to read. Their shortness is purchased by a condensed form of presentation.

Let us look from a different point of view at books with a low density of new technical terms. Since the overall density of technical terms is the same in all books, one should not think those books prefer everyday language. Instead they use those terms already introduced more intensely: the average frequency of a technical term is considerably higher (Figure 28.3).

Figure 28.3 is a good example of the advantage the usage of several linguistic parameters has, compared with one single readability value. Additional information about the text is provided.

Adapting the Language to Age and Ability

The fifteen biology books looked at have in common that they all are approved for the lower secondary years by the same regional Minister of Education. They address, however, within this common framework different subgroups in two aspects.

On one hand, they all appeared as a pair:

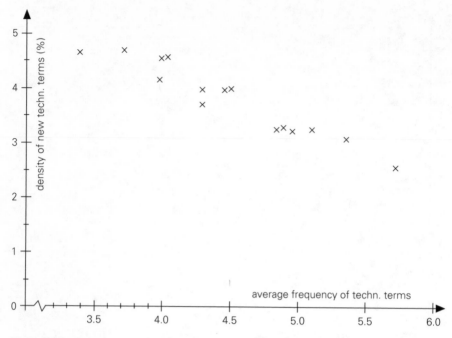

Figure 28.3: Books with a low density of new technical terms have a high frequency

- volume 1 for grades 5 and 6 (age 10 and 11 years),
- volume 2 for grades 7 to 10.

On the other hand, since most secondary schools in Germany have external differentiation, they address:

- either the *Hauptschule* and the *Realschule*,
- or the *Gymnasium* and the *Realschule*.

One can, therefore, bring twelve of the fifteen books into an array:

Grade 7/10	**Hauptschule/RS**	**Gymnasium/RS**
Grade 5/6	BIO 1* LW 1 WDB 1	BH 1 CVK 1 KDL 1
Grade 7/10	BIO 2 LW 2 WDB 2	BH 2 CVK 2 KDL 2

(*) Abbreviations of the twelve book titles as in Graf (1989)

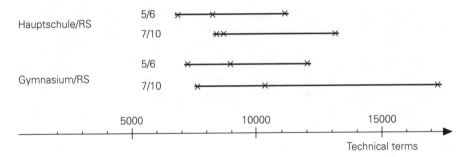

Figure 28.4: Number of technical terms in books by school type and age
Note: Each book is represented by a cross (x).

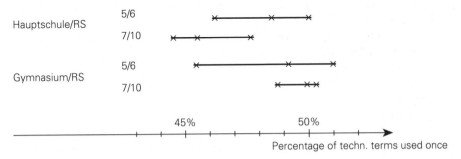

Figure 28.5: Percentage of technical terms used only once in books, by school type and age

Looking at these twelve books one can study under ideal conditions how the authors adapt their text to different ages and abilities.

One would expect large differences in their linguistic properties. The four groups of pupils belonging to the four fields of the array certainly have rather different willingness and ability to read scientific texts. The expected order by increasing ability should be:

Hauptschule/Realschule, grade 5/6
Hauptschule/Realschule, grade 7/10
Gymnasium/Realschule, grade 5/6
Gymnasium/Realschule, grade 7/10.

All of the linguistic variables considered so far are indicators of differences in difficulty. A higher density of new technical terms, for instance, is a feature of a more difficult text. Each of the linguistic variables, when calculated for the twelve books, should thus show four clusters, if the authors managed to adapt their language appropriately to their audiences. Especially the differences between the *Hauptschule* and the *Gymnasium* should be distinct.

However, Figures 28.1 to 28.3 indicated that there are not four such clusters. This result is even more evident in Figures 28.4 and 28.5 and four corresponding diagrams for the other four parameters (not shown here) which all look very similar.

The horizontal axis in both diagrams is oriented such that the difficulty increases to the right. The four bars belong to the four different groups of pupils.

It is true that, as expected, the average difficulty increases in the books of the *Gymnasium*, grade 7/10. More remarkable, however, is the considerable overlap the four bars have with each other. The most difficult book for the *Hauptschule*, grade 5/6, with respect to technical terms, for example, contains more technical terms than the easiest book for the *Gymnasium*, grade 7/10. It has been shown that, for some variables, the books differ only slightly from each other. In addition, in several cases, the less able pupils are confronted with higher linguistic demands. Thus this misadaptation affects the pupils of the *Hauptschule* grade 5/6 most strongly.

What are the reasons for this lack of variance, these deficits in adaptation to the demands? One reason has already been mentioned. Presumably the long tradition of textbook writing acts as a unifying force on the style and prevents the authors from adapting their mode of presentation. Another reason may be that, in German education, there is a tendency to reduce the differences between the types of secondary school, thus neglecting more and more the differences which still exist between the pupils.

Conclusion

The aim of this paper was to show that among the various methods of investigating the language of science education statistically based ones looking at the technical vocabulary have some advantages:

- It is possible to include some information about the very many concepts that appear only once or a few times in a text but nevertheless influence the language strongly by their large number.
- Suitable chosen parameters give considerably more information than conventional readability studies. The diagrams shown clearly indicate that these parameters are by no means random-distributed or insignificant.
- Since modern textbooks are fed into computers for printing, the authors can use a linguistic analysis without major effort to improve their texts in an early stage of production.

It is evident that statistically based considerations of the technical vocabulary reflect only very few properties of the language of science education. Language is by far too complex to be investigated only by one or a few methods.

Nevertheless our results were sufficient to show an inadequate adaptation of the author's language to different ages and abilities. They were sufficient to show that the thick books suffer from an abundance of information whereas the thin books suffer from an extremely condensed way of presentation. The observations gave good reason to raise questions about the influences that act on textbook authors during their work. They gave good reason to remind us of the saying of the philosopher Karl Jaspers (1981): 'In science as a profession everything is worth or could be worth knowing. For education, however, science is full of knowledge not worth teaching.'

References

ALTHAUS, H.P., HENNE, H. and WIEGAND, H.E. (1980) *Lexikon der germanistischen Linguistik*, 2nd Ed, Tübingen, Niemeyer, entry 'Fachsprachen'.

BARNES, D. (1971) *Language, the Learner and the School*, 2nd Ed, London, Penguin.

BLEICHROTH, W., DRÄGER, P. and MERZYN, G. (1987) 'Schüler äußern sich zu ihrem Physikbuch', *Naturwissenschaften im Unterricht — Physik*, **26**, pp. 262–4.

BORSESE, A. (1994) 'Il problema della communicazione linguistica a scuola', *Ens. de las Ciencias*, **12**, pp. 333–7.

BRÄMER, R. and CLEMENS, H. (1980) 'Physik als Fremdsprache', *Der Physikunterricht*, **14**, 3, pp. 76–87.

BULMAN, L. (1985) *Teaching Language and Study Skills in Secondary Science*, London, Heinemann.

CASSELS, J.R.T. and JOHNSTONE, A.H. (1985) *Words that Matter in Science*, London, Royal Society of Chemistry.

CURTIS, F.D. (1938) *Investigations of Vocabulary in Textbooks of Science for Secondary Schools*, Boston, Finn.

DIGISI, L.L. and WILLETT, J.B. (1995) 'What high school biology teachers say about their textbook use', *Journal for Research in Science Teaching*, **32** pp. 123–42.

EVANS, J.D. (1976) 'The treatment of technical vocabulary in textbooks of biology', *Journal of Biology Education*, **10**, pp. 19–30.

FUCKS, W. (1968) *Nach allen Regeln der Kunst*, Stuttgart, DVA.

GARDNER, P. (1972) *Words in Science*, Melbourne, ASEP.

GRAF, D. (1989) *Begriffslernen im Biologieunterricht der Sekundarstufe I*, Frankfurt/M, Lang.

HOWE, R.W. (1990) *Trends and Issues in Science Education*, Columbus, OH, ERIC.

JASPERS, K. (1981) *Was ist Erziehung?*, München, DTV.

KELLY, P.J. and MONGER, G. (1974) 'An evaluation of the Nuffield O-level biology course materials and their use', *School Science Review*, **55**, 192, pp. 470–82.

MERZYN, G. (1987) 'The language of school science', *International Journal of Science Education*, **9**, pp. 483–9.

MERZYN, G. (1994) *Physikschulbücher, Physiklehrer und Physikunterricht*, Kiel, IPN.

PRESTT, B. (Ed) (1980) *Language in Science*, Hatfield, Herts, ASE.

SUTTON, C.R. (1974) 'Language and communication in science lessons', in *Science Teacher Education Project: The Art of the Science Teacher*, London, McGraw Hill, pp. 41–53.

VOLLMER, G. (1980) *Sprache und Begriffsbildung im Chemieunterricht*, Frankfurt/M, Diesterweg.

WHITE, J. and WELFORD, G. (1988) *The Language of Science*, London, DES.

29 Imagery of Information: Prerequisites and Consequences

Elke Sumfleth, Hans-Dieter Körner and Andrea Gnoyke

Abstract

The hypothesis that mental representations are of great importance to the learning process was investigated by collecting descriptions of learners for properties of terms like concreteness, imagery, meaningfulness and understandability. In particular, the question of the role of imagery of information was explored. The assessment results gave high scores, even for terms which stand for imperceptible entities. Therefore it is assumed that perception plays no relevant part in assessing imagery and the other properties. Instead, it is suggested that the reasons for the assessments are conceptual definitional knowledge of the concreteness and practical or mental applicability of imagery. All students attach great importance to visual conceptions. Above all practical experiences and actions lead to direct imagining. Students emphasize that information visually represented is used mentally and is processed as if it were knowledge of actions. Consequently the question is posed, how is learning influenced by pictures? The first results can be summarized as follows: Generally the linguistic parts of a picture attracts attention and the procedure of looking at a picture depends obviously on relevant preconceptions.

Introduction

Das Hindernis, das mich von der Chemie fernhielt, hat etwas zu tun mit der weiten Kluft zwischen wahrgenommener Wirklichkeit und Symbol. Das Wasser, das ich trinke und in dem ich bade, und die Formel H_2O schienen mir keine direkte Beziehung zu haben. (The barrier which keeps me away from chemistry has to do with the wide gap between perceived reality and symbol. The water which I drink and in which I bath and the formula H_2O do not have a direct relation for me.) (Born, 1964)

This quotation, as many others, points out that a division into three worlds (Popper, 1974) — the world of physical states, the world of mental states and subjective ideas, and the world of theories and objective ideas — only has a symbolic character because they depend on each other. The world of mental states has a conciliatory function because sense perception and experience and handed down knowledge combine here. Visual conceptions are assigned to this world of mental states. Imagery means developing in the mind an image of the information unit. These images are not mental pictures, not copies of the environment, but mental representations that are mentally treated images. The hypothesis that these mental representations are of great importance was investigated by collecting introspective descriptions of learners about properties of terms like concreteness, imagery, meaningfulness and understandability. In particular, this research attempts to explain the role of imagery in mental processing.

Assessment of Term Properties

Procedure

To determine the basic relations between these properties of terms, numerous terms have to be investigated. In addition, the number of students assessing these properties must not be too small. Therefore a seven-step scale rating procedure was chosen. Other investigations which have been conducted in a comparable context use different rating procedures, mainly multistep scales (Paivio, Yuille and Madigan, 1986; Baschek, Bredenkamp, Oehrle and Wippich, 1977; Wippich and Bredenkamp, 1977; Offe, Anneken and Kessler, 1981; Krampen, Jirasko, Martini and Rihs-Middel, 1990). They generally provide very high agreement on the results (Westermann and Hager, 1984; Borg, Müller and Staufenbiel, 1990). Moreover rating scales meet in a more general sense the interval scale necessary for a range of statistical operations (Westermann and Hager, 1983; Westermann 1984). The properties of the terms are considered as unipolar (Hager, Mecklenbräuker, Möller and Westermann, 1985; Günther and Groeben, 1978; Anderson, 1968; Schwarz *et al.* 1991).

Thirty-two chemistry terms which were different in respect of the four properties (concreteness, imagery, meaningfulness, understandability) were chosen by a preliminary investigation. Prospective chemistry teachers were asked to name a chemistry term which they considered as very concrete, distinctly visual or having great meaning in chemistry. In addition, each of them was asked to mention one term with the opposite property. A majority of the frequently named terms was then put into the term list for the main investigation. Terms which are probably unknown by younger students were excluded. The final list contained the following terms: acid, ammonia, atom, base, bonding, bromine, compound, combustion, copper, electrolysis, equilibrium, indicator, cadmium, metal, non-metal, periodic table, oxygen, polarity, reaction, reaction-equation, redox-reaction, salt, structure, structure-formula, titration. Terms like experiment, periodic table, reaction-equation and structure-formula mark ways to represent, recognize or interpret chemical contents and are therefore very important in the learning process. The terms were listed at random avoiding a direct grouping of antagonistic and content-related terms. All students were given lists with the same term arrangement, because no statistically relevant effect of different arrangements has been shown (Krampen, Jirasko, Martini and Rihs-Middel, 1990). The terms and their assessment scales were printed on four sheets of paper, one for each term property, and stapled together in random order. The students' instructions were based on those of Baschek and co-workers (1977) and of Westermann and Hager (1984). About twenty minutes were needed to complete the test.

Elke Sumfleth, Hans-Dieter Körner and Andrea Gnoyke

Table 29.1: Correlation of assessments of term properties

	Imagery	Concreteness	Understandability	Meaningfulness
Imagery				
Concreteness	0,45*			
Understandability	0,47*	0,45*		
Meaningfulness	0,29*	0,27*	0,33*	

* = p < 0,001

The subjects were 311 students comprising three groups: Eighty-six students aged about 17 to 18 years, from grade 11 in German gymnasiums and drawn from six courses at three different schools, eighty students beginning university study of the English language and 145 students beginning a university study of chemistry. Differences in age are unimportant in respect of the assessment of these term properties (von Eye, von Eye, and Hussy, 1980; Schwibbe, Räder, Schwibbe, Borchardt and Geiken-Pophanken, 1981; Möller and Hager, 1991). The argument for uniting the different courses in these three groups is supported by the work of Offe and co-workers (1981). Group homogeneity was good. According to Offe *et al.* this investigation can be regarded as reliable (Körner, 1994).

Results

Examining the correlations between the properties themselves there was a strong interrelation between imagery, concreteness and understandability, however meaningfulness was much less correlated (Table 29.1). Factor analyses verify this result and indicate that the assessments concerning the three inter-related properties were based two-thirds on a common factor and one-third on specific criteria. The assessments of meaningfulness were based nearly on the whole on the second factor.

The results showed mainly high scores. The only terms gaining low scores were 'sodium aluminium silicate', 'orbital' and 'titration', especially in the assessments of younger students (grade 11), who do not know these terms. Comparing these results with those of another study concerning everyday terms (Baschek, Bredenkamp, Oehrle and Wippich, 1977) showed that nearly a quarter of the everyday terms but only a few chemistry terms were assessed as very concrete (mean scores between 6 and 7). One example is the term metal which might be seen as a term from daily life. Baschek's study contained this term too and its mean score of 6.1 was nearly the same as in our study. There were more scores between 4 and 6 in the chemistry terms which was higher than that for everyday terms. Assessments below the scale mean (3.5) were more frequently associated with everyday terms. The same was true for imagery. In contrast meaningfulness was scored very differently. There were nearly no everyday terms assessed as very meaningful while a fifth of the chemistry terms belonged to this category. It is especially striking that the assessment of meaningfulness led to high scores when all other properties had low scores. Perhaps this fact reflects the opinion that a professional

372

Table 29.2: Correlation of term assessments concerning imagery (students grade 11)

	Salt	Metal	Non-metal	Copper	Com-bustion	Atom	Bond	Equilib-rium	Polarity	Orbital
Salt	—									
Metal	0.32	—								
Nonmetal	*0.43	0.21	—							
Copper	*0.49	0.33	0.31	—						
Combustion	0.17	0.04	*0.41	*0.37	—					
Atom	0.03	−0.08	−0.04	0.01	0.17	—				
Bond	0.05	0.02	0.07	0.10	−0.01	0.29	—			
Equilibrium	0.15	−0.03	0.20	0.11	0.18	−0.01	0.23	—		
Polarity	0.00	0.13	−0.06	0.20	−0.08	0.14	0.05	0.11	—	
Orbital	0.03	0.01	−0.09	0.09	−0.25	0.13	0.11	0.06	0.04	—

* = $p < 0.001$

language has only a few unimportant terms. A comparison with understandability is not possible because Baschek *et al.* (1977) did not investigate this property.

Standard deviations were partly very high, especially for the assessment of concreteness. Consequently the assessment is very heterogeneous. This comment applies in particular to the terms 'atom' and 'energy' which were distinctly characterized by large standard deviations. Buck (1979) refers to an atom as an 'abstract reality'. This statement does not say whether Buck would assess concreteness by 1 or 7. Reliability of mean assessments was determined following the methods of Paivio *et al.* (1968) and Baschek *et al.* (1977) using a split-half method. The groups were divided into two subgroups randomly and the means of these subgroups were correlated. Reliability values about 0.9 obtained indicate a very high agreement.

The high scores are surprising, particularly for terms which stand for imperceptible entities like 'atom' or 'equilibrium'. Therefore it is assumed that perception plays no relevant part in assessing imagery and the other properties. This is corroborated by correlating the term assessments of perceptible and imperceptible entities. If perception plays an important role, the correlations have to be high within each group of perceptible or imperceptible entities and they have to be low between these groups. For 16-year-old students and university students of English language clear differences were found in the assessments of both groups (Tables 29.2 and 29.3). This is shown by low correlations between these groups. Although the correlations within each group are a little bit higher, one cannot differentiate them clearly from the others. There is no reason to believe that perception has an important influence. In contrast chemistry students judge much more systematically. The corresponding correlations are higher and mainly significant (Table 29.4).

The correlations of the students' assessments of the terms shows that the systematic estimation of university students of chemistry is moulded by contents. The figures in Table 29.4 clearly show that the students differentiated between terms which are used for theoretical explanations and those standing for practical experiences. It is less important if the named entities are perceptible or imperceptible. Overall, the assessments of younger students (grade 11) and of university students of English language are without uniformity and hardly categorizable.

Secondly, the students were asked to give the reasons for their assessments.

Table 29.3: Correlation of term assessments concerning imagery (English language students)

	Salt	Metal	Non-metal	Copper	Combustion	Atom	Bond	Equilibrium	Polarity	Orbital
Salt	—									
Metal	0.27	—								
Nonmetal	0.18	0.24	—							
Copper	*0.43	0.23	*0.43	—						
Combustion	*0.14	0.30	0.25	0.30	—					
Atom	0.20	0.14	0.03	0.04	0.19	—				
Bond	−0.03	0.21	−0.02	−0.15	0.30	0.30	—			
Equilibrium	−0.06	0.25	−0.13	−0.06	*0.40	0.13	0.47	—		
Polarity	−0.01	0.23	−0.14	−0.05	0.15	0.13	0.33	*0.56	—	
Orbital	−0.06	0.21	0.16	0.01	0.08	0.12	0.13	0.20	0.15	—

* = $p < 0.001$

Table 29.4: Correlation of term assessments concerning imagery (chemistry students)

	Salt	Metal	Non-metal	Copper	Combustion	Atom	Bond	Equilibrium	Polarity	Orbital
Salt	—									
Metal	*0.58	—								
Nonmetal	*0.62	*0.56	—							
Copper	*0.38	*0.47	*0.52	—						
Combustion	*0.39	*0.42	*0.55	*0.57	—					
Atom	0.05	0.23	0.22	0.25	0.20	—				
Bond	0.18	0.22	0.25	0.12	0.17	*0.51	—			
Equilibrium	0.12	0.25	0.04	0.14	0.21	0.08	*0.37	—		
Polarity	0.19	*0.38	0.24	0.28	0.17	*0.40	*0.48	0.23	—	
Orbital	0.13	0.16	0.15	0.10	0.06	*0.39	*0.41	0.18	0.48	—

* = $p < 0.001$

For imagery as well as concreteness, only 10 per cent of the students based their arguments on perception. Possibly perception is important when knowledge is constructed and becomes less important when a concept is attained. The relevant criterion for assessing concreteness is conceptual definitional knowledge of the term: 'Very concrete, because it is defined exactly.' Nearly half of the students argued this way although this aspect was not mentioned in the instruction which focused on tangibility. The latter played a subordinate role in students' answers. In respect of imagery, the statements refer to practical or mental applicability of the terms. Knowledge played a subordinate role. Only 6 per cent argued using this aspect. 'Terms are always concrete if you know them. They are visual if you can mentally do something with it.'

Interviews As Teaching–Learning Situations

Thirdly, there is the question how visual conceptions affect chemical problem solving. Therefore interviews were conducted with twenty-four students, twelve chemistry students and twelve English language students. An experimental phenomenon was presented to initiate situations of teaching and learning. First of all the students as learners had to predict the experimental result, and then they had to explain the phenomenon. Afterwards they took the part of a teacher and helped the next student to solve the problem. For that they were provided with different media (texts, pictures) which were criticized by the students during a final interview.

For forecasting the outcome of the experiment half of the participants referred to daily life, nobody argued using theoretical scientific theories. The same was true for the realization of problem. All students referred to everyday experiences or described situations which were connected to the phenomenon presented. Their visual and episodic imagination was to the fore. Hence in explaining phenomena, English language students used knowledge from everyday situations, keeping their interpretations to the macroscopic level, while chemistry students discussed the microscopic structure of matter. For the English language students, only the 'teacher' would call their attention to the microscopic level. Asked how they would explain the experimental phenomenon to a fellow-citizen, all students chose situations from daily life to represent the problem and to enable the learner to interpret it for themselves. This way of behaving makes it clear that the use of everyday experiences, remembered in a visual and episodic form, is an important element at the beginning of a problem solving situation.

The 'teachers' often used illustrations which showed the matter evaporating as discontinuous particles. They regarded these illustrations, or the ideas constructed from them, as connecting links between experiences and applications on macroscopic level and theoretical interpretations on microscopic level (Sumfleth and Körner, 1994).

All students attached great importance to visual conceptions. Above all practical experiences and actions lead directly to imagined entities. Students emphasized that information represented visually is used mentally and is processed as if it were knowledge of actions. It would appear that visual ideas in outline are the first abstraction of a practical action. 'I think it does not work without constructing a mental picture. It may be possible without illustrations and graphics, but not without constructing a model in mind.' Or: 'If you work out an image, understanding comes, otherwise there is knowledge only without understanding.'

Figurative descriptions impart meanings to statements. Each alternative metaphor points to a different way of understanding by somewhat changing the meaning (Sutton, 1992). There is an interrelation between the word and the image created by the word. This is important both for the discovery of new ideas and for learning science, since the learning of science means seeing things in a new way.

Statements are built up by a logical core (the scientific meaning) and by associative surroundings (additional meanings, personal meanings, feelings) (Schäefer, 1984). These surroundings exist before any scientific use of the term and they cause change in meanings and they may form the origin of the mental representations. These mental representations are the base to which new information is related and are deepened by discussions or practical experiences.

Pictures and Learning

Consequently the question whether pictures support the construction of images has to be answered. Three different groups of students were given different kinds of learning material: one group was provided with a text, a second group with a series of pictures, and a third group with a series of pictures and a text. If students know that they have to retell the information taken from the material, then they neglect pictures and confine themselves to the text (Sumfleth and Gnoyke, 1994). To get further information concerning processing of visual learning material we investigated the movement of the field of fixation whilst viewing a picture.

There is a difference between eye movement and movement of the field of fixation. To register eye movement the actual movement of the eye is recorded by the six ocular muscles, while measurement of the movement of field of fixation records the objects looked at, and the duration of fixation. It provides a quantitative description of reading and looking at pictures and allows for qualitative interpretation. The eyes of a reader do not pass continuously over the lines but the gaze jumps from passage to passage. When looking at pictures these jumps occur too, but are very non-directional. In reading one jump takes 10 msec, whilst when looking at pictures it is in the order of 300 msec, depending on the size of the picture (Ballstaedt, Mandl, Schnotz und Tergan, 1981). Then fixation occurs again which takes at least 200–300 msec. Only during fixation is the information grasped. Thus reading and looking at pictures consist of a sequence of jump, fixation, jump, fixation etc. and regressive fixations are possible.

Measurement of movements of the field of fixation (horizontal and vertical) are done by infrared-reflectometry. This method uses differently reflecting areas of the eye. The position of the fovea centralis is determined by a residual quantity of reflected light between the surface moisture, the crystalline lens, iris and pupil. Variation of intensity of light occurring by eye movement is determined by four photocells fixed lateral in a pair of glasses. The number of fixations, the periods of fixation, the number of regressions and the local ways of fixation and triple sequences by gaze-analysis can then be determined (Menz and Groner, 1986). The first three variables supply information on how the viewer searches for a special picture area, how long he or she stays there and how often he or she fixes the same points. By analysing local ways of fixations and triple sequences, different picture areas are defined which are characterized by accumulation of fixation points (Figures 29.4.1 and 29.4.2). This gaze-analysis leads to an arrangement of picture areas within which fixations are not further differentiated. In this way areas with many fixations are separated from those with only few (Figure 29.4.2). Several changes in turns form a path which can be subdivided into triple sequences.

Figure 29.2 shows the attribution of numbers to special picture areas. For instance: field number 3 marks the terms 'Minuspol–Pluspol', field number 21 the electrode on the left, and field number 22 the formulas on the left side. So movement of field of fixation can be followed from area to area. At the beginning all probationers follow the same sequence: The first five points of fixation are the same, all others are then different. The probationers start by using the normal

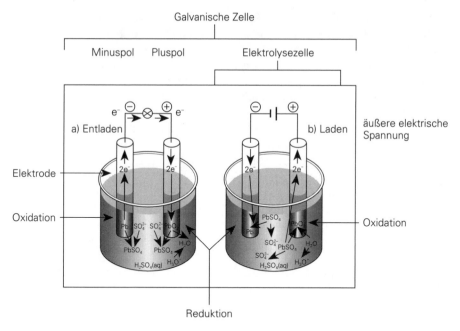

Figure 29.1: Picture shown to the probationers

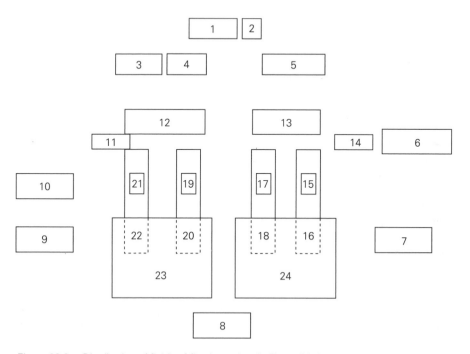

Figure 29.2: Distribution of fields of fixation points in Figure 29.1

reading process and then terms are searched for in a clockwise manner determined by the picture. In the absence of any picture the mode of reading dominates and is characterized by a proceeding line to line.

The group of probationers consisted of chemists and non-chemists in equal proportions. The two subgroups differed in their fixation times on different areas of the picture. Non-chemists looked much longer on the individual terms than the chemists. This is because the terms are to great extent unknown to the non-chemists. The chemists concentrated for more than half of the time on the reaction symbols. Contrary to the non-chemists they compared the areas of the symbols (23 and 24 in Figure 29.4.2) on the right and on the left. There was a second characteristic difference. Almost all chemists identified the terms with details of the picture directly. There was a characteristic search for the whole context. Many of the non-chemists perceived terms and picture details separately. On the whole the linguistic parts attracted attention and the procedure of looking at a picture depends obviously on relevant preconceptions. The information contained in the picture is modulated by the viewer.

After looking at the pictures the candidates had to give a report of the contents of the picture. First results show a considerable congruence between the picture areas which were looked at (ways of fixation, periods of fixation, regressions of fixations) and the information they recalled. The number of mistakes rose when the difficulty of the contents was increased, regardless of periods and regressions of fixation.

Conclusion

This investigation supports the hypothesis (Isfort, 1986) that processes of thinking and coding can be made 'visible' by movement of the field of fixation. The internal representation of the cognitive net determines how information is recognized and remembered. In addition different details in the picture influence this process. Therefore features of the picture (object characteristics) and of the viewer (subject characteristics) are to be seen as manageable variables for learning from pictures.

Until now imagery of information has hardly been investigated by science education research. On the contrary it is taken for granted that pictures are appropriate for learning. The results of this investigation lead to the question of how pictures should be constructed in order to facilitate learning for beginners. Nowadays many pictures made by experts do not support learning by beginners. More investigations in this field are needed to obtain detailed information about suitable presentation of scientific facts in a visual form.

References

ANDERSON, N.H. (1968) 'Likeableness ratings of 555 personality trait words', *Journal of Personality and Social Psychology*, **9**, pp. 272–9.

BALLSTAEDT, S.P., MANDL, H., SCHNOTZ, W. and TERGAN, S.-O. (1981) *Texte verstehen — Texte gestalten*, München, Urban Schwarzenberg.

BASCHEK, I.-L., BREDENKAMP, J., OEHRLE, B. and WIPPICH, W. (1977) 'Bestimmung der Bildhaftigkeit, Konkretheit und der Bedeutungshaltigkeit von 800 Substantiven', *Zeitschrift für experimentelle und angewandte Psychologie*, **24**, pp. 353–96.

BORG, I., MÜLLER, M. and STAUFENBIEL, T. (1990) 'Ein empirischer Vergleich von fünf Standard-Verfahren zur eindimensionalen Skalierung', *Archiv für Psychologie*, **142**, pp. 25–33.

BORN, M. (1964) 'Symbol und Wirklichkeit', *Physikalische Blätter*, **20**, p. 554.

BUCK, P. (1979) 'How real are atoms really? Wie wirklich sind "Teilchen" eigentlich?', *Chimica Didactica*, **5**, pp. 181–94.

EYE, VON, D., EYE, VON, A. and HUSSY, W. (1980) 'Zur Bedeutung der kognitiven Komplexität für die Einschätzung von semantischen Eigenschaften von Substantiven', *Zeitschrift für experimentelle und angewandte Psychologie*, **27**, pp. 534–52.

GÜNTHER, U. and GROEBEN, N. (1978) 'Abstraktheitssuffix-Verfahren: Vorschlag einer objektiven ökonomischen Messung der Abstraktheit/Konkretheit von Texten', *Zeitschrift für experimentelle und angewandte Psychologie*, **25**, pp. 55–74.

HAGER, W., MECKLENBRÄUCKER, S., MÖLLER, H. and WESTERMANN, R. (1985) 'Emotionsgehalt, Bildhaftigkeit, Konkretheit und Bedeutungshaltigkeit von 580 Adjektiven: Ein Beitrag zur Normierung und zur Prüfung einiger Zusammenhangshypothesen', *Archiv für Psychologie*, **137**, pp. 75–97.

ISFORT, A. (1986) *Verbalisierung und komplexes visuelles Material*, Münster, Lit.

KÖRNER, H.-D. (1994) *Vorstellen und Verstehen*, Frankfurt/M, Lang.

KRAMPEN, G., JIRASKO, M., MARTINI, M. and RIHS-MIDDEL, M. (1990) 'Semantische Merkmale vier vielverwendeter politischer Begriffe in fünf Nationalitätsstichproben', *Zeitschrift für experimentelle und angewandte Psychologie*, **37**, pp. 459–85.

MENZ, CH. and GRONER, R. (1986) 'Blickpfade bei der Bildbetrachtung', in ISSING, L.J., MICKASCH, H.D. and HAACK, J. (Eds) *Blickbewegung und Bildverarbeitung*, Frankfurt/M, Psychologie.

MÖLLER, H. and HAGER, W. (1991) 'Angenehmheit (p), Bedeutungshaltigkeit (m'), Bildhaftigkeit (I) und Konkretheit (C) von 452 Adjektiven: Ein Beitrag zur Normierung', *Sprache und Kognition*, **10**, pp. 39–51.

OFFE, H., ANNEKEN, G. and KESSLER, E. (1981) 'Normen für die Konkretheits- und Vorstellbarkeitseinschätzung von 234 Substantiven', *Psychologische Beiträge*, **23**, pp. 65–85.

PAIVIO, A., YUILLE, J.C. and MADIGAN, S.A. (1968) 'Concreteness, imagery and meaningfulness for 925 nouns', *Journal of Experimental Psychology*, **76**, Monograph Supplement, No. 1, Part 2, pp. 1–25.

POPPER, K.R. (1974) *Objektive Erkenntnis*, 2nd ed., Hamburg, Hoffmann Campe.

SCHAEFER, G. (1984) 'Information und Ordnung — Zwei mächtige Begriffe unseer Zeit' in SCHAEFER, G. (Ed) *Leitthemen Information und Ordnung*, Köln, Aulis, pp. 9–45.

SCHWARZ, N., STRACK, F. and HIPPLER, H.J. (1991) 'Kognitionspsychologie und Umfrageforschung: Themen und Befunde eines interdisziplinären Forschungsgebietes', *Psychologische Rundschau*, **42**, pp. 175–86.

SCHWIBBE, M., RÄDER, K., SCHWIBBE, S., BORCHARDT, M. and GEIKEN-POPHANKEN, G. (1981) 'Zum emotionalen Gehalt von Substantiven, Adjektiven und Verben', *Zeitschrift für experimentelle und angewandte Psychologie*, **28**, pp. 486–501.

SUMFLETH, E. and GNOYKE, A. (1994) 'Die Bedeutung bildlicher Symbolsysteme für

Theoriebildungen in der Chemie', *Der Mathematisch-Naturwissenschaftliche Unterricht*, **48**, pp. 14–21.

SUMFLETH, E. and KÖRNER, H.-D. (1994) 'Zur Bedeutung bildhafter Vorstellungen im Lehr-Lern-Prozeß', *Naturwissenschaften im Unterricht — Chemie*, **5**, pp. 182–6.

SUTTON, C. (1992) *Words, Science and Learning*, Buckingham, Open University Press.

WESTERMANN, R. (1984) 'Zur empirischen Überprüfung des Skalenniveaus von individuellen Einschätzungen und Ratings', *Zeitschrift für Psychologie*, **192**, pp. 122–33.

WESTERMANN, R. and HAGER, W. (1983) 'Eine empirische Untersuchung zum Skalenniveau von Normwerten für die Bildhaftigkeit von Substantiven', *Psychologische Beiträge*, **25**, pp. 112–25.

WESTERMANN, R. and HAGER, W. (1984) 'Zur subjektiven Repräsentation und direkten Erfaßbarkeit der Verständlichkeit, des Informationsgehalts und der Bildhaftigkeit von Informationsmaterial', *Zeitschrift für experimentelle und angewandte Psychologie*, **31**, pp. 328–50.

WIPPICH, W. and BREDENKAMP, J. (1977) 'Bestimmung der Bildhaftigkeit (I), Konkretheit (C), und der Bedeutungshaltigkeit (m') von 498 Verben und 400 Adjektiven', *Zeitschrift für experimentelle und angewandte Psychologie*, **24**, pp. 671–80.

Part VII
Science Education Research in Europe

Introduction
Geoff Welford

The final section in this publication is fittingly concerned with research in science education in Europe. Two of the three papers included were part of the input to a symposium which addressed this area. Solomon bases her paper on a comparative study setting school education in the context of a possible future, European, scientific culture. She proposes a way of applying the term 'scientific culture' to the study of curricula. The findings, perhaps unsurprisingly, point to diversity of science experiences at school and beyond, but highlight features across the countries which may contribute to a future popular science culture. Lijnse has as his focus the European PhD summer schools of 1993 and 1994, pointing to differences in several European countries in the research programmes at this level.

These two are joined by Sjøberg who takes us into the variation in understanding across Europe of some terms of fundamental importance to the science education researcher. While he explores meanings to show a difference in language usage between continental Europeans and the Anglo-Americans, he celebrates this diversity and breadth as a real strength in the development of open dialogue within the newly formed European Association for Research in Science Education. His paper is a 'natural' through which to bring this volume to a close, exhorting colleagues to decrease isolation and to take up together the research challenges ahead.

30 Science Education Research in Europe and Scientific Culture: What Can Be Done?

Joan Solomon

Abstract

A project on school science and the future of scientific culture in Europe will be completed in December 1995. It has collected data about science education in fourteen different European countries. Some like France, Germany and Greece are full members of the EU. Others like Sweden, Norway and Poland are likely to enter the EU at some time, and others like Switzerland are fully European, but never likely to join the Union. However scientific culture and knowledge do not respect political treaties although, as this paper shows, they do reflect in many more or less subtle ways, the national cultures. This paper proposes a way of looking at the term 'scientific culture' which can be applied to a study of scientific curricula. The picture obtained of school science, and of science for children beyond the classroom, was not uniform, but it contained pointers to the significant features in the different countries which seemed likely to contribute to the future of popular culture with respect to science.

Science in the EU: Invisible College or Popular Culture?

This paper reports on a study which might well become background reading for those wondering what specifically European research might look like. It is a comparative study of the science education systems with a European purpose. The EU already has reports from organizations such as EURYDICE (The Education Information Network in the EC) and CEDEFOP (European Centre for the Development of Vocational Training) e.g., Structures of the Education and Initial Training systems in the member states of the European Community Luxembourg (1990). In contrast to those, this study sets school education in the context of a possible future, European-wide, popular scientific culture. Hence this paper also attempts to examine what such a culture might be like.

Science, much more than, for example, literature, might be expected to have an academic commonality across different countries and cultures. Since the days of the formation of the Royal Society in the seventeenth century, there has been a notion of the invisible college to which all scientists belong by virtue of the common but esoteric knowledge that they hold. Although meta-scientists like Turnbull and Collins might well cast some doubt on the acultural and universalistic (Merton, 1973) nature of scientific knowledge, it nevertheless remains true today, as it has for almost the complete life-time of European science, that scholars can move from university to university across national boundaries in pursuit of their studies without cultural impediment. Indeed both the historical figure of Erasmus, and the EC's own modern ERASMUS project, bear witness to this.

Only the very tiny proportion of any nation's population which becomes either a research scientist or a science teacher can resonate to such a vision. For the vast majority of people the research perspective is not personal enough to prove attractive or relevant. In schools where the full range of ability study science, what is on offer may need to be much closer to everyday problems if it is to be interesting and meaningful to students, and so needs to take into account the local culture. This is a far more illusive quality to include in our European report than scientific knowledge.

School pupils can 'read' a public view of the perceived local importance of science in the organization of their science education: its funding, its local standing and relevance, and the status accorded to science teachers. These economic and sociological aspects of science education also have a place in this project.

Communications across Europe

The project on which this article is based is directed by Professor José Mariano Gago from Lisbon University, and set up in 1993 under the auspices of the European Community, prior to the Treaty of Maastricht. At that time no advice could be given about education to the member states. Indeed under the treaty of Rome there was a strong recommendation to respect the 'linguistic and cultural diversity' of European countries, which is still upheld. On the other hand, any project designed to open new channels of communication across the community, such as improving language teaching (LINGUA), was strongly supported. No previous report specific to science education in the EC had been made.

The project began by commissioning national reports from every EU country, and from a few still outside the EC, to a common format which would cover the science content in each National Curriculum, the local and national politics of its delivery, informal education through museums and science centres, the training and status of science teachers, and the reality of equal opportunity for all pupils. It also contains some comparative figures on educational spending collected from published data, and a summary European Report. All of this will be published before the end of 1995 by the EC.

A case could be made for seeing popular scientific knowledge as a vehicle for the communication of popular concerns on scientific matters. Whatever we mean by 'scientific culture' it is clear that communications of all sorts move far more surely and fluently within a culture, than across cultural boundaries. Indeed the comfortable comprehension of messages about popular themes, and the intersubjectivity that this indicates, may be taken as the surest test of cultural commonality. Such a popular culture would show the kind of 'habitus' (Bourdieu, 1977) we use in daily life as we engage with natural phenomena or technological artefacts, and also the life-world knowing (Schutz and Luckmann, 1973) which we all accept without question as 'just common sense' (Larochelle and Desautels, 1992). Such a taken-for-granted scientific culture would enable European citizens to exchange views and attitudes about such contentious issues as genetic manipulation and environmental protection. Lay people do not 'exchange views' about scientific knowledge: to do that would be to adopt the role of specialist which would be false to their self image. Most simply forget what they have learnt at school, or just dredge out remnants for trivial pursuits or popular quizzes. However people do need some familiarity with words such as 'DNA' or 'ecology', however conceptually thin or personal, for use in the expression of their views and concerns.

Thus scientific understanding needs, not so much to be conceptually correct, as to be comfortably familiar. Previous educational research on the Public Understanding of Science programme (Solomon, 1992a) demonstrated some of the features and practice of this kind of knowing by monitoring the informal exchange of views on science-based social issues in a large number of groups of 17-year-old students. The issues involved were controversial and emotional, as general release television programmes about science so often are, and the students reacted in a similar fashion during their discussions. Very explicitly they brought their own experiences and values to bear on the subjects; but the role of scientific knowledge in their talk was less obvious. Nevertheless there was evidence that it was the existence of a common base of scientific understanding which made the discussions possible.

A Basis for Scientific Culture

The fundamental purpose of this study is to explore education as the basis on which the population in each country builds up its scientific picture of the world. The contents of this picture will range widely from an understanding and feel for different materials, or enough knowledge of the human body to appreciate some of the new advances in medicine, to an educated sense of wonder about the stars and the universe. In this European Report, as in the reports from each individual country on which it is based, attempts were made to highlight those out-of-school activities (e.g., visits to hands-on or field study centres, the inclusion of technology or the history of science) which might have an important effect on this scientific image of the world received by the school students.

Culture is, as Clifford Geertz remarked in his influential book *The Interpretation of Cultures* (1977), a very complex matter. His definition, however broad the brushwork, manages to encompass popular, national and scientific culture, as well as different families' ways of living in the home (Solomon, 1994). It is all the more useful because it lays special emphasis on the way in which culture can best be explored and analysed.

Believing with Max Weber, that man is an animal suspended in webs of significance he himself has spun, I take culture to be those webs and the

analysis of it, therefore, to be not an experimental science but an interpretative one — in search of meaning. (Geertz, 1977, p. 5)

For our purposes this cultural web will include the meaning and value of
education itself as well as that of science. This is a reminder that cultural influences
will not operate in just one direction. Up till some fifty years ago a 'cultured'
person in Britain, and most of the countries of Europe, was one learned in the
classics rather than in science. This exerted a strong influence on what was taught,
and also on the 'hidden curriculum' which indicated what was most valued by
society. The government and diplomatic service, for example, was populated by
classics scholars, and only a tiny proportion of headteachers of schools had an education in science. To some extent this classical tradition still exists in Italy as we
shall see in the next section. In other countries where 'science and technology'
conjure images of high scholarship, and facility in mathematics, this too will affect
the school system. This will have important effects on the future EU citizen's
scientific culture.

The project did not ask directly for information about attitudes and national
culture. Instead it put together a picture of science education, looking for trends,
similarities and differences in what science was taught to children. It also enquired
about recent public concerns and controversies, if any, about the country's science
education. Out of this material impressions of the general drifts in education might
emerge which could show what the future citizens of Europe might build into their
scientific culture. This involved not only the substance of what they learnt, but also
how it was presented, both overtly and in a more hidden way.

Science in the National Curriculum

The first section in the National Reports begins with a survey of the number and
duration of science lessons in both the primary and secondary schools. Some generalizations can be drawn from this part of the national reports:

1 All EU countries now have a National Curriculum which stipulates the
science content and recommends the number of lessons in which it is to be taught,
in its secondary schools. In Germany, as in Switzerland, the curriculum is stipu-
lated region by region (Laender or Canton). In the UK and Belgium the situation
is similar, whereas in Spain the National Curriculum is only 40 per cent of what
is taught. The rest is determined by the autonomous regions.

2 Science is differently defined in the different member states of the EU. In
some countries the social sciences are all included under the heading 'science'. In
others, social studies, history and geography are also considered a part of science.
Technology or business studies may be presented as a part of physics (in Italy).
It seems likely that students' attitudes towards 'science' will be strongly though
subtly affected by its broader definition.

3   No country stipulates exactly how science should be taught although official recommendations are often made. In almost every country these include exhortations to have their students carry out practical work in the school laboratory. This is compulsory in the UK where it becomes a part of the assessment of students, and that is bound to affect the students' image of science and its significance. Often they are also urged to make the context of what they teach as relevant as possible to the student's everyday life. Clearly this affects what we might call the 'webbishness' of the cultural message received — the connections between everyday life-world knowledge and science knowledge.

4   Science has recently become a compulsory subject in primary schools in most member states of the EU, or else is just about to be so. This level of schooling, far more than secondary teaching, is likely to be reported back to the family and shared with them as 'what I did today in school'. The content of the primary curriculum is expected to be thoroughly accessible and may comfortably interact with home attitudes. In this way too science may enter the popular culture, possibly for two different generations, the parents and their children, at the same time.

An integrated topic approach fits in very easily into the common primary school practice of having one teacher deliver all the subjects to her or his class. In most of the European countries this is what happens in most of the primary schools of Europe. This is another way of making the connections which is so essential for the development of culture. (Only in the UK are there separate sections in the primary part of the compulsory National Curriculum which can be read as physics, chemistry and biology.)

5   In most EU countries (e.g., Sweden, Spain, Denmark, Portugal, Germany, Greece and Poland) the emphasis is now on 'the environment' in primary science teaching where previously it was on 'nature study'. A range of arguments and concerns surround this integrated delivery of primary science. It can be argued that the integrated environmental approach is more appealing to the sensitivities of young children and their concerns with animals and plants. It does therefore have clear implications for the formation of attitudes towards science. However this integration may sometimes mean that it is hard to tell exactly what science content is covered, or indeed, whether the subject might be presented exclusively as personal, moral or social education. There are worries in some countries (e.g., in Spain, Norway, Ireland and Sweden) that the physics and chemistry content, in particular, may have been 'integrated away'.

6   There is enormous diversity in the teaching of technology and in its meaning. Sometimes it is based within the science curriculum e.g., Sweden, Greece, Denmark and the Netherlands. In other countries, such as Poland, Spain, Portugal and the UK, it is a separate subject in the National Curriculum. In Switzerland, Ireland, France, Denmark and Germany technology does not figure at all in the compulsory school curriculum, and in many countries it is an option available only at higher levels (e.g., Germany). 'Technology' may imply any one of the following:

- a study of the main implements of production as, for example, in Italy, with or without a discussion of social effects and civic implications;
- information technology and the use of computers; and
- acquisition of workshop skills for the designing and making of technological artefacts (e.g., the Netherlands, Denmark and the UK at junior secondary level). In Sweden there has been a recent move away from technology as applied science, and it now has all three of the meanings above, but as a separate subject in its own right.

7   The use of stories and controversies from the history of science is mentioned in many national curricula but rarely at the mandatory level. (The exceptions to this are Denmark in physics at higher levels, and the UK at junior secondary level.)

In terms of the development of a scientific culture in Europe this last aspect of school science could be important and valuable for two reasons. In the first place it could make connections, once again, with other fields of knowledge. Secondly it enables the pupils to use their usually highly developed understanding of how real people are, to illustrate the scientific concepts that they learn. Scientific 'breakthrough' is never achieved without the sort of persistent endeavour and heart-warming delight that is easy for anyone to recognize. When science becomes more 'human' in this way it is not only easier to learn, it also descends from the heights of esoteric knowledge, to a level where it links with other people-based enterprises within the local culture.

The Organization of Science Education

Control of education is a recognized way of affecting the affiliation and national consciousness of the next generation. In countries such as Belgium, Germany, Spain and the UK, there are regions which have considerable control over their own curricula. Indeed the whole subsidiarity (diversity) debate becomes especially heated when control of education seems to be passing out of the hands of the dominant cultural community.

In many countries there is a movement underway to decentralize power either to the schools themselves or to a local board of parents and industrialists and other local influences. In Greece, Spain and Portugal where the school-leaving age has only recently been raised to 16, control is still exercised centrally, but reform is being discussed. Poland is emerging from a centralized political system and trying to organize a more local structure of control, and Ireland is moving in the same direction. Even French education, once a model of central control, is allowing schools more autonomy in the delivery of their curriculum and its timetable. Sweden and the Netherlands have removed some of the detailed state regulation and transferred it to local authorities and schools. However it is important to note that decentralization may result in local groups exerting just as much control over the management of schools as the central government once did.

The fairest way to assess the level of a country's funding of education, and the one which most clearly shows the level of a government's commitment to education (although not perhaps the way most obvious to its students), is to measure the expenditure on education as a percentage of GNP. However, the EU will be interested in the effects of such government provision on the experience of individual students, and indeed may see this as an area for action. These more realistic data depend not only on the absolute wealth (GNP) of a country but also on the size of its student population and the age-range of compulsory education.

Actual practice may depend on another power structure — that of pedagogic judgment. Sweden, for example has decided that methods of implementing educational policy should remain entirely in the hands of teachers. The most recent version of the UK science curriculum leaves 20 per cent of time free for teacher choice. If recommendations for new practice derive from the science teachers themselves, as they often do, the teachers will have ownership of both the idea and its practice. This gives them a powerful professional position. The European countries also differ in how they train teachers and how much innovation they expect from them. Ireland, the UK, the Netherlands, and all the Scandinavian countries have active Science Teachers' Associations affiliated to ICASE, which are often a source of new teaching ideas. Greece positively encourages teacher-led projects while other countries, such as Poland, are more likely to keep the reins of innovation in the hands of the university staff.

Going Beyond the School

Apart from organized visits to museums there was very little contact between school science and the outside world before the 1960s. But there has been rapid change. Prizes and competitions, like the international physics, chemistry and biology Olympiads, or industry-based competitions, are becoming more common and exert some influence on school science teaching. At a deeper level, the question of the 'relevance' of science (sometimes in opposition to its supposed 'validity') began to be argued out during the late 1970s and 1980s. In school education this debate originated in the Netherlands, and to a lesser extent in Britain, where the terms 'in society' began to be found in the title of new science syllabuses.

The National Reports record the use of out of school resources. In some cases there may be organized science expeditions which teachers fit in with their science teaching. Field centres for biological work are one of the best established of these resources and the most recent addition to the list is the Hands-on Science Centre. These differ from museums in being interactive so that students can perform quasi-experiments. These centres aim to amuse as well as to educate and occasional voices have been raised questioning whether any real learning can take place alongside such hilarity. Interactive centres are now part of a worldwide movement (EXCITE) and considerable international research (e.g., Haury and Rillero, 1994) shows that the experiences they provide for school students are memorable, and they are also beginning to show that science can be shared with the family. In this sense they provide rather direct access to that subset of national culture which is the home environment.

Quite different out-of-school influences may enter the classroom because the science course calls for the discussion of public issues which are science-based, such as preserving the environment or sanctioning biotechnological processes. The prevalence of this varies very much country to country and follows cultural forces. Where public debate on these issues is encouraged, as in Sweden, Germany, the Netherlands, Denmark and (more recently) Poland, discussion will figure in the school syllabus. Where scientific controversy occurs more rarely amongst the public, as in France, Ireland and Italy, it hardly figures at all in the schools. In this way it is national culture which drives the science curriculum. It should be added that if discussion of controversies is apparently encouraged, but not included in the national examinations, as in Sweden, the UK and the Netherlands, there will be little incentive to teach it. There is also considerable evidence that teachers may resist its introduction because of the extra time and new skills needed to run balanced classroom discussions.

Student Interests and Motivation

The topic is crucial to our estimation of scientific culture and yet it is a field of research which is rarely visited. Compulsory education in science achieves little if it leaves students with a lack of interest or even an active distaste for the subject.

Both age and gender affect the magnitude and range of interest in science. The designation of physics as a boy's subject, and biology as a girl's subject is widespread and has been the object of much discussion, especially in feminist circles. However it is not uniform across Europe. The divide scarcely exists at all in Portugal, Spain and most Eastern European countries. It is strongest in Britain, Sweden, Italy and the Netherlands, in the same way as it is in the USA and Australia. Clearly then it is a cultural artefact and one which the EU might do well to take steps to diminish. There is also some evidence to suggest that practical work, either in the school laboratory or in the field, increases student enjoyment of science, especially for the younger students and for boys (Pell, 1985). As we have seen in most European countries, primary science is set in an environmental context for this very reason. A recent study of pupils' interest in biology conducted in Germany and Poland indicates that it decreases with age for both genders as does enthusiasm for most school subjects.

Another factor which may be important is the prominence of mathematics in the science being learnt. Potential employers often use mathematics as a criterion for ability in commerce as well as in science. This practice serves to keep up a level of difficulty which is not helpful for their public image. It has also served to lure able students to specialize in commercial or business studies where the rewards are very much higher. The science considered to be the most difficult — physics — involves the most mathematics. It is also the science in which there is most gender differentiation, and falling recruitment at university level. Although cause and effect may be hard to tease out, in the context of the familiarity that popular culture demands, such an image of intellectual demand is not helpful.

The 'hidden curriculum' transmitted to students, just as surely as the definition of force or photosynthesis, is composed of the meanings (Geertz' significances) not found in any curriculum. In general they are about the worthiness of learning and other less academic matters like school chauvinism. Bourdieu and Passeron (1977) have found this aspect of school culture overly oppressive, and called it 'aggression' against the home cultures of working class students. Within any one subject, such as science, the hidden curriculum also transmits messages about the meaning and nature of the subject itself. Research has shown that how the teacher teaches, the questions asked, the answers given, the stories and even jokes told about scientists all combine to construct a kind of mini-culture, together with its mini-epistemology in the minds of the pupils. Painstaking classroom research by Brickhouse (1989) and Lederman and Zeider (1987) and others has shown how strongly this affects pupils' understanding of the nature of science. A recent questionnaire analysis of nearly 800 students at Key Stage 4 (age 15) showed that, even in these days of a formal centralized curriculum for science, the single most important influence on pupils' answers to questions about the nature of science and what real scientists do, is the science teacher (Solomon, 1995), even when no lessons at all are overtly dedicated to describing the 'nature of science'.

The second cultural aspect of school science is more deliberately taught; this is the scientific way of problem-solving. In science lessons it appears as full of rigorous objective definitions of terms, and overarching decontextualized principles, both of which are much at odds with the life-world way of thinking and all its richnesses of verbal meanings, and local explanations of phenomena (Pepper, 1942; Solomon, 1992b; Geertz, 1993). It takes considerable acculturation, what Schutz and Luckmann (1973) call a secondary process of socialization, to get students to think in this foreign way, and indeed many never achieve it at all (Solomon, 1984; Costa, 1995). To be able to operate in two different domains of thought, to cross over, as it were, between one culture and another, is hard, although quite essential if one is to be a professional scientist.

Interest and motivation based on science-as-a-career seems to be variable, even in this time of high unemployment. Science-based industry is reducing its technical labour requirements at the same time as raising their entry qualifications. Most young people do not even recognize the common careers which involve science at a lower level, such as hospital or food technician. On the other hand names like 'doctor', 'engineer' and 'technologist' summon up the image of a stable career with high prestige and rewards. Our data suggest that, in some countries, this produces a strong, but possibly unrealistic, motivation for learning science, accompanied by intensive coaching programmes in out-of-school hours, especially in the less affluent EU member states such as Greece and Ireland.

Joan Solomon

European Science Education Research for the Future

Will science educational research really change so as to have a genuine European dimension? Or will it be no more than what has gone on before, but with a comparative flavour?

This report has shown that there is a much deeper quandary to explore. Europe, it is said, contains more deep-seated cultural diversity than any other comparable block of countries, so the challenge to develop an education which is peculiarly European, and yet does not destroy local and national culture, is a hard nut to crack. It is not difficult to imagine the fury and indignation that would arise if a Stalinist pan-European science education were to be imposed from Brussels! Nevertheless it could be argued that the European Union will be no more than a squabbling set of competing economic units unless it acquires some measure of common culture. EU projects, like Socrates, are trying to knit together the learning experiences of university students; this is a sound way to begin fashioning this culture. School education reaches far more of our young people. The future of scientific culture in Europe, even if it continues to exhibit national variations, is built on school foundations. Exploring and strengthening these is an appropriate challenge to science education researchers.

References

BOURDIEU, P. (1977) *Outline of a Theory of Practice*, Cambridge, Cambridge University Press.

BOURDIEU, P. and PASSERON, J-C. (1977) *Reproduction in Education, Society and Culture*, London, Sage.

BRICKHOUSE, N. (1989) 'The teaching of the philosophy of science in secondary classrooms: Case studies of teachers' personal theories', *International Journal of Science Education*, **11**, 4, pp. 437–49.

COSTA, V. (1995) 'When science is another world: Relationships between worlds of family, friends, and science', *Science Education*, **79** (in press).

DUSCHL, R. (1988) 'Abandoning the scientistic legacy', *Science Education*, **72**, 1, pp. 51–62.

GEERTZ, C. (1977) *The Interpretation of Cultures*, New York, Basic Books.

GEERTZ, C. (1993) *Local Knowledge*, London, Fontana Press.

HAURY, D. and RILLERO, P. (1994) *Perspectives of Hands-on Science Teaching*, ERIC Clearinghouse for Science, Mathematics and Environmental Education, Columbus, Ohio.

LAROCHELLE, M. and DESAUTELS, J. (1991) ' "Of course, it's just obvious": Adolescents' ideas of scientific knowledge', *International Journal of Science Education*, **13**, 4, pp. 373–89.

LEDERMAN, N. and ZEIDER, D. (1987) 'Science teachers' conceptions of the nature of science: Do they really influence teaching behaviour?', *Science Education*, **71**, 5, pp. 721–34.

MERTON, R. (1973) *The Sociology of Science*, University of Chicago Press.

MILLER, J. (1983) 'Scientific literacy: A conceptual; and empirical review', *Daedalus*, Spring Issue, pp. 29–48.

PELL, A. (1985) 'Enjoyment and attainment in secondary school physics', *British Journal of Research Journal*, **11**, 2, pp. 123–32.

PEPPER, S. (1942) *World Hypotheses*, University of California Press, Los Angeles.

SCHUTZ, A. and LUCKMANN, T. (1973) *Structures of the Life-world*, New York, Heinemann.

SOLOMON, J. (1984) 'Prompts, cues and discrimination: The utilisation of two separate knowledge systems', *European Journal of Science Education*, **6**, pp. 277–84.

SOLOMON, J. (1992a) 'The classroom discussion of science-based social issues presented on television: Knowledge, attitudes and values', *International Journal of Science Education*, **14**, 4, pp. 431–44.

SOLOMON, J. (1992b) *Getting to Know about Energy*, Lewes, Falmer Press.

SOLOMON, J. (1994) 'Towards a notion of home culture', *British Education Research Journal*, **20**, 5, pp. 565–77.

SOLOMON, J. (1995) 'Large scale exploration of pupils' understanding of the nature of science', *Science Education* (in press).

31 First Experiences with European PhD Summer Schools on Research in Science Education

Piet Lijnse

Abstract

In 1993 and 1994, two trials have taken place with summer schools for PhD students working on research in science education. This contribution describes the format of these schools and how participants have experienced them. Then some differences in PhD research in science education in several European countries are discussed, and their significance for further European cooperation.

Introduction

1992 was meant to be an important year on the way to a more united Europe. For me, it functioned as a trigger to write to several groups and centres working in science education, to ask their opinion about summer schools for PhDs as a first step to wider European cooperation. As a result, in September 1992, a small number of people came together in Utrecht to discuss the matter. It was decided that we would organize an initial, small scale trial to take place in Utrecht in July 1993, and a second somewhat larger one in August 1994 in Thessaloniki, to be organized by Dimitris Psillos. Subsequently it was suggested, summer schools could be organized, once every two years, by the European Association for Research in Science Education, which would be established in the meantime.

As planned, in July 1993, about fifty people (thirty PhDs and twenty staff) came together for a week in a small conference centre in the Netherlands. In August 1994, about sixty-five people (forty-five PhDs and twenty staff) came together in Greece. For both meetings, an application for EEC-Erasmus funding was sent in by the participating institutions. However, only the second time was the bid successful. Since then, two experimental summer schools have taken place and we may now ask ourselves what we have learned, and whether the undertaking has been worthwhile and should be continued.

Format

The scope of the schools was not restricted to a particular subdomain, but to research in science education in its broadest sense. The schools were not a collection of lectures by lecturers, but, as agreed, *for* and *by* the PhD students. The aim was to provide an opportunity to learn from their peers who are facing similar problems. Most of the time was spent on allowing students to report and discuss their own research. Working groups were formed of about twelve students, each coordinated by two or three staff members. A safe working environment was thereby created, in which the students could report and discuss freely. They were asked, not so much to talk about their results, but rather about the questions they had and the problems they were facing. Being somewhere in the second year of PhD research was considered an optimal point of time for their participation, although this was not used as a strict criterion. Each student was given about 1.5 hours to report and discuss his or her work with the group. Apart from the main activity, some plenary contributions were given by staff members at both schools. In the first school any research groups represented were also given an opportunity to describe their programmes of work.

Evaluation So Far

In general it can be said that both schools have been very successful in the sense that the PhD students valued highly the opportunity to meet and discuss with their peers their research. However, the staff involved were also very positive about the schools. The working groups appeared to provide a good and supportive environment, which often prompted intensive discussion. English was the official language. However, the different mother tongues constituted a problem, though much effort was put into overcoming it. Nevertheless, the language barriers call for further consideration. Students said they had learned much from the experience. In several cases, lasting working contacts have been made, which is a very important result. Above all, many students valued the summer school highly because they felt that, to a certain extent, it broke down their feeling of isolation as research workers. Some expressed surprise that so many people were doing similar work. For many it appeared to be the first international occasion in which they had participated. All these aspects contributed considerably to the success of both weeks, and we may conclude that the schools largely fulfilled their main educational task.

However, plenary sessions and discussions appeared to present problems. It was often not clear what particular function the general lectures were supposed to serve for the PhD students. Plenary discussions mostly took place between staff and did not sufficiently involve the students. Also, the presentations of research programmes, which formed part of the first school, were not suited for a meeting of this kind.

Future organizers, therefore, will need to think more deeply about the role of plenary sessions. Such talks should probably be minimized, and replaced by workshops on special carefully chosen methodological or theoretical topics. This would reflect the general feeling that the active climate of being at work together was essential. Nonetheless, despite these defects, both PhDs and staff were very enthusiastic about this new kind of activity. For almost all participants, 'Summer schools are here to stay'!

What Else Did We Learn?

It became obvious that the concept of a PhD student is not very clear. One similarity between all PhD students is that all are working on some piece of original research. Another is that almost all have a background in one of the sciences, at least to the level of a UK first degree or beyond. However, almost all other aspects appear to differ across countries. Some research students are very experienced teachers, others have just followed a teacher training course, or not even that having just completed their science degree.

Related to these differences in teaching experience, of course, are large differences in age and in ways of being appointed. In France, for example, most 'students' are experienced teachers who are working on a PhD without payment, in their own free time. In the Netherlands, we could have such teachers studying for PhDs as well (but almost nobody does), but it was more usual that part-time or full-time university staff members are working on a PhD. Nowadays, this possibility is largely replaced by the category of young 'research assistants in training' who are specially appointed at a rather low salary in return for a rather intensive research training during their PhD work. This arrangement comes closer to full-time British PhD students who are working on special grants, sometimes secured from external sources.

There are also large differences in research training. In France, research students follow a part-time one year course of research (DEA). In the UK, many may have a science education-related MEd. In the Netherlands, some students may have pre-graduate research training in science education, while others will have no educational research background at all.

As well as differences in background at the start are the differences in required training during the course of their PhD work. As noted in the Netherlands, students have to undergo a rather extensive training, while in other countries no special training seems to be required.

To provide such training is, of course, at the heart of the purpose of the summer schools. Apart from very few centres, the numbers of people at a particular institution working toward a PhD are rather small. As a consequence, students often work in isolation. If they have to get research training, this is often not a problem as far as general training in research methodology and educational theory is concerned. However, it is not usually possible for the students to follow special 'science education research' courses, as the small number of students simply does not support the provision of such courses.

Another point that has not yet become clear is that there probably are considerable differences in the required quality, duration and depth of the research theses. What did emerge, however, was an important difference in the ways in which the research topics are chosen. In the UK, for example, PhD students largely seem to choose the topic of their research themselves, more or less according to their own interests, while, in the Netherlands it is the research group which decides and advertises for someone who is willing to do the job. Hence research programmes, which are essential in the Netherlands to be able to plan successive PhD studies, appear to play a much less prominent role in other countries.

As a consequence, the topics which formed the subject of PhD research seemed to be very diverse and often involved 'one-shot' studies. Therefore the problem of how to learn from and build upon the experience of others, came clearly to the fore. What constitutes a productive research programme and how do we make sure that not every topic has to be worked on again in its own national context? Is the teaching of energy in, for example, France and England and Greece, so different that it is has to be a matter of similar research in each country? Or is it that we do not communicate at a detailed enough level to profit from each other's studies? In my opinion, the current research literature and publications do not promote sufficient European communication. Maybe, there should be a new European Journal, or mechanisms to exchange PhD theses on a much broader scale. The last suggestion then raises the problem of which language these should be written in.

Apart from the choice of topics and the existence and content of research programmes, another factor is the institutional background within which science education research is being embedded. Two main positions appear to be present. In most of the UK, for example, science education research seems to be predominantly part of the work of Schools or Departments of Education. This means that the research that is done is, in my opinion, strongly influenced by more general educational theories and research problems. In other countries, like Germany and the Netherlands, science education research is, however, predominantly embedded in the science faculties. As a result, the research done is often much more content oriented and domain specific. Although there is a role to play for both approaches, I think that the need for more communication applies particularly to the 'content-specific branch', as the other framework already has its networks.

What Does All This Imply?

Although the above describes a number of differences in the situation of PhD students, we should not forget some major similarities. For example, in all countries, PhD students are present, their number seems to be increasing, and they express a willingness and a need to make contacts internationally, to undergo further training and to exchange experiences. Therefore, my conclusion is that summer schools should become a regular activity. The proposed European Association should secure their organization and provide a means of getting the necessary support from the European Union.

The summer schools have also been useful in bringing staff together and making them think about further cooperation. In this respect, the format of the summer schools, though certainly not yet ideal, has been very stimulating as it makes people actively do research together for a week, instead of only hearing people talk about their results. Not surprisingly, therefore, the wish has been expressed that the organization of similar working meetings for experienced researchers would also be a very welcome activity, the more so as productive research programmes still deserve closer attention. It should also be a main task for the Association to organize such working meetings, so that a European community of science education researchers will come alive.

References

LIJNSE, P.L. (Ed) (1994) *European Research in Science Education*, Proceedings of the first PhD Summer school, Utrecht, CD-ß Press, p. 359.
PSILLOS, D. (Ed) (1995) Proceedings of the second PhD Summer school (in preparation).

32   Science Education Research in Europe: Some Reflections for the Future Association

Svein Sjøberg

Abstract

This paper explores the variation in understanding across Europe of the basic terms, 'science', 'science education', 'didactics' and 'pedagogy' to demonstrate a division between understandings common in continental Europe and the Anglo-American tradition. However, it argues that such diversity is not a point of weakness, but a point of strength for a future European Association for Research in Science education which will enable the development of breadth and open-minds.

At a time when researchers in science education in Europe are establishing their own association for research in science education, it may be fruitful to raise some critical questions about what we share and what might be our possible differences. Also, when many European countries are moving towards a union, it is easy to consider that uniformity and similarity are our aims. My own underlying views run contrary to this perspective. Rather, my main point will be that our strength lies in a large diversity and variation in, geographically speaking, a very small area of the world — namely Europe. I will argue the great cultural diversity combined with small geographical distances is the strength that should be utilized, not something that we should try to smooth out and remove. In fact, we should learn to treat both similarities and differences as a strength to be utilized.

'Education' and 'Science': Sources for Misunderstanding?

The contrast between European cultural traditions is well manifested through the concepts we use to describe our own field of research — indeed a point which was brought out with strong emotions many times at the Leeds conference. Whereas the Anglo-American tradition uses the concept 'science education' to describe our common field of research, and this communicates well, other European languages use other words. Consequently, any word-by-word translation of 'science education' is indeed very misleading.

Let us start with the English concept of 'education'. If 'education' is the most general and embracing term for all kinds of studies relating to schooling and up-bringing, the best translation would in many European languages be versions of

'*Pedagogik*', as in the German and Nordic languages. The similar sounding English term, 'pedagogy', is rather different and narrow, which translates more as 'methods of teaching'. And, while the most embracing term for professional academics working in the field of education in many European countries are variations of '*pedagoger*', this translates very badly into English; e.g., the American Heritage dictionary: 'Pedagogue: One who teaches in a pedantic and or dogmatic manner.'

The issues that science educators address, would, in most European languages fall under variations of the concept '*Didaktik*'. This concept is found in languages like German, French, Italian, Spanish, Portuguese, Dutch, Swedish, Danish, Norwegian etc. A translation of '*Didaktik*' to English 'didactics' is very misleading, since 'didactics' in English seems to be used as a rather pejorative term to describe a particular method of teaching, more or less telling, preaching and moralizing. The meaning of '*Didaktik*' (and its variations in different European languages) is very different, although this term is used somewhat differently in different European countries. The term '*Didaktik*' has a long history in Continental European thinking. Many see John Amos Comenius' major work *Didactica Major* (1628, p. 32) as a foundation for the *didaktik* tradition. It was not until 1896 that this work was translated into English, and then with the rather unfortunate title *The Great Didactic*. Although there are large variations between the different continental *Didaktik*-traditions, the basic questions of educational concern are 'Why?' and 'What?' questions. These key questions are therefore connected with purposes, legitimization and selection of worthwhile knowledge. The 'How?' question is also present to varying degrees, and when included, is referred to as the 'wide definition' of *Didaktik*. It is therefore important to note that questions of teaching methods are *not* the key issues of *Didaktik*. Much Anglo-American research seems to centre around teaching and learning, often connected with the learning of particular concepts, and research questions often stem from the fact that educational psychology (often in a behaviourist tradition) is seen as more or less synonymous to educational theory. Questions of what is the best, or the most efficient way of teaching particular content are rarely found in the *Didaktik* tradition. I think it is fair to say that the European Continental *Didaktik* tradition is more philosophically rooted around questions of what constitutes an 'educated' person and what knowledge may be seen as worthwhile. In recent years, educational researchers from both Anglo-American and '*didaktik* tradition' have started to meet to explore similarities and differences. A workshop was hosted by the Institute für die Pädagogik der Naturwissenschaften in Kiel in Germany in 1993, and in August 1995 a follow-up was hosted at the University of Oslo. Some of deliberations are published in Hopman and Riquarts (1993). Major actors from both traditions have taken part in these discussions, which are not directly linked with science education. It is also interesting to note that many of the key writings in, for instance, German *Didaktik* (like Weniger, 1926, 1965; Blankertz, 1970; and Klafki, 1983) are not available in an English translation. Continental theory and empirical research are rarely referred to in the dominant English-language journals in science education. The opposite does not seem to be the case, as most continental science educators seem to be well versed in the English literature.

This leads to another key concept in the European Continental tradition, the concept of *Bildung* (in German and Swedish), translated to Formation in French and Latin languages and '*dannelse*' in Danish and Norwegian. The underlying considerations address basic philosophical issues relating to the meaning of life, what constitutes an educated (*Gebildet*) person etc. Concepts like personal growth, autonomy and independence are major concerns. In many countries, the debate is also politically grounded and connected with visions about prerequisites for a democratic society. Pragmatists like James and Dewey raised similar issues in the American debate, but they offer two rather different lines of thought.

My concern is here to point out that the two key concepts in the Continental European tradition, *Didaktik* and *Bildung*, do not translate easily into English. I think that this is not only a matter of differences in terminology, but rather of deep-rooted differences in how problems are conceived, conceptualized and approached in the different cultures. It is also interesting to note that several of the European research institutes and university departments have the word *Didaktik* in their titles, and most of the researchers present at the Leeds conference probably have some variation of '*Didaktik*' in their academic title and also award university degrees involving the term '*Didaktik*'. This is certainly the case for my own institution and its academic degrees. It should therefore be no surprise that many people from non-English speaking countries (i.e., the whole of Europe except the UK!) feel strongly about this issue. Therefore, my contention is that we have rather different European traditions in our thinking about fundamental issues in education, and that these could be an asset for us to explore in the context of *science* education.

However, confusion may also stem from the other word in science education, i.e., that of 'science'. English language dictionaries or encyclopaedias provide definitions like: 'Science may be broadly defined as the development and system-atization of positive knowledge about the physical universe' (Grolier CD-ROM encyclopaedia) and 'Branch of systematic study especially of the physical world' (Encarta CD-ROM encyclopaedia). Therefore, the word science, without any quali-fier, in English implies the *natural* sciences. Indeed, there are indications that many children in English-speaking countries, mainly think of the *physical* sciences (and *not* the life sciences), when 'science' stands alone (Baker and Leary, 1995).

The use of such concepts in other European languages is very different. When one in German uses the term '*Wissenschaft*' (in the Nordic languages and in Dutch etc.: '*Vitenskap*') or the French '*la science*', the meaning is more embracing, including all forms of systematic study, and without an implicit priority given to the study of the physical world. In most languages one has to add a qualifier, like *Naturwissenschaft* in German. But although this seems very simple, it may add new confusion. In my own language, Norwegian, the common name for all 'natural sciences' is *naturvitenskap* (very similar to German). But the corresponding school subject is called *Naturfag*. Although the school subject is supposed to cover the same subject matter, there is evidence that for pupils and many teachers, *Naturfag* means 'The study of Nature', often concretized to the study of flowers, plants and animals, i.e., descriptive biology. The result may be that while 'science' in some countries may be interpreted as mainly physics and chemistry, in other countries it is interpreted first as *all* forms of systematic enquiry, and even when a qualifier is added, it may become nearly the opposite of what is implied by the 'original' English term.

It is evident that these differences in the meaning of key words in 'our field' may have profound consequences. An obvious example would be any comparison of 'attitudes to science'. Depending on the choice of translation, the outcomes may reflect more the differences between languages than anything else. Even more striking are the differences found in the use of words like 'technology' — but that lies outside the scope of this article. For a brief discussion, see Sjøberg (1995). Great care should therefore be taken in all comparative research in science education, even within the rather limited context of Europe.

Both our key words, 'science' and 'education' are therefore problematic. More confusion is added when two are put together as 'science education'. In some languages, this may be understood as in French: *'La science de l'Education'*. This means 'the science of education', i.e., meaning the social science that has education as its object of study. These terms then refer more or less to the whole field of educational research, a field that is often removed from the content of teaching — in our case: science! If, therefore, our association has the key words 'science education' in its title, the immediate meaning conveyed may be very different for each country.

At first sight, one may think that the issues I have raised here may be clarified with care in the choice of proper translations. This may of course help, but I do think that the issues are more deep-rooted and that some of them may stem from different cultural and philosophical traditions. At a very superficial level, one may use labels like an 'Anglo-American tradition' which is empirical and pragmatic versus a continental European tradition that is more analytical, philosophical and rational.

A possible example from science education is this: Much of the debate about the use of the theories of Jean Piaget may stem from the fact that Piaget's own writings have been interpreted within an Anglo-American context that is very different from his own. The philosophical and epistemological aspects of his theories have often been neglected when the focus has become stage-wise classification and how best to 'accelerate' children through these.

Equal Science Contents: Reality or Illusion?

Many surveys indicate great similarities in the content of science in school curricula across cultures, not least in Europe. Such evidence can be found in previous IEA (International Association for the Evaluation of Educational Achievement) studies like the SISS-study (Second International Science Study) and the comparisons of science curricula in twenty-three countries published in Rosier and Keeves (1991). Also, the ongoing TIMSS (Third International Mathematics and Science study) study indicates many similarities. The science education group in UNESCO has also made a curriculum grid based on the TIMSS classifications and gathered data from a host of countries. In such contexts, one often notices the great similarity of science curricula across cultures. However, there are reasons to question such a conclusion. What most curriculum analysis does, is to take apart the curriculum into smaller units, basically units derived from the separate sciences as academic disciplines. In doing so, the analysis removes the curriculum from its context. Some science curricula, especially at the lower levels, may be organized around topics or issues. In Norway, for instance, science is taught as an integrated subject for all pupils up to the age of 16–17. In other countries, the organization is like that of the corresponding academic discipline even at an early age, or there are different ways to organize science for different groups of pupils, often with the implicit message that the 'real' way to organize high status science is through its academic disciplines. When curricula are taken apart in a curriculum analysis grid, such major differences are often blurred. The result may be that two curricula, that in reality are very different, may look very similar. The very method of research may create both similarities or differences that may be mere artefacts of the analysis. A more qualitatively oriented study, where the researchers actually bring the relevant material to share and discuss will probably be more illuminating than the paper-and-pencil analysis.

Conclusion

In my opinion our future strength as a European association for researchers in science education is not the uniformity of Europe, but rather that we have a 'laboratory' with great cultural diversity within a small area. It is the variation, and not the uniformity, that should be seen as our strength. Indeed, we may also question the attempts to 'harmonize' school systems and curricula in Europe. When diversity is seen as an advantage in Nature, why should it not also be considered a virtue in culture and education?

In the present move towards a European Union by many (but not all!) European countries, it may also be appropriate to warn against Eurocentrism and isolation. The narrow nationalism that has previously dominated many European countries, for instance as manifested in history textbooks, should not be replaced by a new form of extended nationalism — Europeism. Our future cooperation should be coupled with an open mind towards other cultures — an issue that should provide a challenge also in science education.

Additionally, part of our comparative strength is that, in spite of the variations, there is a common cultural heritage that may enhance our communication. Coupled with short geographical distances, this could indicate more personal and face-to-face cooperation between science education researchers in Europe. In any case, our association will be most welcome. For, in the future, we will not need to go to the NARST and similar conferences in the US in order to meet our colleagues next door!

Svein Sjøberg

References

BAKER, D. and LEARY, R. (1995) 'Letting girls speak out about Science', *Journal of Research in Science Teaching*, **32**, pp. 3–27.

BLANKERTZ, H. (1970) *Theorien und Modelle der Didaktik*, München, Juventa Verlag.

HOPMAN, S. and RIQUARTS, K. (1993) *Didaktik and/or Curriculum*, Institut für die Pädagogik der Naturwissenschaften an der Universität Kiel, Kiel.

KLAFKI, W. (1983) *Studien zur Bildungstheorie und Didaktik*, Weinheim, Julius Belz Verlag.

ROSIER, M.J. and KEEVES J.P. (Eds) (1991) *The IEA Study of Science I: Science Education and Curricula in Twenty-three Countries*, Oxford, Pergamon Press.

SJØBERG, S. (1995) 'Technology education, diversity or chaos?', *Studies in Science Education*, **25**, p. 289.

WENIGER, E. (1926, 1965) *Didaktik als Bildungslehre*, Weinheim, Julius Beltz Verlag.

List of Contributors

Prof René Amigues, IUFM d'Aix-Marseille, 2 Avenue J. Isaac, 13621, Aix-en-Provence, cedex, France.

Dr Annemarie Møller Andersen, Royal Danish School of Educational Studies, Dept of Biology, Geog and Home Economics, Emdrupvej 101, DK-2400, Copenhagen NV, Denmark.

Dr Björn Andersson, Dept of Educational Research, University of Göteborg, Box 1010, S-43126, MOLNDAL Sweden.

Dr Mike S Arnold, 171 Sapgate Lane, Thornton, Bradford, BD13 3DY, UK.

Mrs Hilary Asoko, CSSME, School of Education, The University of Leeds, Leeds, LS2 9JT, UK.

Mr Frank Bach, Dept of Educational Research, University of Göteborg, Box 1010, S-43126, MOLNDAL, Sweden.

Mr Montserrat Benlloch, Faculty of Psychology, Autonoma Univ. of Madrid, 28049 Madrid, Spain.

Prof Alfredo Bezzi, Dipartimento di Scienze della Terra, Università di Genova, Viale Benedetto XV, 5, 16132 Genova, Italy.

Mr Richard Boohan, Institute of Education, University of London, 20 Bedford Way, London, WC1H 0AL, UK.

Dr Carolyn Jane Boulter, Luicers, 47 High Street, Theale, Reading, UK.

Dr Bob Campbell, Dept of Educational Studies, University of York, Heslington, York, YO1 5DD, UK.

Mr Justin Dillon, CES King's College London, Waterloo Road, London, SE1 8TX.

Dr Wolff-Gerhard Dudeck, Institute for Physics Education, University of Bremen, Box 330440, 28334 Bremen, Germany.

Prof Reinders Duit, IPN-Institute for Science Education, University of Kiel, Olshausenstr 62, 24098 KIEL 1, Germany.

Dr Richard A Duschl, DIL, University of Pittsburgh, 4C12 Forber Quadrangle, Pittsburgh, PA 15260, USA.

Dr Sibel Erduran, Peabody College, Vanderbilt University, Nashville, USA.

Prof John Gilbert, Education and Community Studies, University of Reading, Bulmershe Court, Earley Reading RG6 1HY, UK.

Dr Shawn Glynn, University of Georgia, Athens, USA.

Dr Andrea Gnoyke, Department of Chemistry, Institute of Chemical Education, Essen University, Essen, Germany.

Prof George Ionnidis, The School of Humanities and Social Sciences, Department of Education, The University of Patras, 261 10 Patras, Greece.

Dr Kim Issroff, School of Education, The Open University, Milton Keynes, MK7 6AA, UK.

Dr Barbara Hodgson, School of Education, The Open University, Milton Keynes, MK7 6AA, UK.

Dr Hans-Dieter Körner, Department of Chemistry, Institute of Chemical Education, Essen University, Essen, Germany.

Dr Koos J Kortland, Centre for Science and Mathematics Education, Ornstein Lab, P O Box 80.008, 3508 TA Utrecht, Netherlands.

Mr John Leach, CSSME, School of Education, The University of Leeds, Leeds, LS2 9JT, UK.

Dr Michael Lichtfeldt, Freie Universität Berlin, Zentralinstitut für Fachdidaktiken, Habelschwerdter Allee 45, D-14195 Berlin, Germany.

Prof Piet Lijnse, Centre for Sci and Maths Education, University of Utrecht, P O Box 80.008, 3508 TA UTRECHT, Netherlands.

Dr Roger Lock, School of Education, University of Birmingham, Birmingham, B15 2TT, UK.

Mr Fred Lubben, Dept of Educational Studies, University of York, Heslington, York, YO1 5DD, UK.

Mr Azam Mashhadi, Dept of Education Studies, University of Oxford, 15 Norham Gardens, Oxford, OX2 6PY, UK.

Prof Gottfried Merzyn, Universität Göttingen, Waldweg 26, D-37073 Göttingen, Germany.

Dr Robin Millar, Dept of Educational Studies, University of York, Heslington, York, YO1 5DD, UK.

Mrs Patricia Murphy, School of Education, The Open University, Milton Keynes, MK7 6AA, UK.

Mr Mick Nott, Centre for Science Education, Sheffield Hallam University, Sheffield, S10 2BP, UK.

Dr Jonathan Obsorne, Centre for Educational Studies, King's College London, Cornwall House, Waterloo Road, London SE1 8TX, UK.

Dr Juan Ignacio Pozo, Faculty of Psychology, Autonoma University of Madrid, 28049 Madrid, Spain.

Dr Judith M Ramsden, Dept of Educational Studies, University of York, Heslington, York, YO1 5DD, UK.

Mrs Mary Ratcliffe, School of Education, University of Southampton, Highfield, Southampton, SO9 5NH, UK.

Dr Eileen Scanlon, IET, The Open University, Milton Keynes, MK7 6AA, UK.

Prof Hannelore Schwedes, Institute for Physics Education, University of Bremen, Box 330440, 28334 Bremen, Germany.

Mr Phil Scott, CSSME, The University of Leeds, Leeds, LS2 9JT, UK.

Ms Gaynor Sharp, The Open University, 91 Palmerston Road, Earlsdon, Coventry, CV5 6FH, UK.

Mr Steve Sizmur, NFER, The Mere, Upton Park, Slough SL1 2DQ, UK.

Prof Svein Sjøberg, Science Education SLS, University of Oslo, PB 1099 Blindern, 0316 Oslo, Norway.

Dr Ros Smith, Homerton College, Cambridge, UK.

Dr Helene Sørensen, Royal Danish School of Educational Studies, Dept of Biology, Geog and Home Economics, Emdrupvej 101, DK-2400, Copenhagen NV, Denmark.

Dr Joan Solomon, 27 Little London Green, Oakley, Aylesbury, Bucks, HP18 9QL, UK.

Dr Vassiliki Spiliotopoulou, The School of Humanities and Social Sciences, Department of Education, The University of Patras, 261 10 Patras, Greece.

Dr Else-Marie Staberg, Dept of Education, Umea University, 90187 Umeå, Sweden.

Prof Elke Sumfleth, Chemistry Education Dept, University of Essen, Schützenbahn 70, D-45127 Essen, Germany.

Dr Clive R Sutton, Science Education Group, School of Education, University of Leicester, 21 University Road, Leicester LE1 7RF, UK.

Dr Andrée Tiberghien, CNRS-IRPEACS Equipe COAST, ENS, 46 Allée d'Italie, 69364 Lyon, Cedex 07, France.

Dr Rod Watson, Centre for Educational Studies, King's College London, Cornwall House, Waterloo Road, London SE1 8TX, UK.

Mr Geoff Welford, School of Education, The University of Leeds, Leeds, LS2 9JT, UK.

Dr Jerry Wellington, Division of Education, Sheffield University, Sheffield, S1 4ET, UK.

Ms Liz Whitelegg, Centre for Science Education, Open University, Milton Keynes, MK7 6AA, UK.

Index

ability 172, 240, 361, 363, 365–8
Abrahams, M.R. 249, 251
abstract 22, 25, 87–8, 95, 97, 168, 173, 218, 226, 271, 305, 375
academic history of teachers 284, 290–1
Acevedo Diaz, J.A. 312
action research 302–4
activities, out-of-school 385, 389–90
activities:
    air 11–13
    biotechnology 231, 234
    CASE interventions 305–7
    change/difference 87–9, 95, 98
    concept development 37, 40–1, 45
    decision making 128, 138–9
    electricity 54–9
    GFK 159
    heat/temperature 25–31
    investigation tasks 192–8
    learning 155
    methods of science 270–2, 274, 278–9
    organization 184
    story-telling 149–50
    teaching/learning 115–21
Adey, P. 139, 305
age:
    analogy use at secondary 22–3, 27, 33
    combustion at secondary 243, 245–52
    compulsory science at primary 387, 390
    computer based problem solving at secondary 64–5, 67
    concept change in lower secondary 200–2, 208, 210
    concept of energy at secondary 100–13
    concept mapping at primary 74–84
    concept of matter at secondary 10
    controversial issues at secondary 229–30, 241
    decision making at secondary 128
    developing concepts at primary 36–47
    electricity at secondary 59
    gender and primary science 297–9, 301–4
    gender and secondary science 305–6

models in primary 177–87
particle matter 212, 216
personal meaning making and secondary science 325–36
of PhD students 396
physics at junior secondary 115–23
practical investigation and lower secondary 191, 194, 198
quantum physics and secondary 255, 258
student interest 390–1
teaching sequences on gases for secondary 7, 19
term properties and secondary 372–3
textbook adaptation to 361, 363, 365–8
thermodynamics at secondary 85–6, 95, 98
understanding nature of science at secondary 272, 275, 277–8
Agostinelli, S. 65, 67
Aikenhead, G.S. 127, 128, 312, 322
air 9, 11–17, 89, 200–10, 244–7, 249, 326–7, 331, 333–5
Alexander, P. 184
Althaus, H.P. 363
Alting, A. 303
Alvarez, M. *et al.* 312
Amigues, René 64–73
analogy 22–33, 36–9, 41–2, 46–7, 50–63, 122, 166–74, 179, 184–5, 254–5, 259, 343
Andersen, Annemarie Moller 303
Anderson, N.H. 371
Andersson, Björn R. 7–20, 243, 244, 245, 249
Anglo-American tradition of science education 399–400, 402
Anneken, G. 371, 372
argumentation 137–8, 182–4, 186–7
Arnold, Michael S. 22–33
Arsac, G. *et al.* 101
Arthur, R. 313
Artigues, M. 104
Asoko, Hilary M. 36–47, 215, 327

Assessment of Performance Unit Science
    Project 306
assisted performance, teaching as 325,
    328–9, 331–2, 335–6
Atkins, P.W. 94
Atkinson, P. 291
atmosphere 8, 12, 144
atom 216–17, 219–24, 256–8, 373
attainment targets 115, 192, 239–40
attitudes:
    European scientific 385–7
    gendered science 297–300, 305
    geoscience students 313
    learning 65
    science teachers 36, 289
    teacher and classroom behaviour 337
    to biotechnology 229–41
    to educational research 347, 352
    to science and language difference 402
attributes/structure of analogical
    representation 167–8
audio-taping, use in research of 27, 40, 68,
    74, 77, 126, 130, 205, 274, 327
Aufschnaiter, St.V. 54
Ausubel, D.P. *et al.* 75

Bach, Frank 7–20
Bachelard, S. 103
Baker, D. 305, 401
Ballstaedt, S.P. 376
Bannister, D. 313
Bar, V. 244
Barlex, D. 86
Barnes, D. 77, 362
Baschek, I.-L. 371, 372, 373
behaviour 224, 299–303, 307–8
Bell, B. 178
Benlloch, Montserrat 200–10
Berkheimer, G.D. 10, 11
Berkovitz, B. 10
Bethge, Th. 217, 255, 256
Beyer, K. 308
Beyth-Marom, R. *et al.* 128, 139
Bezzi, Alfredo 312–23
Bildung concept 401
biology 193, 361–5, 387, 390
biotechnology 229–41
Bishop, A. 286
'black box' view of science knowledge
    286–7, 291
Black, D. 25, 51
Black, M. 145
Blakeslee, T.D. 10, 11
Blankertz, H. 400

Bleichroth, W. 361
Bliss, J. *et al.* 341
blood 144
Bodmer, W. 126
Bohr model 255–6, 259
Boo, H.K. 251
Boohan, Richard 85–98
Borchardt, M. 372
Borg, I. 371
Bormann, M. 256
Born, M. 370
Borsese, A. 362
Boujaoude, S.B. 243
Boulter, Carolyn Jane 177–87
Bourdieu, P. 384, 391
Bragg, Sir William 254
Brämer, R. 362
Bransford, J.D. 154
Bredenkamp, J. 371, 372, 373
Brickhouse, N. 271, 283, 391
Bridges, D. 238
Briggs, H. 9, 216
Brinkman, F. 338
Britton, B. 167, 171
Brook, A. 9, 23, 86, 216, 245
Brousseau, G. 102
Brown, A. 153
Brown, D. 9
Brown, J.S. 171
Bruer, J.T. 153, 155, 156
Bruner, J. 328, 329, 361
Buck, P. 217, 340, 373
Buckley, B. 179
Bullock, D. 327
Bullough, R. 337
Bulman, L. 361
Bunge, M. 166

Caillot, M. 65
Calvin, W.H. 215
Campbell, Bob 351–7
Campione, J. 153
canonical/noncanonical circuit diagrams
    64–72
Carey, S. 19, 201, 215, 273
Carmichael, P. *et al.* 74
Carrascosa, J. *et al.* 312
Carré, C. 36, 87
Carretero, M. 201
Carroll, J.S. 116
Cassels, J.R.T. 362
Cassens, H. 255, 256
Chalmers, A.F. 326
Chambers, D.W. 312

change/difference 85–98
charge 53, 257
chemical change 93–4, 244–8, 250–2, 327
chemistry 193, 212–13, 217, 251–2, 303, 308, 371–3, 387
Chevallard, Y. 101, 102
Chi, M.T.H. 201, 209, 252
Children's Learning in Science (CLIS) project 10–11, 23, 215
circuits, electrical 36–47, 50–7, 166, 169, 174
  diagrams 64–73
citizenship 126, 229, 269
Clark, C.M. 154, 337
Clemens, H. 362
Clement, J. 9, 25
Clift, P. *et al.* 177
Closset, J.L. 65
cluster analysis 208, 312, 315–16, 320–1, 367
Cockburn, C. 308
cognition:
  adaptive 170
  conflict 10, 51
  development 64, 139
  models 255
  science learning 153
  situated 171
  structure 212, 215
  teacher 338
  theory 200–1, 210
Cognitive Acceleration in Science Education (CASE) project 297, 305–7
cognitive psychology 148, 156–9, 171
cognitive sciences 154–63
Cohen, D.K. 153
Cole, M. 328
collaboration 64–73, 75–84, 181–2, 184
  gendered 297–301, 303–4, 307–8
Collaborative Learning and Primary Science (CLAPS) project 297–302, 304
Collet, G. 111
Collins, A. 171
Collins, H. 290
combustion 8–9, 93–4, 243–52
Comenius, John Amos 400
commentary, organizing 333–4
common-sense 42, 86, 146, 362, 385
commonality, scientific 383–4
communications across Europe 384–5, 397

community:
  modelling 169–70, 179
  research 146–8, 352–3, 357
  scientific 36–7, 52, 101, 112, 144, 149, 271–2, 274, 326
computer based problem solving 64–73
concept:
  change 50–63, 157, 178, 215
  change: ideas to theories 200–10
  combustion 243, 252
  cosmology as 337, 340, 349
  development 22, 24–30, 36–47, 117–21, 192
  education/science 399–402
  electricity 65
  exploring 353, 355
  gases 9–10, 19
  gendered 301
  mapping 74–84, 212
  memorized 194, 196
  mental models and models of 169–72, 174
  movement between planes 325–8, 335
  of PhD student 396
  picture and 87, 94, 98
  quantum physics 254–62
  teacher scientific 284
  in textbooks 361, 368
  use of term 166
  visual 370, 374–6
concreteness as term property 370–4
confidence 36, 305–6
conjuring experiments 288–9
Connelley, F.M. *et al.* 161
conservation:
  energy 106
  mass 256
  matter 9, 122, 205–8
  matter/mass 244–5
constructivism 9, 14, 74, 116–17, 143, 148–51, 170, 178, 183, 215, 338
content 8, 76, 177, 191–2, 239, 270–2, 278–9, 303, 308, 402–3
context 101, 106, 229, 235, 238, 271–2, 274, 276–7, 279, 285, 289
  and texts 177–87
continuity/intensity of current 39, 44–6, 53–61
controversial issues 229–41
conversation, science lessons and 148–51
Cosgrove, M.M. 169, 244
cosmologies 337–49
Costa, V. 391
Coulthard, R.M. 77

coupling of change 90, 92, 94
Cowie, H. 137
Craig, J. 303
critical incident analysis 186–90
Cross, R.T. 230
Crystal, D. 183
cultural development theory 326
culture:
 influence on science 149
 influence on theory 178–9, 186–7
 scientific citizen 270, 274, 278–9
 scientific/national 383–92, 399, 401,
 403
curriculum:
 controversial issues 239–41
 cross contexts 238
 European science 402–3
 science 192
 science epistemology 269–72
 Science-Technology-Society (STS)
 126–7, 130
 society and 297
Curtis, F.D. 362

Daintith, J. 24
Dalton, P. 313
Darwin, Charles 145, 146
Dawson, C.J. 273
de Leeuw, N. 201
decentralization 388–9
decision making 115–23, 126–39, 154–5,
 286
Delamont, S. 291
democracy 229, 269–70, 278–9, 401
density of technical terms in texts 363–4,
 366–7
Desautels, J. 385
description 65, 117–18, 120–1, 245–9,
 276–7, 375
Dewey, John 401
diagnostic testing 11, 14–18, 194, 198,
 274–6, 328
Dickens, P. 229
didactic/dialogic exchange 183–4
didactical transposition 102, 112
didaktik concept 400
DiGisi, L.L. 361
Dillon, Justin S. 243–52
disappearance as concept 244–5
discourse 226, 297, 301, 301–2, 307, 325,
 327–9, 331, 334–5
discussion:
 concept development 51
 gendered 301, 304, 309

group 40, 47, 74, 76–80, 83–5, 87,
 95–8, 117, 119, 121, 131, 134–9,
 159, 182, 186, 231, 238, 327
 language differences in 362
 teacher 288
diSessa, A. 201
displacement as concept 244–5
disputes, scientific 271–2, 274, 277–8,
 390
dissolving as concept 88, 94
Dixon, B. 230
Dobson, K. 126
Doyle, W. 157
Dräger, P. 361
'Draw-a Scientist Tests' 312
Driver, Rosalind 9, 11, 19, 22, 37, 86, 88,
 103, 149, 177, 215, 216, 243, 245,
 270, 272, 273, 275, 277, 326, 327
Dudeck, Wolff-Gerhard 50–62, 174
Duguid, P. 171
Duit, Reinders 23, 25, 86, 105, 166–74,
 179, 201, 216
Duncan, J. 178
Duncker, K. 25
Dunnett, G. 313
Dupin, J.J. 39, 51, 52
Durant, J.R. 229, 269
Duschl, Richard A. 153–63, 272

earth 81, 146
Ebbutt, D. 352
Edelman, G.M. 215
education 385–6, 388–9, 399–400, 402
Edwards, D. 177, 329
Eger, M. 74
Eichinger, D. 11
Eichinger, J. 322
Eijkelhof, H.M.C. 116
Einstein, Albert 254, 255
electricity 25, 50–63, 64–73, 145
 see also circuits, electrical
electrons 53, 255–61
elicitation methods 312, 314–15, 318,
 321–2
ELIN project (Pupils Learning in Nature/
 technique) 302–4
Elliot, J. 352
Ellse, M. 86
Emberton, J. 327
emphasis of ideas 330, 334
enculturation 36–7
energy 24, 38–9, 41–6, 51, 85–98, 100,
 104–8, 110–12, 122, 256–7, 373
Engel Clough, E. 88

enquiry:
    nature of 275–8
    scientific 191–8
entropy 85–7, 92, 94–5
environment/nature study 387, 390
epistemology in curriculum 269–79
equality of opportunity 303, 353, 355, 384
equilibrium 22–33, 89, 95, 373
Erduran, Sibel 153–63
Erickson, G.L. 23, 353
ethics 231, 238–41, 286–7, 289
Europe:
    PhD summer schools in 394–8
    science education research in 399–403
European Association for Research in
    Science Education 394, 397–403
European Union 383, 386, 389–90, 392,
    397
evaluation:
    controversial issues 238
    data 161, 277–8
    decision making 137
    scientific 289
    student investigation data 191, 194,
        196–8
Evans, J.D. 362
Evans, R. 19
evolution 145–6
exogalactic model of universe 344, 346
experience:
    everyday 19, 24, 38, 46, 50–1, 86, 104,
        107, 117–21, 174, 184, 217, 221,
        325, 327, 329–31, 333–5, 370, 372,
        375, 387
    gendered science 306, 309
    learning and 170
    practical 148, 151, 193
    scientific 22, 36, 274, 278
    of surrounding world 313
    teacher 285–7, 291
explanations, scientific 103, 287, 289, 291
explicit/implicit knowledge 273
exploratory talk 80, 301–2
Eye, D. and A. von 372
eye movement 376

facilitator, teacher as 153–4, 156–7
factor analysis 206, 372
Falkenmark, Malin 8
Faraday, Michael 147
feminism 307–8, 390
Fensham, P.J. 14, 305, 353
fermentation 145
Feynman, Richard 254

field:
    centres 389
    of fixation 376–8
    of reference 103, 106–7
Fine, Arthur 254, 255
finiteness 341
Fischer, K.W. 327
Fischler, H. 217, 255, 257, 258
Fleming, R. 138
flu vaccination 130, 132
fluoride 130–4
FOCUS programme 315–16, 321
food irradiation 274, 277
Fransella, F. 313
free energy 86, 94
French, J. 291
frequency of technical terms in texts
    363–6
Fucks, W. 363
functional relations 103–4, 110–12

Gago, José Mariano 384
garbage in physical science 115–23
Gardner, H. 153, 154, 215, 216, 224
Gardner, P. 362
gases 7–20, 200, 202, 246–7
gaze-analysis 376
Geddis, A.N. 39
Geertz, Clifford 385–6, 391
Geiken-Pophanken, G. 372
gender:
    biotechnology 229, 231–7, 241
    effects in science classrooms 297–309
    student interest 390–1
generalizability 77, 262, 271, 276, 331
generic/investigation specific knowledge
    194, 196–7
genetic engineering 229–41
Gentner, D.R. 25, 167
geography 312–13, 315–16, 320–1
geology: a science, a teacher, or a course
    312–23
germ theory 270
Gick, M.L. 25, 31
Giere, R.N. 110, 157, 160
Gilbert, G.N. 290
Gilbert, John 141, 177–87, 279
Gitomer, D. 156
Glaser, R. 155–6, 159
Glasersfeld, E. von 117, 171, 343
Glynn, Shawn M. 25, 166–74
Gnoyke, Andrea 370–8
goal states of cognitive psychology science
    156–9

Goldberg, F. 51
Goodfield, J. 341
Gough, N. 178
Gowin, D.B. 75, 160, 162, 171
Graf, D. 361, 362, 363, 366
Grannott, N. 182
Green, P. 216
Greenacre, M.J. 206
Griffin, P. 328
Grob, K. 51
Groeben, N. 371
Groner, R. 376
Grosheide, W. 51
group:
    decision making 126–39
    gendered 298–9, 301, 303–4
    work 11, 40, 119–20, 148, 155
    see also discussion, group
Growth in Knowledge Frameworks (GFK)
    153–4, 156, 159–62
Gruber, H. 215
Guasch, E. *et al.* 312
Günther, U. 371

habitats 78–80
Hager, W. 371, 372
Halliday, M.A.K. 77
hands-on:
    experience 36
    Science Centres 389
Happs, J.C. 243, 313, 321
Harding, S. 307
Harlen, W. 177
Harrison, A.G. 172
Harrison, R.E. 339, 340
Harvey, William 144
Haury, D. 389
Hawking, W.S. 343
heat 22–33, 101, 144–5, 201, 203–10,
    246, 250–1
heating and insulation 195–6
Heisenberg Uncertainty Principle 255–7,
    259
Hennessey, S. 171
Hesse, M.B. 145
Hewson, P. and M. 283
Hickman, M. 205
hidden curriculum 386, 391
Hiele, P.M. van 117
Hirokawa, R.Y. 128
history of science 149, 156–7, 162, 340,
    346, 349, 388
Hodson, D. 157, 284
Hofstein, A. 116, 127

Holyoak, K.J. 25, 31
Honda, M. 19
Hopman, S. 400
Horton, P.B. *et al.* 75
Howe, R.H. 362
Hughes, S. 230
humanization of science 147, 388
Hunt, J.A. 126
Husén, T. 230, 271
Hussy, W. 372
Huxley, T.H. 146
hydrogen 93–4

ideational exchange 77–83
image:
    of geosciences 312–23
    of information 370–8
    scientific 22, 31, 312, 322
    of scientist 313
Imsen, G. 308
In-depth Focus in teacher training 353–7
individual:
    decision making 127–34, 139
    knowledge 101
    participation in discussion 81–2, 84
indoctrination 237
information:
    processing 66–7, 69–70, 154–5, 158,
        160
    vigilence 128, 134–9
infrared-reflectometry 376
Ingendahl, W. 216
integrated science 387, 403
integration of knowledge 213, 215, 221
interactions, learning and social 298, 302
interactive centres 389
interest, student 231, 240, 390–1
internal/external personal constructs 314–16
internalization 326–7, 329, 335
intersubjectivity 332
interventions, CASE and gender 297,
    305–7
interviews, use in research of:
    concept development 27, 31, 33, 40
    decision making 126, 130–4, 138
    electricity 59
    gendered science 297, 299–300, 305,
        308–9
    images of science 312, 322
    meaning making 328
    nature of science 283–4, 288
    particles 212, 214, 224–6
    student nature of science 274
    as teaching-learning situations 375

intonation 330–1, 334
intuition 23, 25, 201, 216, 362
intuitive/school/expert learner 216, 224–6
Ioannidis, George 337–49
Irwin, A. 230
Isaacs, A. 24
Isfort, A. 378
isolation of researchers 395–6
Izquierdo, M. 252

Jabin, Z. 46
James, W. 401
Janda, M. 183
Janis, I.L. 128
Jaspers, Karl 368
Jay, E. 19
Jenkins, E.W. 230, 278
Jirasko, M. 371
Johnson, E.J. 116
Johnson, M. 179
Johnston, D.D. 128
Johnston, K. 11, 216
Johnstone, A.H. 362
Jordan, R.R. 313
Joseph, Sir Keith 239
Joshua, J. 51, 52
Joshua, M.A. 104
Joshua, S. 39, 104
Joslin, P. 169
Judson, H. 177
Judy, J. 184
Jung, W. 171

Kahney, H. 25
Kass, H. 284
Keen, T.R. 313
Keeves, J.P. 402
Kelly, A. 308
Kelly, A.E. 338
Kelly, G.A. 312, 313, 314
Kelly, P.J. 130, 362
Kesidou, S. 23
Kessler, E. 371, 372
key ideas 330–1
Kilborn, Brent 159
Kitcher, P. 157
Klaasen, C.W.J.M. 117, 118
Klafki, W. 400
Klopfer, L. 153
knowledge:
    action and 191–8
    biotechnology 229, 231–4, 237
    as communication 384–5
    concepts as 200–2, 208, 210

conversation and 149
cosmology as 337, 340
decision making 128
everyday/scientific 215–16, 218, 224, 226, 387
evolution 339
experience and 217
in-action 288
language and 177–8, 181–2
models and growth of 153–63
nature of science 326, 331
pedagogical content 7, 37–8, 47, 153–4, 284–91
personal/scientific 169–70
scientific 19
social issues 115–16
status of scientific 269–73
student representations 275–8
teacher scientific 36–8
teaching and learning 100–9, 112
universal scientific 383
use 322
Koliopoulos, D. *et al.* 105
Körner, Hans-Dieter 370–8
Kortland, Koos J. 115–23, 131
Kouladis, V. 285
Krajcik, J. *et al.* 153
Krampen, G. 371
Kruger, C. 38
Krugly-Smolska, E.T. 126
Kuhn, D. 138, 182, 272
Kuhn, T.S. 103, 106, 156

labour, social division of 304, 307
Lakin, S. 284, 285, 313, 322
Lakoff, G. 179
Landers, R. 167
language:
    barrier to European research 395, 397
    collaboration 70–3
    combustion 250–1
    concept mapping 74–84
    everyday/subject 213
    gender and scientific 308
    importance of 325–6, 329
    knowledge and 215–16
    pictures and 85–98, 370
    primary science 177–80, 182–3, 185
    quantum physics 258–9
    science education and 353, 355, 361
    scientific 22, 25, 27, 32
    scientific understanding 143–7
Lantz, O. 284
Larochelle, M. 385

Latour, B. 286, 290
Lavoisier, A. 144–5
Laws, P. 278
Layton, D. *et al.* 272
Leach, John 19, 178, 269–80
learning:
  analogy and 170–2, 174
  concepts 335–8
  as conscious process 216
  conversation and 148–51
  experience and 193
  gendered attitudes and 298, 301, 305
  knowledge and teaching 100–9, 112
  styles 307–9
  teacher training and barriers to 356
  teaching and process of 115–22
Leary, R. 401
Leatherdale, W.H. 166
Lederman, N.G. 271, 284, 285, 312, 391
Lee, O. 10, 11
Lemeignan, G. 105
Lemke, J.L. 14, 179, 326
length of text 363–5, 368
Leontiev, A.N. 326
Lesh, R. 338
Levy-Strauss, C. 105
Ley, D. 178
Licht, P. 338
Lichtfeldt, Michael 212–26, 255, 257, 258
light 146
Lijnse, Piet L. 116, 216, 394–8
Lind, G. 166
Lindauer, I. 169
linguistic variables in science texts 361–9
Linn, M.C. 23, 272
Liston, D.P. 352
literacy, scientific 126, 269–70, 278
Lock, Roger 229–41
Löffler, G. 217
Longden, K. 244
Lowe, R. 177
Lubben, Fred 177, 191–8
Lucas, A. 284
Luckmann, T. 385, 391
Lumb, P. 312
lunar eclipse 182–5
Lütken, H. 303

McClelland, V.A. 352
McGuigen, L. 245
Mach, E. 173
McKnight, C. 313
McLaughlin, M.W. 153
Madigan, S.A. 371, 373

Makensen, M.V. 217
Malpighi, Marcello 144
Mandl, H. 215, 376
Mann, L. 128, 139
mapping 26, 31, 33, 74–84, 166–7,
  217–19, 221–3
market-place image 186
Martin, E. 24
Martin, Jane R. 307
Martini, M. 371
Marx, G. 86
masculinization of science 300, 307–9
Mashhadi, Azam 254–62
Mason, J.H. 347
mathematics in science 390–1
matter 10, 13, 85, 87, 89–95, 201–2, 210,
  255
Matter and Molecules (MAM) 10–11
Matthews, M. 157
meaning:
  language and 362
  personal 178–9, 182, 186, 325–36
  shared 71–2, 74–7, 82–4
  surrounding world 313
meaningfulness as term property 370–2
measurement 191, 197–8
mechanistic concept of atom 258–9, 262
Mecklenbraüker, S. 371
Medawar, P. 290
Mehwut, M. 243, 244
Menz, Ch. 376
Mercer, N.M. 177, 301, 329
Merton, R. 383
Merzyn, Gottfried 361–9
Mestre, J.P. 215
meta-understanding 19–20
metacognition 155, 159, 309
metal 372
metaphor 38, 145–6, 148–9, 166, 168,
  171, 179, 186–7, 255, 313, 328,
  347–8, 375
Meyer, J. *et al.* 177
Miles, C. 230, 231, 235
Millar, Robin 19, 22–33, 86, 177, 191–8,
  270
Miller, J. 270
mineralogy 312, 314–16, 320–1
modelling/engineering/scientific frames
  196
models:
  classical quantum physics 254–5
  combustion 243, 246
  concept development 41–2, 44–5
  decision making 128

growth of scientific knowledge and 153–63
mental 22–6, 65, 143, 145, 149, 151, 166–74, 195–6, 210, 370–8
particle 9, 19–20
physical 143
physics 103, 112
in primary classroom 177–87
science as process 284
scientific 216–19, 226
scientific as form of speech 143–51
thermal processes 23–6
of universe by teachers and children 337–49
Models in Science and Technology — Research in Education (MISTRE) 179
modification as concept 244–7
Möller, H. 371, 372
Moll, L.C. 327
Monger, G. 362
morals 231, 238, 240–1, 286–7
Morris, N. 179
motion 92, 271
motivation 117–18, 173, 215, 390–1
Müller, M. 371
Mulkay, M. 290
multiple correspondence analysis 200, 206–9
Munby, Hugh 159, 162
Murphy, Patricia 298, 304, 308
museum education 384
Muth, K.D. 167

narrative 179, 181, 185–7
of introduction 325, 328–35
National Curriculum England and Wales 36, 126, 128, 177, 192–3, 230, 239–40, 278
science in 384, 386–8, 391
teachers' understanding 284, 287
nature of science 9, 19, 130, 255, 269–80, 283–91, 322
Nature/technique integrated science 302
Neale, D.C. 38
Nesbitt, J.E. 312
Newell, A. 154
Newman, D. 328
Newton, D.P. and L.D. 312
Niedderer, H. 51, 217, 255, 256
Norman, D.A. 169, 185
Northfield, J.R. 353
Nott, Mick 279, 283–91
Novak, J.D. 75, 160, 162
Novick, S. 87, 216

nucleic acid 146
Nussbaum, J. 9, 87, 216

observation, use in research of 40, 95, 224–6, 283, 303, 329–32, 335
Oehrle, B. 371, 372, 373
Offe, H. 371, 372
Ogborn, J. 86, 95, 285
Ojala, J. 341
orbit, concepts of atomic 256–60, 262
Orpwood, G. 159, 162
Osborne, R. Jonathan 40, 74, 244
ownership of learning 302, 389
oxygen 93–4, 244–5, 247–9, 251

packaging 119–20, 130–2, 134
Paivio, A. 371, 373
palaeontology 312, 314–17, 320–1
Paris, M. 252
particulate theory 10–20, 23, 212–13, 216, 249, 251, 255–6, 258–62
Passeron, J.-C. 391
Pathways to the Atom-idea 19, 212–26
Payne, J.W. *et al.* 128
Pedersen, G. 303
Pell, A. 389
Pepper, S. 391
perceptions 62, 115, 145, 298, 322, 370, 373–4
persistance of concepts 23, 200, 209–10
personal construct psychology 312–16, 318–21
personality 304, 317, 319–20, 322
persuasion 143, 146–7
Perutz, M. 290
Petasis, L. 156
Peterson, P. 154
Petri, J. 217
petrology 312, 314–17, 320–1
Pfundt, H. 172, 216, 243, 245
pharmaceuticals 236, 238
phenomena:
  analogy and 172, 179
  based reasoning 276–7
  concepts of 9
  cosmologies 339–41
  everyday 325–6, 329, 331
  experimental 216, 375
  language about 361
  natural 36
  re-description 143–5
  revolutionary 254, 256
  science as representation of 184–5
  thermal 23

philosophy 159, 162–3
physical/metaphysical/symbolic models of
    universe 342–8
physics 50–1, 100–7, 212–13, 217, 303,
    308, 387, 390–1
Piaget, Jean 19, 170, 171, 172, 177, 215,
    327, 343, 361, 402
Piburn, M. 305
pictures:
    language and change process 85–98
    and learning 376–8
    metaphors as 168
    of universe 337, 341–8
Pimenoff, S. 351
play-oriented approach 54
pollution 119–20
Pope, M.L. 313
Popper, K.R. 370
Posner, G.J. 215
Post Graduate Certificate in Education
    (PGCE) 351–5
pot-filling model 151
Potari, D. 349
potential energy 92–3
Pozo, Juan Ignacio 200–10
practical work 11–12, 37, 67–70, 103–4,
    108–10, 112, 117, 149–50, 159,
    286–9, 387, 390
pragmatism 23, 401–2
preconceptions 10, 22, 37–8, 41, 76,
    116–17, 119–20
preferences:
    group collaboration 299–301
    subject 297, 303–7
pressure 52–60
pressure groups 230
Prestt, B. 361
Price, R.F. 230
Prieto, T. 88, 243, 245, 246
probabilistic concept of atom 259, 261
probability 94, 257
problem solving 10–13, 18, 51, 110,
    115–18, 121, 153, 156, 215, 375, 391
    computer based 64–73
Procedural and Conceptual Knowledge in
    Science (PACKS) project 191, 193–8
processes to understanding 273, 275
professionalism 291, 317–20, 322, 357,
    389
progression 11, 39, 98, 110, 116, 146,
    217, 219, 222
    understanding 243–52
prototypical situations 65, 87, 97, 100–13
Psillos, Dimitris 394

psychology:
    cognitive 148, 156–9, 171
    constructive 170–1
    educational 400
public understanding 146, 148, 229, 241,
    269–72, 278–9, 325
Public Understanding of Science
    programme 385
pupil as scientist 148–9
purpose:
    of enquiry 191
    of science 269–73, 275

quality:
    of experimental testing 70–2
    of teaching 116, 312, 315, 317–19, 323
quantum physics 217, 254–62
questionnaire, use in research of:
    biotechnology 229, 231, 234
    classroom research 351, 354–5
    combustion 235–7, 243, 249
    decision making 130
    gendered science 297, 299, 304–6
    nature of science 283–5
    particulate theory 212–14, 217–18
    quantum physics 258, 263–5

radiation 25
Räder, K. 372
Ramsden, Judith M. 351–7
rapport, personal 317–19
Ratcliffe, Mary 118, 126–39
rating scales 371
re-descriptions 143–9
readability values 361, 363, 365, 368
reality 19, 75, 126, 154, 177–8, 184,
    217–19, 224, 226, 271, 273, 277,
    313, 348
reasoning 9, 22, 30–2, 50, 52, 58, 61, 65,
    106, 127, 131, 137–9, 155–6, 159,
    169, 171–3, 340
    causal 249, 251
    relation-based 276–7
Reay, D. 301
recycling 119–20
reflection 37, 69–70, 73, 104, 119–22,
    178, 238, 285, 288, 305, 309, 329,
    337–9, 352
rehearsing ideas 332, 334–5
Reiner, M. 252
Reinman-Rothmeier, G. 215
reinterpretation of activity 196
Reisby, K. 303
relationship of concepts 200–2, 207, 210

relevance of science 389
reliability 197–8, 234, 271, 291, 322,
    372–3
Renkl, A. 215
Renstrom, L. 244
repertory grid technique 184, 312–15,
    317–18, 320–2
repetition:
    experimental 289
    of scientific ideas 330, 332, 334–5
rephrasing ideas 330, 334
representations:
    of geology and its teachers 312–13
    models as 166–7, 169–70, 180, 184–5
    symbolic 106, 109
research:
    networks 146, 148
    summer schools for 394–6
researcher-scientists 290–1
resistance 53, 55–60
Resnick, L.B. 153, 154, 171, 215, 252
responsibility 102, 104, 108–9, 112
retention, conceptual 10–11, 17, 19, 42–5,
    59–60, 98, 194, 196, 216, 217
Rhöneck, Chr.V. 51
Richards, S. 290
Richert, A. 286
rigging experiments 288–9
Rihs-Middel, M. 371
Rillero, P. 389
Riquarts, K. 127, 400
Roberts, D.A. 127, 138
Roberts, Doug 159
Roberts, G.G. 230
role:
    of analogy 171
    of collaboration 66
    of gender in science curriculum 298,
        300–1, 303–4
    of imagery of information 370–1
    of models 173
    of school science 272, 279
    of teacher 36, 102, 153, 312, 325,
        328–9, 335, 338, 391
    of university staff 352
Root-Bernstein, R. 158–9, 160, 161
Rosier, M.J. 402
Ross, J.A. 131
Ross, K.A. 86, 93, 243
Roth, W.-M. 75
Rouse, W. 179
Rowell, J.A. 273
Royal Society 383
Roychoudhury, A. 75

Rozier, S. 106
Rudduck, J. 137, 351, 352, 357
Russell, Tom 159, 162, 244
Ryan, A.G. 312, 322

Sainsbury, M.J. 75
Salter Science 128
Saltiel, E. 243
Samarapungavan, A. 274
Sanmartí, N. 252
Scanlon, Eileen *et al.* 298
Schaefer, G. 215, 375
Schaffer, S. 287
Schauble, L. *et al.* 196
Schibeci, R.A. 312
Schilling, P. 52, 174
Schmid, G.B. 86
Schmidt, D. 51
Schnotz, W. 376
Schollum, B. 243
Schon, D.A. 145, 288, 352
Schrödinger, E. 92
Schrödinger wave equation 255–6
Schuell, T.J. 215
Schutz, A. 385, 391
Schwab, J. 163
Schwarz, N. *et al.* 371
Schwedes, Hannelore 50–62, 174
Schwibe, M. and S. 372
science, definition of 401–2
Science-Technology-Society (STS)
    curriculum 126–7, 130
Scientific Questions/Closure Probes 275–8
Scott, Phil H. 5, 19, 174, 215, 216, 218,
    325–36
Screen, P. 279
Searle, J.R. 77
Second International Science Study (SISS)
    402
Seeley Brown, J. *et al.* 279
Seiler, Th.B. 216
selection:
    of analogy 39
    of scientific ideas 330, 332
Selley, N. 291
semiotics 73, 179
Semrud-Clikeman, M. 167
sequence:
    analysis 66–72
    learning 52, 54, 61
    teaching 7–20, 22, 24, 40–7, 100–13,
        118–22
Séré, M.-G. 9, 198, 200
Shapin, S. 287

Sharp, Gaynor 297
Shavelson, R. 154
Shaw, M.L.G. 313, 315
Shayer, M. 305
shell of atoms concept 257, 259
Sheppard, C. 279
Shipstone, D.M. 40, 50, 106
Shulman, L.S. 7, 37, 153, 286
Simon, H.A. 154
Sinclair, J.McH. 77
Sizmur, Steve 74–84
Sjøberg, Svein 308, 399–403
skill:
    communication 153–4
    decision making 127, 138–9
    development 304–5
    manipulative 196
    process 38
    research 353, 355–6
Slotta, J.D. 201
smeared charge, electrons as 260–1
Smith, D.C. 38
Smith, Ros 46, 289, 297
Snow, Richard E. 148
social construct, science as 307–8
social interactions and personal meaning
    making 325–36
social issues 85, 104, 115–23, 126–39,
    229–30, 237–9, 269
social relations of science 270–2, 278–9
social sciences 386
social settings and language 178–81,
    183–4, 186
society and science 353, 355
socio-cultural approaches 154, 160
Solomon, Joan 25, 51, 86, 105, 148, 149,
    273, 279, 285, 383–92
Songer, N.B. 23, 272
Sørensen, Helene 303, 304
source/target domain of model 52, 54, 58,
    166–8, 172–3
Spiliotopoulou, Vassiliki 337–49
Spoken Language and New Technology
    (SLANT) project 301–2
Staberg, Else-Marie 297, 307
state, change of 9, 217, 247
status:
    of hypothetical theory 103
    of science 177–8
Staufenbiel, T. 371
Stavy, R. 9, 10, 23, 200, 215
Stead, K. 313, 321
Stein, B.S. 154
Stenhouse, L. 352

stereotyping 238, 306, 313
Stern, P. 154
Stoddart, T. 37
Stofflett, R.T. 37
story-telling 22–4, 149–50, 328
stratigraphy 312, 314–16, 320–1
Strauss, S. 23
Strike, K.A. 215
string analogy for circuits 39, 41–2, 46
Strube, P. 312
Stubbs, M. 77
Students' Conceptions of Quantum Physics
    Project (SCQP) 255, 258–61
Stylianidou, F. 98
Sumfleth, Elke 370–8
summer schools for PhD students 394–8
Summers, M. 36, 38
Sutton, Clive R. 14, 22, 143–51, 168,
    177, 361, 362, 375
Swatton, P. 278

Talbert, J.E. 153
Tasker, R. 40
teacher:
    cosmologies of primary 337–49
    development in Denmark 302–4
    image of science 271, 278
    images of geosciences 312–23
    personal meaning 325
    PhD student as 396
    as researcher 351–3, 355–7
teaching:
    analogy and 173–4
    as assisted performance 325, 328–9,
        331–2, 335–6
    collaborative 181
    concept change and 200–2, 208
    controversial issues 229–41
    decision making skills 138–9
    electric circuits 36–47
    electricity 50–63
    epistemological approach 279
    gender and 298–9, 302–3, 305–6, 308
    learning process and 115–22, 153
    model 154–6, 286
    quality of 116, 312, 315, 317–19, 323
    science concepts 327–8
    science in EU 387
    scientific enquiry 191–3
    sequence 7–20, 22, 24, 40–7, 100–13,
        118–22
    strategies 9–10
Teaching about Why Things Change 86–7
technology 126, 229, 307–9, 387–8

television/school science 241
temperature 22–33, 58, 88–92, 95–6, 196,
    201, 203–10
Tenney, Y.J. 167
Tergan, S.-O. 376
terms:
    concept mapping 76–81, 83
    properties of 370–4
    technical 361–8
terrestrial/celestial models of universe
    344–6, 348
text 37, 117, 158, 172, 213
    contexts and 177–87
    linguistic variables in 361–9
    'Ludvig, Lisa and the air' 11–14, 17
Thagard, P. 157
theory:
    concept 118, 121
    data 154, 158–61, 326
    evidence 19, 274, 276–7, 289
    implicit, and knowledge 200–2, 205–6,
        208–10
    practice 153, 155, 192–3, 351–3, 355,
        373, 375
thermodynamics 23, 33, 85, 90, 92–3
theses, research 396–7
Thiele, R.B. 171, 173
thinking 46, 54, 61, 138, 153–6, 160, 226,
    337–41, 348, 378
Third International Mathematics and
    Science Study (TIMSS) 402–3
Thomas, G. 269
Tiberghien, Andrée 23, 100–13, 243
time, lack of 305, 309
Tobin, K. 148
Todd, F. 77
Todtenhaupt, S. 217
Torricelli, E. 144
Toulmin, S. 160, 161, 341
training:
    researchers 396
    school-based 351–2
    scientists 269–72, 279
    teacher 7, 37–8, 139, 156, 279, 286,
        289, 303, 306, 337–9, 384, 389
    value of classroom research for
        351–7
transmission 9–10, 37, 116, 151
transmutation as concept 244–52
Travis, A.S. 244
Treagust, D.F. 25, 47, 169, 171, 172
Trumper, R. 105
truth 215
Turney, J. 230

Tyndall, John 145
Tytler, R. 278

uncertainty 237, 274
understandability as term property 370–3
understanding:
    analogy and 168–9, 171, 173
    circuits 50–4, 61–2
    concept development 36–8, 44, 46–7
    education and 230, 238, 240
    electricity 65, 72
    familiarity 385
    GFK and 159
    knowledge 101–4, 107–12
    language and 177–9, 181, 185
    modelling and 143, 146–7, 149, 151
    particulate theory 212–26
    scientific enquiry 191–8
    shared 147–8, 325, 329, 331–2, 335–6
    social issues 116
    STS and 127
    student and nature of science 269–80
    teacher 284–5, 288, 290
    textbooks 361–2
    thermodynamics 85–6, 98
Unger, C. 19
universe, models of 337–49
utilitarianism 269–70, 278–9

validity 104, 112, 116, 271, 322
value judgements 127, 137–8
van den Berg, Ed. 51
Veje, C.J. 303
veterinary products 236
videotaping, use in research 59–60, 110,
    212, 214, 299, 302–4
Views on Science-Technology-Society
    (VOSTS) 312, 322
visualization 144–5, 173, 260–1
Vollmer, G. 361
Vonk, H. 338
Voorde, H.H. ten 117
Vosniadou, S. 171, 201
Vygotsky, L.S. 177, 215, 325, 326, 327,
    328, 335, 361

Wagemans, C. 303
Wagenschein, M. 171, 216
Walker, D. 126
Walkerdine, Valerie 308
Warren, J. 86
water 8, 23, 90, 93–6, 217–18
    analogy 22, 25–33, 41, 50–63, 166,
        169, 174

Watson, J. Rod 243–52
Watts, D.M. 86, 178
wave phenomena 255–6, 259–61
Webb, P. 36
Webb, R. 352
Welford, Geoff 189, 267, 295, 359, 381
Wellington, Jerry 279, 283–91, 313, 322, 352
Weniger, E. 400
Wertsch, J.V. 332
Westbrook, S.L. 249
Westermann, R. 371
Whisonant, R.C. 323
White, J. 361
Whitelegg, Elizabeth 297–309
Whitfield, R.C. 286
Wightman, T. 216
Williamson, V.M. 249
Wilson, S. 286

Wippich, W. 371, 372, 373
Wittgenstein, L. 75
Wittrock, M.C. 74
Wolpert, L. 290, 329
Wong, E.D. 169
Woolgar, S. 290
Woolnough, B.E. 193
working together 300, 302–4, 307–8
Wright, E. 154
written analysis, use in research of 40, 95, 110, 126, 130, 132–3, 135
Wynne, B. 270

Yuille, J.C. 371, 373

Zeichner, K.M. 352
Zeider, D. 391
Ziman, J. 157, 229
Zohar, A. 138